Contemporary Systems Thinking

Series Editor:
Robert L. Flood
Norwegian University of Science and Technology
Trondheim, Norway

More information about this series at http://www.springer.com/series/5807

Janet McIntyre-Mills

Planetary Passport

Re-presentation, Accountability and Re-Generation

Springer

Janet McIntyre-Mills
School of Social and Policy Studies
Flinders University
Adelaide, SA
Australia

ISSN 1568-2846
Contemporary Systems Thinking
ISBN 978-3-319-58010-4 ISBN 978-3-319-58011-1 (eBook)
DOI 10.1007/978-3-319-58011-1

Library of Congress Control Number: 2017939321

Printed on acid-free paper

This Springer imprint is published by Springer Nature
The registered company is Springer International Publishing AG
The registered company address is: Gewerbestrasse 11, 6330 Cham, Switzerland

'Indigenous people are all about place…land is our mother. This is not a metaphor. The natural world is in constant dialogue with us, although we do not always listen or respond …' (Walker cites Manulania Meyer, 2014:306)

We are the land

Our history and choices are written in the landscape

South Africa and Indonesia include urbanisation and a growth in city environments where 65–75% of the population will live by 2030.

Expanded Pragmatism dedicated to sustainable social and environmental justice is more important than ever. In the wake of the elections in UK, Europe and USA people have demonstrated the depth of their dissatisfaction that Rorty (1998) summed up in 'Achieving Our Country'. He stressed that 'something will crack' as follows:

> "The non suburban electorate will decide that the system has failed and start looking around for a strongman to vote for—someone willing to assure them that, once he is elected, the smug bureaucrats, tricky lawyers, overpaid bond salesmen, and postmodernist professors will no longer be calling the shots."
>
> Rorty said 'nobody can predict' what such a strongman would do in office, but painted a bleak picture for minorities and liberal causes. 'One thing that is very likely to happen is that the gains made in the past forty years … will be wiped out…Jocular contempt for women will come back into fashion'." (Rorty 1998: 87–91 and cited by Helmore 2016).[4]

Rorty criticises the lack of agency in academia and the tendency to emphasise criticism rather than contributing to practical transformation. This is summed up in the pithy statement by Whitman (1998) when reviewing 'Achieving our Country', as follows:

> "Old Leftists sought to effect a just distribution of wealth and opportunity. New Leftists 'specialise in what they call the 'politics of difference' or of 'identity or of 'recognition' (citing Rorty 1998:76–77). Old Leftists worry about minimum wages, adequate housing, and universal health care. New Leftists worry about how to … recognize otherness. (79)"

Policies of recognition and identity whilst very important need to be applied in the practical transformation of praxis to protect country and the people living in the run down city areas or isolated regional areas. Thus the book makes a plea for:

- Making a difference and to demonstrate in coalitions spanning social and economic sectors how we can support social and environmental justice.
- Demonstrating transformation through practical interventions in 'hands on' demonstration projects that bring hope.
- Providing the basis for learning what works why and how.

Pragmatism and agency remain important ways forward for the future of democracy. The volumes in this series propose agency for both social and

(Footnote 3 continued)
cites sections from 'Wall Street to Wellbeing' and the chapters sketched in volume 1 of this series. Chapter 5: extends a paper delivered at the Annual Islamic Studies and a version published as part of the Participatory Education Research Journal.

[4]Helmore, E. 2016 https://www.theguardian.com/us-news/2016/nov/19/donald-trump-us-election-prediction-richard-rorty.

environmental justice through practical interventions[5] that support so-called Blue Economy initiatives (Pauli 2010, 2016)[6] and that focus specifically on social and environmental justice for the voiceless, in response to the votes of 'no confidence'.

When democracy fails to enable credible candidates to stand then voluntary voters do not bother to vote. Hannah Arendt also predicted this in 'Crisis in the Republic' (1972). Perhaps if the Democrats had not silenced Sanders there may have been a different outcome?

Adelaide, Australia Janet McIntyre-Mills

[5]The ideas in this book prefigure the founding of a centre. The focus is on the development of a social and environmental justice approach and addresses a closed development, production, construction and evaluation cycle to underpin social, economic and environmental decisions. And results from a recent meeting of minds with David Stanley (a visionary thinker and practitioner) and Rudolf Wirawan (CEO of Wirasoft and President of the Indonesian Diaspora).

[6]https://m.youtube.com/watch?v=SGcol_4nqyU Accessed 28/01 2017.

Acknowledgements

My thanks to my husband, Michael for taking the photographs that appear on pages 80–82.

I acknowledge that the images from Wikipedia are from the creative commons. The other photographs are my own.

Contents

About the Author

Janet McIntyre-Mills is an Associate Professor at Flinders University, Adjunct Professor at the University of Indonesia and Islamic National University in Indonesia, and Honorary Professor at the University of South Africa. Her research focuses on systemic representation, accountability and regeneration applied to social and environmental justice concerns such as health, housing and social inclusion and the mitigation and adaptation to climate change. She addresses complex needs by exploring the meanings and 'what if' questions with diverse stakeholders. [email:janet.mcintyre@flinders.edu.au]

Her books include: 'Global Citizenship and social movements' (Routledge), 'Critical Systemic Praxis for social and environmental justice' (Springer), 'User-centric design to address complex needs' (Nova Science). 'Transformation from Wall Street to Wellbeing: Joining up the dots through Participatory Democracy and Governance to Mitigate the Causes and Adapt to the Effects of Climate Change'. (Springer) and 'Systemic ethics and Non-anthropocentric Stewardship: Implications for Transdisciplinarity and Cosmopolitan politics' (Springer).

Glossary

- 'Axiological' refers to the notion that social and environmental justice is central to new participatory architectures for democracy and governance (Mertens 2016; Romm, 2016).
- 'Addiction' refers to extreme forms of consumption that undermines wellbeing and relationships.
- 'Anthropocentrism' refers to a human-centred approach that disregards other living systems. This volume takes the next important step in the research agenda, to link the notion of relationships across humans, animals and the land as a source of indigenous and non-indigenous well-being and the broader societal need for environmental protection and effective ecosystem management of domestic, liminal and so-called wild or natural habitat (Donaldson and Kymlicka 2011).
- 'Boundary' in this work refers to a *priori* norms and *a posteriori* indicators which can be re-framed through different conceptual constructions of the way in which relationships across the material and non-material world ought to be defined.
- 'The Blue Economy' approach of Gunter Pauli (2010) outlines more than a 100 ways to create opportunities through environmental thinking that does not privilege the environment at the expense of people, instead his approach is to find ways to enable the unemployed to benefit through working on environmental challenges. His motto is: 'There is no unemployment in eco-systems' (Gunter Pauli 2016).[7] He stresses the need to provide integrated opportunities through design that taps into the abundant talent and environmental opportunities that can be found and to ensure that the designs protect both people and habitat. This is a systemic approach that could ensure that people come up with solutions that do not create binary oppositions between people and the environment. It is unnecessary to argue that for people to flourish the environment must suffer.

[7]https://www.speakersassociates.com/speaker/gunter-pauli Accessed 20/12/2016.

- 'Consumption footprint' refers to the size of human impact on the environment. The Intergovernmental Panel on Climate Change (IPCC) formula shows that the excessive consumption of energy resources impacts on the size of our carbon footprint; this footprint is defined in terms of: E (Emissions) = Population X Consumption per person X Energy Efficiency X Energy Emissions. This suggests that the privileged lives of some could lead to 'existential risk' for all forms of life on the planet (Bostrom 2011).
- 'Commodification' refers to the process of treating people, plants and animals as sources of profit.
- 'Country' refers to interconnected living rooted in integrated knowledge systems rooted in oral history, protected by law and a sense of awe and spirituality.
- 'Cultural memory' refers to a practical record of what works, why and how to support living systems. It is rooted in respect for norms and values. First Nations remember their stories by referring to landmarks, rather than ink marks on a page. The plants, animals, rocks and rivers hold stories and act as memos for our thinking. Unlike societies who think only in abstract and record their ideas on paper in books, the Aboriginal sacred text is the land. They are its caretakers and in ideal terms according to Rose (1992) they see some life forms 'as family'[8] which makes all the difference as to how they relate to living systems. Memory can be coded in the landscape to help oral cultures to retain knowledge to enable them to survive and to thrive (Kelly 2017).
- 'Consciousness' is explored in terms of the role it plays in how we think and how we value the social and environmental fabric of life. This shapes the conceptual, political and spatial landscape which in turn shapes our physical and mental wellbeing. Can we train the mind to become more flexible and adaptable through public education and a sense of the sacred? Just as we aid thinking through mental exercise or we aid physical agility through Pilate's exercises perhaps we can work on the mind and body and increase our plasticity and flexibility across the life span?
- 'Complexity' refers to the number of interrelated variables facing policy makers that can be viewed very differently depending on one's assumptions and values.
- 'Critical Systems Thinking' refers to the capability of human beings to think about living systems and to consciously make decisions about how they choose to live. This has implications for personal and public ethics, democracy and governance.
- 'Democracy' is flawed and current forms are in need of revision, because voting is insufficient as a means to hold those in power to account. Participation in thinking through the consequences of choices in regular public engagement processes is important so that people think through the implications of their choices in the short, medium and long term.
- 'Flows' refer to the way in which human choices shape the landscape of our daily lives and need to be understood as part of the situated knowledges that are

[8]https://www.youtube.com/watch?v=qp3Ktlfy0Hw&app=desktop.

valued or discarded. What we choose to value is translated into being through our thinking and practice. The scope of the new architecture for democracy and governance is to make a plea for a planetary passport to enable us to rethink boundaries and relationships at multiple levels of democracy and governance through applying multiple forms of intervention to protect living systems. The capabilities approach on which this approach rests is non-anthropocentric and it provides a basis for addressing better governance to prevent discrimination against the majority of this generation and the next. Alternative ways of framing human–nature relationships in legal, ethical and spiritual terms are illustrated by the Bolivian[9] and Ecuadorean constitutions in line with their belief in Mother Earth or Pachamana. Despite the fact that the rhetoric is undermined by the political reality of 'business as usual', it provides a first step. The next step is providing the necessary 'planetary passport' that aims to inspire loyalty to the planet as well as acting as a monitoring system to protect living systems.

- 'Global commons' refers to earth, water, the air that we breathe and genetic material that is the basis of living matter.
- 'Governance' refers to working across conceptual and spatial boundaries to protect food, energy and water security. This is vital as a first step towards preventing poverty. Governance needs to address the big issues of the day, namely poverty and climate change by protect biospheres, rather than merely protecting national interests in a 'zero sum' approach. Governance needs to be fluid, systemic and organic. The notion of nested overlapping systems that are responsive to needs to use the organic analogy of water as a synecdoche for our hybrid or interconnected lives.
- 'Hybridity' refers to our interconnectedness and the need for better public education on understanding our shared rights and responsibilities to act as stewards of living systems of which we are a strand in the 'web of life' (Capra

[9]The Guardian, Sunday 10 April 2011 18.17 BST http://www.theguardian.com/environment/2011/apr/10/bolivia-enshrines-natural-worlds-rights "Bolivia enshrines natural world's rights with equal status for Mother Earth: Law of Mother Earth expected to prompt radical new conservation and social measures in South American nation" The Law of Mother Earth, now agreed by politicians and grassroots social groups, redefines the country's rich mineral deposits as "blessings" and is expected to lead to radical new conservation and social measures to reduce pollution and control industry. The country, which has been pilloried by the US and Britain in the UN climate talks for demanding steep carbon emission cuts, will establish 11 new rights for nature. They include: the right to life and to exist; the right to continue vital cycles and processes free from human alteration; the right to pure water and clean air; the right to balance; the right not to be polluted; and the right to not have cellular structure modified or genetically altered. Controversially, it will also enshrine the right of nature "to not be affected by mega-infrastructure and development projects that affect the balance of ecosystems and the local inhabitant communities". "It makes world history. Earth is the mother of all", said Vice President Alvaro García Linera. "It establishes a new relationship between man and nature, the harmony of which must be preserved as a guarantee of its regeneration. "The law, which is part of a complete restructuring of the Bolivian legal system following a change of constitution in 2009, has been heavily influenced by a resurgent indigenous Andean spiritual world view which places the environment and the earth deity known as the Pachamama at the centre of all life. Humans are considered equal to all other entities".

1996). As human beings, we have diverse origins and we are dependent on many other forms of life (including domesticated and wild plants and animals) for our survival. Our relationships with other living systems can become increasingly detached in urban environments, where people tend to forget that cities are dependent on ecological systems for survival. Three options underpin human behaviour, namely cooperation, competition and a realisation of our interconnectedness. This book argues for the latter. In the liminal spaces in which we live our shared lives, we need to find ways to live with human and biological diversity (see Donaldson and Kymlicka 2011). This involves creating opportunities to connect with others and the places where we live. Options for living need to be preserved and protected to enable (a) integrated co-located living so that human beings can live alongside other non-human animal species by protecting ecological habitat as spaces within cities and preserving (b) agricultural spaces as well as (c) conservation of separate wilderness spaces for fauna and flora. These three options support non-anthropocentric cooperation and interdependence based on recognising the hybridity of all living systems.

- 'Habitat' refers to the living spaces for human beings and other living systems which we share and on which we are mutually dependent in overlapping biospheres.
- 'Indigenous People' and 'Indigeneity' are concepts that can empower or be used to disempower. Indigenous groups exist in different circumstances, some have political rights in the form of treaties or constitutional recognition, some have land rights (often hard won through years of litigation), some have limited cultural rights and recognised limited political representation. Some have none of the above, some are not minority groups, but they have survived a history of colonisation and prefer to see themselves as Indigenous. This is why the United Nations GA 2007, Resolution 61/295 of 13 September enables self-identification. This is certainly the case in South Africa where the majority African culture describes itself as 'Indigenous' not just the smaller minority groups like the San and Khoi. In South Australia, leadership based on 'speaking as country' is a growing movement in recognition of Indigenous ways of knowing and being.
- 'Inequality' is discussed in terms of the price of escalating inequality—national and global. Consumption based on living simply and ethically and well versus consumerism to express status is based on very different values and they have very different consequences for others and for the environment. Zavestoski (2002) has stressed the voluntary simplicity movements (Alexander 2011) such as post materialism, slow living, eating local food, reducing energy usage, reusing, recycling only tend to occur when status needs have already been met. Paradoxically some of the changes become status driven consumerism that can appear to be simple but often lead to change for the sake of appearance (Binkley 2008) and do not lead to greater levels of happiness. The gap between rich and poor has grown wider than ever before in human history (Oxfam 2016). Now the 1% own more than the bottom 50% of people on the planet. But empirical research shows paradoxically that more equal societies do better in terms of

achieving a range of social, economic and environmental indicators (Wilkinson and Pickett 2009). Higher levels of unhappiness lead to higher levels of addictive habits spanning food choices and what people tend to buy and why. When people have a sense of low status they tend to consume status items for appearance sake and to alter the mind through drugs and food choices (Wilkinson and Pickett 2009; McIntyre-Mills 2008, 2011 and 2014). Furthermore, addiction to meat (Donaldson and Kymlicka 2011), due to increasing population and the lack of agricultural land is unsustainable.

- 'Planetary passport' refers to a new architecture for democracy and governance informed by non-anthropocentric Indigenous perspectives to promote an appreciation that protection of people and the planet requires working across national boundaries and ensuring that the social contract is extended to protect non-citizens and the environment on which we depend.
- 'Policy' refers to who is included and excluded in the decision-making on who gets, what, why, how and to what effect. Policy makers need to work with stakeholders in ways that explore their assumptions and values based on local, lived experiences of what works, why and how. This user-centric policy approach starts with the axiological assumption that local knowledge is relevant and that the intention of policy needs to privilege the perceptions of service users to the extent that their diverse ideas do not undermine the rights of others.
- 'Re-generation' in this context refers to decision-making that fosters biodiversity and living systems within shared and protected habitats.
- 'Species' as a concept is a central concern for this book which is discussed in relation to the issue of categorization, membership, displacement and decision-making (in terms of state sovereignty, territory, colonisation and its implications for human, animal and plant life). As urbanisation encroaches on the wild spaces and displaces other forms of life relationships that are anthropocentric need to be re-framed to enable regeneration and sustainable living that is non-anthropocentric. Thus the book makes a case for recognising our hybridity and interconnectedness. The attitudes to animals have been shaped by the 'us/them' and 'zero sum' approach and it is now applied to human beings who are unwanted. The contributions made by Donaldson and Kymlicka (2011) to animal rights through exploring our relationships with other animals need to be given centre stage in redressing current political impasse in animal rights. Donaldson and Kymlicka (2011, 253) cite scholars who think that the collapse of habitat and food resources will occur first and then the ethical choices will change. They re-emphasise that by 2025 there will be insufficient water and land to support meat eating and that meat diets are unhealthy. They could have added that meat has been linked (along with sugar[10] as two of the most carcinogenic substances). Surely if all that separates us from other primates is a small

[10]Ironically sugar trade and slavery are linked. Slavery came to an end when it became too difficult to maintain economically and because of the work of Wilberforce in bringing about changes through political pressure and finding ways to manipulate a corrupt system of law (McIntyre-Mills 2014).

percentage of the genome—then we need to respect our connections with other creatures and we need to take seriously the contribution we should be making as stewards of the environment and other creatures by recognising and appreciating the biodiversity of which we are a strand?

- 'Sustainability' refers to a sustainable local community is determined by a sustainable region in which food, energy and water supplies are considered as major determinants for wellbeing (see McIntyre-Mills 2014 and McIntyre Mills, De Vries and Binchai 2014).[11] This work makes a plea for leadership to support regeneration based on a recognition of who we are and goes beyond sustainable development (Girarde 2015).[12] We are human animals, who have rights and responsibilities to care for humanity and other species as we are one strand of a living system.
- 'Social contract' refers to protection of citizens within the boundaries of a nation state. According to the United Nations High Commissioner for Refugees, Guterres (UNHCR 2014), for the first time since the Second World War, the global figure for displaced persons has now passed 50 million and, by 2050, this figure could be as high as 150 million (Rusbridger 2015, 13). And yet the needs of the displaced are not addressed through the current architectures of democracy, governance and education. Surely it is time to reframe the social contract and to support public education to enable people to join up the dots?
- 'Spirituality' refers to the sense of the sacredness of living systems and that human beings are part of a continuum of life.
- 'Systems view' refers to the notion that human beings are a strand in the 'web of life' (Capra 1996).
- 'Transformative approaches' (Mertens et al. 2013; Mertens 2016) refer to assumptions about power, the value of Indigenous knowledge, the need to make life chances and the lack of life chances visible and the need to develop trusting relationships that are responsive to complex needs. In this book, the approach is

---

[11]No community can be expected to transform from a high carbon lifestyle (or aspiring to this lifestyle) without feeling part of the design process and owning the decisions as to how resources should be used. Young people (Osler and Starkey 2005); the disabled, asylum seekers and sentient beings (Nussbaum 2006) along with future generations live 'precarious lives' (Butler 2005). Those perceived as different are not protected (Young 2011). The ability to show compassion underpins cosmopolitanism (Butler 2011). Butler's work stresses 'the need to rethink the human as a site of interdependency'. She emphasises that humanity needs to be able to ask for assistance and we need to be able to anticipate that we will be heard and that people will respond with compassion. Do we wish to live in a world where we do not want to help one another and in which we deny the pain of sentient beings? (Butler 2011). If we are prepared to recognise not our resilience, but our mutual vulnerability, it provides a basis for stewardship. We are all reliant on others and need to be able to depend on our connections with others. Held et al. (2005) proposed that the core challenges of the day are to address the vast differences in the standard of living between the rich and the poor. The problem is not only one of externalities that are not factored into calculations of the degradation to the environment, it is a way of thinking and 'being in the world' that shifts the extraction of profit to where labour is cheaper and where governments and citizens are less likely to complain about degradation of environment. Short-term profits are made at the expense of future generations.

[12]http://www.theguardian.com/sustainable-business/blog/sustainability-unhelpful-think-regeneration.

developed to incorporate 'Wellbeing stocks' a concept adapted from Stiglitz et al. (2010: 15) to refer to a multidimensional measure of wellbeing spanning: '1. Material living standards (income, consumption and wealth), 2. Health, 3. Education, 4. Personal activities including work, 5. Political voice and governance, 6. Social connections and relationships, 7. Environment (present and future conditions), 8. Insecurity, of an economy as well as a physical nature'. The aim of the concept is to enable people to re-evaluate economics and to become more aware of the way in which we neglect social and environmental aspects of life. The pursuit of profit at the expense of people and the environment is a central problem for democracy and governance.

- 'Zero sum approach' is expressed as competition across species, classes or sovereign states needs to be set aside in recognition that we are part of one web of life—that we are interdependent and that all sentient beings have rights. As stewards—human beings have the additional responsibility to care for the land on which we are all depend.
- 'Zero Waste approach' is based on a systemic approach to recognising sources of abundance through reusing, repurposing and recycling, in order to regenerate living systems.

Summary

Human security is threatened through the growth in the gap in consumption between the wealthy and the rest. The life chances and quality of life of the few are at the expense of the majority of human beings and other living systems in this generation and the next.

There is another way to do things based on re-framing existing markets and by valuing living systems of which we are a part. Instead of controlling them we need to find ways to mimic and work with them. Currently, many of the powerful control what constitutes knowledge and so increasingly corporatised organisations in the public, private and volunteer sectors reward those who operate successfully within the existing failing market.

They have created grand narratives and myths that have resulted in excessive consumption by some and the desire by others to emulate an unsustainable way of life.

The desire to emulate the rich and powerful is often linked with a desire to be respected, hence the market success through the symbols of excessive consumption expressed in upmarket street codes, luxury food and fashion and the ability to hide the bulk of capital in off shore accounts and undeclared tax.

The case is made in this book for human beings to take responsibility for the social, economic and environmental changes that human beings have made.

We have invented and created concepts such as nation states and the value of certain commodities (based on economic myths about what is valued), thus human beings have the power (if not the will) to *re-frame these concepts* to protect living systems—of which we are a part.

Because these legal architectures were constructed they can be reconstructed in ways that regenerate living systems and which protect the fabric of life. Human beings have the unique capability to shape the planet for a sustainable future or to continue down the current path.

A case is made for protecting local knowledge based on the evidence built up over generations of caring for country.

Prologue: Hunger and Thirst: Learning From History, Experience and Place

December 2014, Christmas Day was one of the coolest days in the past 50 years at Innamincka. We were following the trail of the explorers Burke and Wills. The rare cool day of around 23 degrees followed an unusual month of rainfall. And so the time I spent in the Channel Country with my husband, (a third generation Australian) was cool, unseasonal and unusually pleasant.

I reflected that like Burke and Wills Michael and I remain strangers in country despite our identification with the land. I am a South African and an Australian with a settler history. In South Africa, my relative George Thompson did research for the London Geographical Society and another strand of family were French Huguenots in the Cape, while Michael's family settled in the Barossa area in Australia.

I return many times a year to South Africa as a dual citizen and mobile volunteer. As we move across landscape we both long for a deeper recognition of our hybrid relationship with the land on which we depend. Cities encroach on territory and we lose ourselves to concrete. The story of Burke and Wills provides a cautionary tale for us today. Burke and Wills were keen to explore the interior of Australia. They left Adelaide in a hurry without preparation and without respecting the advice of Aboriginal stewards who offered assistance. They perished in an area where food and water was available.[13] The local knowledge of Aboriginal custodians was rebuffed and they both died of thirst and hunger. 'Being strangers in country' is the leitmotif for this book. We need to know our spaces and have a connection with place.[14]

[13]Some did realise that the land was able to offer a good living to those who could understand how to read it. Lord Grey who also experienced time in colonial service in South Africa said: 'generally speaking, the natives live well; in some instances there may be at particular seasons of the year a deficiency of food, but if such is the case, these tracts are , at those times deserted, it is however, utterly impossible for a traveler or even for a strange native to judge whether a district affords and abundance of food, or the contrary, but in his own district a native is very differently situated; he knows exactly what it produces, the proper time at which the several articles are in season, and the readiest means of procuring them....'(Sahlins, 1972 pg8 cites Grey, 1841, vol 2 pp. 259–262). Grey reflects on conditions of life for Australian Aboriginal people who were living in country. It is instructive to remember that cultures albeit diverse can find surprising similarities. In South Africa, the local African farmers were forced to take part in the formal economy through the British colonial government passing hut and dog taxes. This was a way to force people into the formal economy.

[14]Sahlins stresses that it is important to realise that development is often part of the problem and not the solution. Today cities can become urban deserts where people face the prospect of starvation. The status gained through conspicuous consumption by the world elites includes time to spend in nature in pristine environments along with the ritual potlatch (public consumption of resources) to demonstrate wealth and power. These are evident in so-called traditional societies and high capitalism. So the continuities and differences can be found without oversimplifying patterns. The desire for time to enjoy nature during holidays and early retirement is the excuse for work life imbalance. But often the choices are not available for the most marginalised.

A colleague at the Natural Resources Foundation shared a story of a Kaurna man who wanted to tell stories of country to his son. Because he lives in a house on a very small city block he decided to go to a nature reserve in Adelaide where he could sit around a fire. He arrived at closing time and was told that he could not make a fire. His sense of dispossession is part of this story. The need to create opportunities for families to connect with the earth and water through play and adventure is important for developing ecological citizenship.

Without connecting with the animals and plants, we become disconnected from nature and the diverse creatures living there. We need to create opportunities to connect through adventure play and through communing with nature. Unless we can help children to connect early on they are less likely to become champions of biodiversity.

So children need to connect with the land by playing with sand, identifying lizards, birds, frogs as liminal creatures who deserve protection. The duty to protect bees, the blue tongue lizard, the gecko or the Mallard Duck (Lever-Tracey, Constance, 2015), Koala or possum need to be discussed as part of the heritage for current and future generations. Ecological citizenship needs to be fostered through education programs with natural resource organisations to find ways to live together and to manage potential conflict across species.

The grave site of Burke and Wills

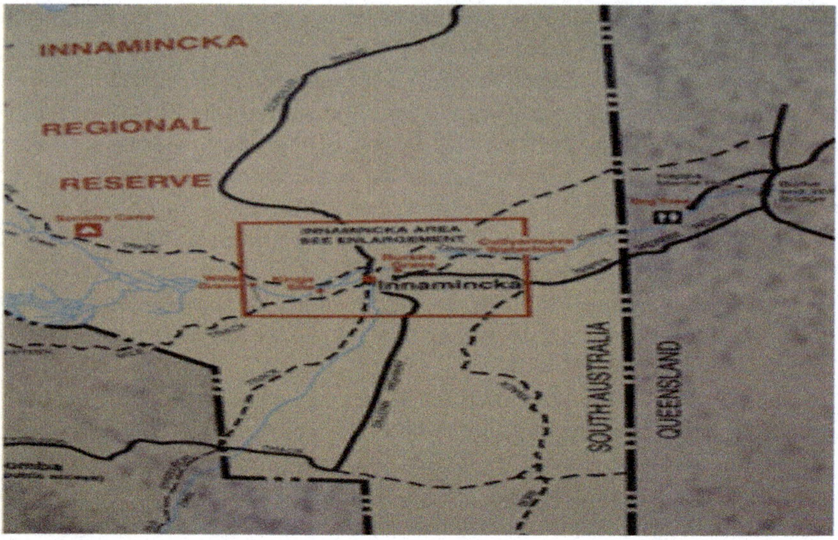

Yesterday I had a coffee with a colleague who told me a story of his teaching and learning career, starting with his life on a mission, spending time in a children's home and not knowing his place, because he was seen as 'too white' to be Aboriginal and 'too brown' to be a so-called white Australian. Fortunately he was accepted by the immigrant Italian communities as 'one of us' and is now confident of his identity in many spaces.

It echoes the story of another friend from Alice Springs who is part of the current ongoing Royal Commission into child abuse in Children's Homes. Over the years he has told me stories in our regular phone calls and visits about being hungry and eating whatever he could find in the garden including so-called 'weeds' as a child. And so learning about 'bush tucker' from others at the children's home and later from his adopted Aboriginal family became an important part of his journey towards healing. He stressed that he was unsure of his identity and his culture as an orphan and believes that he must have been one of the 'Stolen Generation'. His own family did not claim him. His story is of being adopted by an Aboriginal family from the children's home and at last feeling part of an extended family. He now has 'a mob of his own extended family' by association.

Recently, he grieved the loss of his foster mother from Alzheimer's disease and he grieves the incarceration of his nephews for driving under the influence of alcohol and methamphetamines (known colloquially as Ice). He is a veteran of the war in Vietnam and has served Australia. He sees himself as an insider and an outsider in many ways. He writes regularly to politicians and speaks out about what he considers to be the biggest challenges of the day, namely food, energy and water. He is open minded on some issues and less so on others: He sees 'Australia for the Australians' as a result of his experience as an 'old soldier'. He thinks that it is time to stop the discrimination inside our borders against Australian Aboriginals. P.T. stresses that poverty will lead to discontent here in Australia. Listening to the opinions of diverse Australians (including the young and the elderly[15]) is very important for a sense of history and a sense of political concerns. Without listening, cooperation and co-determination become less likely.

Well-being—not the economic bottom line needs to be the focus of our attention for cultural transformation, according to Stiglitz et al. (2010). But this requires re-conceptualising culture.

This is confirmed by some research in another home by my MA graduate, Monei Seduku (2015) who stressed that the business model approach has led to an erosion of caring values by staff responsible for large numbers of patients. The rising number of elderly people according to the traditional economics model is seen as a business opportunity. But the service needs to retain the core values and to ensure that people are not warehoused for profit. Currently more is being done with less which has resulted in lowering standards instead of finding ways to enable elderly people to continue to contribute to society[16] and enabling families to have more time to connect with the vulnerable in society and to care for them in their own homes where appropriate. The need to enable ongoing participation for well-being is supported by the Australian Disability Discrimination Act of 1992 which makes it against the law to discriminate against someone if they have a disability.[17]

---

[15]Respect for the elderly, heritage and learning from the experience of the past is as important as striving in the present and working towards the future. Given the shortage of staff in old age homes and the time pressures that staff are under, they tend to take short cuts that do not support all the capabilities of the elderly. Policies for better governance and risk management across the public, private and volunteer sectors need to apply a priori norms for protecting the capabilities of human and other sentient beings as well as the environment to ensure the well-being of living systems. I have spent time for the past 3 years visiting my husband's relative in a dementia ward. The number of falls I have heard about within the last three years seems excessive and indicative of understaffing by under trained, over worked and underpaid staff.

[16]http://www.abc.net.au/news/2014-06-30/christine-bryden-dementia-diagnosis. The need to continue to be active in order to remain healthy is paramount for those with dementia. Christine-Brydenin 'Keeping brain active to survive' underlines the importance of participation and engagement for all. Ms. Bryden believes that a combination of family, religious faith and medication has helped her survive and continue to function when so many others have not. And she believes that by using her brain she has to some extent rewired what is there....Professor Hodges says there has not been much study into whether mental activity helps rewire the brains of those with dementia, but he does not dismiss the idea. He believes in the future there are likely to be two approaches to treating dementia. 'One is disease specific - drugs that target the protein in that type of dementia,' he said. 'The other is more ameliorating—improving function by general mental stimulation, things that enhance brain function generally'.

[17]https://www.humanrights.gov.au/dda-guide-what-areas-life-does-dda-cover Accessed 7/01/2015 'The Age Discrimination Act 2004 helps to ensure that people are not treated less favourably on the ground of age in various areas of public life including: employment, provision of goods and services education administration of Commonwealth laws and programs. The Act also provides for positive discrimination—that is, actions which assist people of a particular age who experience a disadvantage because of their age. It also provides for exemptions in the following areas: superannuation, migration, taxation and social security laws State laws and other Commonwealth laws,

The extent to which the human rights of non-citizens, the voiceless, including elderly people, young people, sentient beings are discriminated against needs to be addressed across the local, national and post-national level. Do we wish to live in a world where we do want to help one another and in which we deny the pain of sentient beings?

The acts that are covered by the Australian Human Rights Commission provide the lens through which I consider the following current issues in society: employment, provision of goods and services, education and administration of laws and programs. The scope of the book is on rethinking boundaries, relationships, interconnection and flows spanning human choices in our daily lives (being and praxis) at multiple levels and through applying multiple forms of intervention. The capabilities approach on which this approach rests is non-anthropocentric and it provides a basis for addressing better governance to prevent discrimination against the majority of this generation and the next. But human rights also need to be extended to include the rights of sentient beings.

(Footnote 17 continued)
some health programmes. For further information: All about age discrimination. Disability Discrimination Act 1992 makes it illegal to discriminate against the disabled in the following areas of life: Employment. For example, when someone is trying to get a job, equal pay or promotion. Education. For example, when enrolling in a school, TAFE, university or other colleges. Access to premises used by the public. For example, using libraries, places of worship, government offices, hospitals, restaurants, shops, or other premises used by the public. Provision of goods, services and facilities. For example, when a person wants goods or services from shops, pubs and places of entertainment, cafes, video shops, banks, lawyers, government departments, doctors, hospitals and so on. Accommodation. For example, when renting or trying to rent a room in a boarding house, a flat, unit or house. Buying land. For example, buying a house, a place for a group of people, or drop-in centre. Activities of clubs and associations. For example, wanting to enter or join a registered club, (such as a sports club, RSL or fitness centre), or when a person is already a member. Sport. For example, when wanting to play, or playing a sport. Administration of Commonwealth Government laws and programs. For example, when seeking information on government entitlements, trying to access government programs, wanting to use voting facilities'. The Disability Discrimination Act 1992 has as its major objectives to: eliminate discrimination against people with disabilities, promote community acceptance of the principle that people with disabilities have the same fundamental rights as all members of the community, and ensure as far as practicable that people with disabilities have the same rights to equality before the law as other people in the community. For further information visit Disability Rights, Australian Human Rights Commission Act 1986, The Australian Human Rights Commission Act 1986 (formerly called the Human Rights and Equal Opportunity Commission Act 1986) established the Human Rights and Equal Opportunity Commission (now known as the Australian Human Rights Commission) and gives it functions in relation to the following international instruments: International (ICCPR), Convention Concerning Discrimination in Respect of Employment and Occupation (ILO 111), Convention on the Rights of Persons with Disabilities, Child, Declaration, Persons, Declaration, and Belief. In addition, the Aboriginal and Torres Strait Islander Social Justice Commissioner has specific functions under the AHRC Act and the Native Title Act, 1993 to monitor the human rights of Indigenous people. For further information: The Australian Human Rights Commission (information sheet). Australian Human Rights Commission Regulations 1989 (Cth) Section 3(1) of the Australian Human Rights Commission Act 1986 (Cth) defines discrimination. The Australian Human Rights Regulations 1989 (Cth) lists additional grounds which will constitute discrimination under the Act.

The scope of the book is on rethinking relationships at multiple levels and the implications for ethical living. It discusses ways in which to move from competition, towards cooperation and an appreciation of both space for diversity and our interconnectedness.[18] When we look into the eyes of the vulnerable and displaced, surely we can feel empathy for their plight?[19]
The decision made by the UK to leave the European Union occurred as I was writing the final sections of this book. In many ways, this was a vote of no confidence by the most marginalised in UK about how their lives would be impacted by open borders. But the cosmopolitan London voters supported remaining in the EU and elected the new major Saddiq Khan. Partly this is because mobile knowledge workers and the elite have benefited from globalisation and open borders, whereas those in more regional areas feel disconnected from the wider economy. Those who are working class, unemployed or underemployed fear the implications of wider connections and changes over which they have no control.

The book explores perceptions on whether people understand that growth in the size of the ecological footprint of cities, growth in the size of species loss, growth in the size of the gap between rich and poor (as a result of the greed of a few at the expense of the majority in this generation and the next) is unsustainable. The stigmatisation and victimisation of 'the other' needs to be addressed through re-conceptualising the law on sovereignty (Donaldson and Kymlicka 2011:173; McIntyre-Mills and De Vries 2011).

According to Vandana Shiva (2013): 'Growth = poverty'. Growth in the size of the population and cities, growth in the size of the gap between rich and poor as a result of the greed of a few at the expense of the majority in this generation and the

---

[18]As Fiona Stanley said in her 11th Annual Hawke Oration in 2008: the challenges faced by Aboriginal Australians in terms of housing, health and social inclusion are issues that will be felt by many as climate change deepens the impact on social, economic and environmental challenges (McIntyre-Mills 2011, 2014). All life will be threatened.

[19]Silencing and distancing of those who seek asylum raises many long-term issues for social justice. The implications of silencing and ignoring 'the other' at the organisational level or interorganisational national and post-national level can be gleaned by considering the implications at an interpersonal level. A precedent for such an approach is the work on the 'Authoritarian Personality' by Theodor Adorno written as a reflection on the Second World War. The notion that fascism is associated with a particular kind of personality associated with particular types of *interpersonal relations* already exists. Perhaps another step can be taken when considering the implications of silencing and ignoring non-citizens? I make this link in my forthcoming book on Hunger and Thirst: towards a planetary passport for low carbon footprints (McIntyre-Mills 2016 forthcoming). The implications of silencing and distancing at the international relations level can be gleaned by considering the implications at an *interpersonal level* of distancing and the prevention of bonding. See for example: The Image of the Mother's Eye: Autism and Early Narcissistic Injury. The studies of autism also show that bonding between mothers and children can be enhanced through gazing deeply into a child's eyes, Maxson J. McDowell, Ph.D., http://cogprints.org/2593/1/eye22fixed_by_cogprints.html.

next is unsustainable.[20] Unrestrained urban planning is leading to problems associated with congestion in cities, lowing living standards and high risks for women and children who are unable to express their strategic rights within households, communities and local government.

The issues of the day are summed up as follows:

- The growing gap between rich and poor (in particular those who are displaced)[21]
- The growing rise in over reach of the ecological footprints of the Global North.
- The growth in unsustainable food choices.
- The need for a new narrative (see Jakob Von Uexkoll) on a new architecture[22] for participatory governance and participatory democracy and a new sense of relationship with nature and other animals.

The way in which diverse people understand the implications of urbanisation, loss of territory, loss of species and the implications for living systems of which we are a strand (Bird Rose 1996, 2005; Atkinson, 2002; Shiva, 1988, 2005, 2007) needs to be addressed.

---

[20]https://www.youtube.com/watch?v=o2o_QQTYc7I Festival of Dangerous ideas, 2013. Also see Shiva and Barlow (2011) on the need for a new narrative of Earth Democracy and Governance (http://www.democracynow.org/2011/4/22/earth_day_special_Vandana_Shiva_andMaudeBarlow To halt the growth in greed, the growth of the gap between rich and poor and a shift to a demographic transition (from high growth rates and high infant mortality—this is typical of developing countries) we need more participatory democracy and governance not more top-down control of the size of the ecological footprint. Thus I support a participatory communication thesis by critics of the Club of Rome—in other words the Global Agora thesis. This is an ongoing debate that rages between 'the population bomb' supporters and those (like Amartya Sen and Martha Nussbaum who say that literacy and the right to a strategic voice will make a difference to achieving the demographic transition. But she does not voice concerns about the nature of capitalism and remains a social democrat.

The approach is to extend participation, in order to enhance capabilities and to engage the public so that they understand the implications of their choices socially, culturally, politically economically and environmentally. So we also make a case for engagement as a way to enable people to value the environment, social relationships and time which we have so little of in capitalist societies and universities! It advocates a move away from Authority Ranking and Market Pricing to Community Sharing and Equality Sharing and endorses some of the ideas of Fiske (1992), who develops these four categories. All people (and animals) have two basic options—compete or cooperate. Frans De Waal (a primatologist) stresses that human beings evolved through their ability to do both. He stresses that Darwin has been misinterpreted by those who emphasise competition.

Rational decision-making is important to govern the Anthropocene but also an understanding of our perceptions and emotions. This has been emphasised by Johnathan Haidt who stresses that we need to acknowledge the role of emotions and perceptions when we make decisions.

Thus we need to make decisions based on *if then* heuristics or *critical systems thinking*—that is why I still like the work of West Churchman and why his Design of Inquiring Systems Approach has been adapted and extended. It is an approach based on questioning and an awareness of many ways of knowing when planning for the future and governing the Anthropocene.

[21]As stressed elsewhere (see McIntyre-Mills 2011) rising living costs led to food riots and the so-called Arab Spring, culminating in the Occupy Wall Street Movement.

[22]http://www.worldfuturecouncil.org/838.html.

It is a priority for regional neighbours to protect their shared biosphere. We are the boundaries and we can recognise our dependency on the land and that some of the results of our attempts at development are the monsters of our own creation. In this sense I also draw on Hannah Arendt's core notion of the 'banality of evil' to reframe the way in which *we do democracy and governance and nudge people to make the right choices*. New architectures for democracy and governance need to be piloted to support regeneration (rather than merely sustainability) because the current system is so deeply problematic that it requires our being the change in our daily lives.[23] The many people who are moving across Europe have been treated with varying degrees of welcome. The Port of Calais could become a symbol of inhumanity and the breakdown of the nation state system where no formal services are provided to those trying to leave France en route to the UK. The exit by Britain from the EU because of the fears expressed by those who feel marginalised and 'unprotected' is indicative of the concerns of the majority in the UK and the concerns of many in USA.

The violence attributed to recent asylum seekers in Paris and Cologne has resulted in efforts to support Syrians in neighbouring countries and to return illegal migrants to Turkish camps: 'In exchange, the EU would accept up to 250 000 documented asylum seekers a year from Turkish camps'.[24]

Europe's approach to Asylum seekers needs some consistency based on a priori norms and a posteriori measures of service delivery. These need to be put into place to protect displaced people and to ensure that indicators for humane responses are applied consistently.

We are interconnected—we can address differences through 'loyalty' and 'voice', but exit—as described by the development economist Albert Hirschman—is no longer viable because the problems are planetary. According to Evans Pritchard, the Nuer understood that they needed to unite against a common enemy so they set aside their differences. Whereas in the past concerns were at the level of the organisation or nation state, they are now at a post-national, regional level.

The level of analysis needs to be extended by working on transdisciplinary capacity to address the convergent challenges (social, economic and environmental) that do not allow exit from the problem which is in global. Similarly, we need to understand that spaceship earth is the new 'ark of the covenant'—planet earth and

---

[23]This book is based on the idea that democracy and governance needs to be reframed through better representation, accounting and accountability. This involves valuing the human and natural resources and relationships appropriately and appreciating the opportunity costs that are created by not doing the right thing and not monitoring the distribution of resources fairly and equitably. This requires a transformation in our thinking, decision-making and practice to ensure a liveable future for this generation and the next. Accelerated climate change will adversely affect well-being and sustainability (Flannery 2005; Singer 2002; Stiglitz et al. 2010) if we continue to consume at current rates (Davies &World Institute 2008).

[24]Magnay, J 2016, 'Berlin's push to swap failed asylum seekers', Jan 30–31. *The Weekend Australian*, pp 12.

not the ark of 'us versus them'. Thus the clash of cultures (Huntington's 1996) thesis is problematic.

Young people (Osler and Starkeyb 2005); the disabled, asylum seekers and sentient beings (Nussbaum 2006) along with future generations live 'precarious lives' (Butler 2005). Those perceived as different are not protected (Young 1990). The ability to show compassion underpins cosmopolitanism (Butler 2011).[25] Her work stresses 'the need to rethink *the human* as a site of interdependency'. Butler stresses that humanity needs to be able to ask for assistance and we need to be able to anticipate that we will be heard and that people will respond with compassion. Unless this is possible it leads to a life that can be unbearable.

The strong cosmopolitan argument is that the basis for categorical or containerist thinking is fallacious. But we do not live in separate containers; we live in interconnected, overlapping regions. Our fates are interlinked. The onus is upon all nations to rethink the notion of democracy and governance. The economy cannot continue to extract profit at the expense of the well-being of current and future generations.[26]

The participatory democracy and governance potential is discussed in Chap. 4 in terms of prefiguring a means to hold the market to account. Could this approach ensure that the use of local and regional resources that underpin the common good are protected and shared fairly? To what extent could localised living in regions support appropriate technology and use solar or biofuels to run digital systems that monitor from below? Could these enable alternative forms of democracy and governance, based on the principles of subsidiarity (Schumacher 1973)? Could this process avert the 'Asbergerish' future in which people become less willing to engage face to face, because it is too challenging? (Greenfield 2003: 78) How will this affect the ability of human beings to empathise with others (including sentient beings)?

The social movements such as the Arab Spring and Occupy Wall Street Movement provide an example of the way in which citizens within nation states and citizens in a range of contexts can claim a wider public space. But it also raises questions about what will happen in the future. Not only do stocks for the future need to be developed, but the current commodification of people, animals and the environment is simply unsustainable. The current energy choices are also unsustainable. This requires a transformation that is long overdue. The introduction of nuclear energy and the continuing use of coal as an energy choice are equally

---

[25] Butler explores the violence towards the other and other species…in NominalistWay•6 videos.

[26] The strong cosmopolitan approach extends the argument by Kant on hospitality to strangers by stressing the importance of the recognition of transnational agreements and law on environmental concerns and human rights (Nussbaum 2006). The strong cosmopolitan will define human security (Kaldor 2003, 2004) in terms of meeting human rights for all and avoiding a Eurocentric approach when addressing human rights. Cosmopolitanism as a concept has developed through responding to the critics such as Babar and Derrida who emphasised the difficulties associated with universalism and the colonial and imperialist overtones.

fraught. Butler[27] links the need for food and employment; these along with energy and water are the basics of life that need to be protected through transnational movements.[28]

The argument against dialogue across boundaries (Kymlicka 2010) is that people do not have shared languages and meanings and that their perceptions and values would differ to such an extent that they are unlikely to be able to engage in dialogue. The argument is that even if the dialogue does occur the so-called 'territory of the mind' will be very diverse and so creation of shared agreements would be difficult. Strong cosmopolitans argue, however, that by virtue of being human animals we share many quality of life needs, if we are to achieve our capacity to live full lives.

Thus democracy needs to be deepened and widened. This is not a contradiction if it is governed by transnational constitutions that provide space for diversity and freedom, but within the limits of international law that protect future generations of life from poor policy decisions that could undermine the very fabric of life as we know it. A public sphere has already been created through the various forms of digital media that allow for carefully facilitated conversations based on an awareness of our precarious lives[29]. Stanescu (2012: 575) sums it up as follows:

---

[27]http://www.salon.com/2011/10/24/judith_butler_at_occupy_wall_street/ By viewing the video, the performative approach to social change through being the change through re-framing the current socio-economic system. 'If hope is an impossible demand, then we demand the impossible'. 'In brief remarks to the occupiers at Liberty Plaza, Butler offered her take on the continuing "demands" debate: People have asked, so what are the demands? What are the demands all of these people are making? Either they say there are no demands and that leaves your critics confused, or they say that the demands for social equality and economic justice are impossible demands. And the impossible demands, they say, are just not practical. If hope is an impossible demand, then we demand the impossible—that the right to shelter, food and employment are impossible demands, then we demand the impossible. If it is impossible to demand that those who profit from the recession redistribute their wealth and cease their greed, then yes, we demand the impossible'.

[28]Justice needs to be addressed through social movements that are buttressed by international law that is supported by regional courts. The recognition of the EU (despite its many failings) with a Nobel Peace Prize is indicative of the potential of the EU. In a confederation, citizens do not have a direct say in international affairs and an example of confederalist model is the EU. Archibugi (2010) stresses this is different from the 'more rigid constitutional structure' of the United States or Australia, for example. Florini (2003) in her book 'The coming democracy' outlines an argument for the potential of the European Union (and other con federations) to scale up the Aarhus convention to enable all citizens—who are members of the EU to have a say. Currently, the Aarhus convention addresses environmental concerns but it ought to also address social concerns. This could provide the architecture for a balanced approach to involvement by members of a federalist union that respects the identity of sovereign nations and their citizens—to the extent that their freedom and diversity does not undermine the rights of others. Protection ought to be extended to include young people, future generations, the voiceless (including sentient beings) the disabled and those who are not protected by citizenship rights and the social contract.

[29]We need to develop an increased understanding of our interrelatedness with others and the land. We need to become more conscious that we are part of a systemic web of iterations and that human beings are not only dependent on one another, but are also connected with other sentient animals. We are all dependent on the land, air and water (McIntyre-Mills and De Vries 2011, 2012b).

> Butler develops vulnerability and precariousness as an ethic, a social ontology, and a politics. It is because we are beings who can be hurt and killed that we have sociality, that we have a capacity for being-together. Although precariousness seems to refer to an individual life, it is rather a way of thinking connections, of claiming kinship and relations. This is not about beginning with the self's own precariousness, and then expanding that notion to others. …It has to be an understanding of the precariousness of the Other" (Butler 2004a, 134). Precariousness is a place for thinking the ethical because it begins with the Other, rather than with the self.

In corporatised academic environments we are:

> Waging a fight over values in a field in which the market seeks to be the only measure of what we value. My sense is this is one reason people have taken to the streets. For the problem, as you know, is not only that critical thinking risks becoming unfundable within institutions divided by market values, but that basic rights and entitlements are also eroded within such a context….. (JB in in Butler and Athanasiou, 2013:190).

If we understand that so-called 'liquid morality' to cite Baumann's term is no excuse for abandoning the core values of compassion and a realisation of our interconnectedness—then we will refuse to accept the amoral stances of closing borders on homeless[30] people escaping war torn areas as:

> sea arrivals passed the half million mark last week as the UN refugee agency, UNHCR, announced that 'more than 502 000 refugees had entered Greece this year'[31].

This book makes the case that thinking (and policy that results from our values matter) as they are translated into practice that can shape and destroy people and the planet. Turok's (2012) conclusion that 'the future is quantum' needs to shape ethical praxis informed by local Indigenous wisdoms. If we are prepared to recognise not our resilience, but our mutual vulnerability, it provides a basis for stewardship within and across boundaries. Our sense of place is extended when we recognise that we are all reliant on others and need to be able to depend on our connections with others. What if we could recognise our vulnerability and what if we could foster a sense of caring for others that recognises our humanity and our links with others?

---

[30]For if we are beings who can be deprived of place, livelihood, shelter, food and protection, if we can lose our citizenship, our homes, and our rights, then we are fundamentally dependent on those powers that alternately sustain or deprive us, and that hold a certain power over our very survival. Even when we have our rights, we are dependent on a mode of governance and a legal regime that confers and sustains those rights…In other words, we are interdependent human beings whose pleasure and suffering depend from the start on a sustained social world, a sustaining environment….But when someone is born into malnutrition or physical exposure or some other condition of extreme precarity, we see precisely how the deprivation of that sustaining world can and does thwart or vanquish a life." (AA in Butler and Athanasiou, 2013:4).

[31]Oct 21st posting by UNHCR cited by Euractiv.com http://www.euractiv.com/sections/global-europe/western-balkans-route-still-preference-most-refugees-319172 accessed 20/12/2015.

Focusing Thoughts

The use of the less powerful sentient beings and the planet as a resource for plunder is not a new phenomenon, but this book addresses the question: Can the social, economic and environmental deterioration of life chances experienced by the majority on this planet be understood as systemic silencing based on power and control? The gap between the rich and poor has widened in the interests of the wealthy who control governance systems as a 'plunderbund' (see Hay 2013).[32] The issue is how is this possible? Harari (2015) sketches out the idea that human beings evolved through their ability to create grand narratives about what constitutes reality and what can be considered right and legal.

The Berlin Conference of (1884–1885) is held up as an example of complicit cooperation to plunder Africa through colonial exploitation of the Indigenous population under the guise of eliminating slavery, but eventually the role of Belgian King Leopold was exposed through the concerted efforts of social critics such as Edmund Morel. The plunder has continued and escalated to the extent that all life forms on the planet face 'existential risk' (Bostom 2011). The recent exposure through the so-called Panama Papers of the way in which world leaders are conspiring with tax havens underlines the need for a new form of architecture. The papers show how widespread tax avoidance has become[33] by 'having their tax free cake and eating it'. The current politicians are voicing their needs at the expense of the majority. We are unable to exit the planet, we need to protect it for current and future generations by considering both a priori capabilities (Nussbaum 2011) and a posteriori consequences of neo-liberal decisions.

The UN Declaration of the Rights of Indigenous Peoples (2008) affirms that:

> All peoples contribute to the diversity and richness of civilizations and cultures, which constitute the common heritage of humankind…

Furthermore the Declaration recognises that:

> respect for Indigenous knowledge, cultures and traditional practices contributes to sustainable and equitable development and proper management of the environment.

Unfortunately Indigenous knowledge has not been valued or privileged in public and private sector policy decisions.

Inspiration is drawn from the work of Wangari Maathai, for example, who founded the green movement by encouraging the simple act of planting trees in Kenya. This simple act could make a difference through mobilising people to take action through small steps towards green living. She calls this 'the humming bird approach', because these actions although small in themselves can make a collective change. Besides it is better to make a small contribution than to do nothing.

---

[32] The richest 1% had 9% of national income in USA in 1970 but it had increased to 23.5% in 2007 and it had increased to 40% in 2011(Hay 2013).

[33] Chakraboritty, A. 'Hidden money corrupts', *The Guardian Weekly*, 15.0416.

Her argument is that in taking action we can inspire others to join us. In her Nobel Peace prize speech, she invites us all to think about the sort of environment we want for the future, one in which children can play safely in clean water and in which they can watch tadpoles grow.

The magic of being shown how a tadpole grows into a frog is one I associate with my childhood, playing in a river bed in the Karoo when visiting my cousins who lived on a farm. My grandfather showed me how they change and I was fascinated by the process of transformation. This is important for our own human development and learning and the spiritual sense of connection when we understand that transformation is indeed possible! So perhaps this volume can make a small contribution to enabling the people who already live simple lives that do not 'consume the planet to excess' (Urry 2010), to receive the rights that they deserve through demonstrating their responsibility to this generation and the next.

Similarly Vanadana Shiva provides another source of inspiration. We are all seeds for change and we are reliant on water, earth and seeds for our survival. She has also reminded us that we return to the soil and that we are thus connected to the earth worm and the organic life that is rooted in the soil. The work of Vandana Shiva also helps to restore the balance through focusing on the commodification of life. Whilst the work of Donna Haraway is central in explaining how dualistic thinking leads to the creation of boundaries and we need to value people and the land. Haraway[34] blurs the dualisms of the human nature divide by reminding us of our hybrid, co-evolution as human beings with companion species and how these cultural relationships shape both human beings and other species in terms of breeding and interspecies health.[35] This requires us to *do interspecies ethnography*, in order to understand where we have come from and where we are headed. This is perhaps not what Haraway means by the notion: 'we are the boundaries'.[36] But her

[34]https://www.youtube.com/watch?v=lqglzX_y5wM
https://www.google.com.au/search?q=donna+haraway+and+utube&ie=UTF-8&oe=UTF-8&hl=en&client=safari
https://www.youtube.com/watch?v=59N5xwmw5x0, Uploaded on Oct 3, 2007.
https://www.youtube.com/watch?v=59JD2eKIZfQ
http://www.egs.edu/

[35]Dogs and cats chose us as a companion species and have developed mutually beneficial relationships as hunters who assisted in obtaining food for humans. Later as human beings settled in agricultural community's cats kept away pests from storage areas. We have shaped the breeds of our companion animals. Some of the most extreme ways we have shaped dogs has resulted in their being unable to give birth, unless they are given a caesarean operation as their heads are too large https://www.youtube.com/watch?v=59JD2eKIZfQ.

[36]Brian Hare (2013) writes of his research on the co-evolution of dogs and man by studying wolves. He too explains found that dogs sought out human beings, because of their desire to scavenge for bones left over from meals. The friendlier less aggressive animals were invited to come closer to the camps and eventually they slept nearby or with their extended pack providing warmth and sharing the warmth of the fire. They also provided protection for their extended pack as guards. The friendliest dogs were the most successful at cohabitation with human beings. Human beings who bonded with dogs were more successful at defense and hunting. And so the co-evolution was mutually beneficial. Hare studied the biological changes that occur through

work provides a starting point for reworking a new form of governance and democracy based on understanding our hybridity and interconnections.

This book is an attempt to discuss ways to enhance representation, accountability and regeneration. I have been inspired by Albert Hirschman's (1970) three options for action, namely: 'loyalty, voice and exit', but these are no longer sufficient, because the scope of current problems are at the planetary level not merely at the organisational or national level.

In line with 'Transformation from Wall Street to Wellbeing' (McIntyre-Mills et al. 2014) this is a constructivist approach to discuss how it would be possible to live differently through modelling a new architecture for democracy and governance. The idea is to map the actions in a de-identified 'on line' version of a passport which can be linked to a user ID and code. This passport could then be used to demonstrate living in ways that enhance the life chances of future generations and regenerate habitat. The next steps could be to demonstrate ethical decision-making across the production, construction and consumption cycle and to hold people to account for their social, economic and environmental choices.

Assumptions and values are at the heart of a critical systemic approach that privileges the local, lived experiences of people but which makes it possible to communicate across the boundaries of space and culture in ways that explore areas of shared concern and that respect differences to the extent that freedom and difference does not undermine the rights of others in this generation and the next.

Other sources of inspiration include the work of Deborah Rose Bird on multi-species ethnography and extinctions. It provides another way to help us think about the implications of non-anthropocentricism and caring for others or 'for country'. And finally the inspirational recent book by Brown and Strega (2005) on narrative 'research as resistance' has been extended to include archival and media material as a way to foster regeneration by connecting the policy dots and inspiring practical change.

This volume strives to make a contribution by making the simple case that human beings are indeed 'better off' when they live in egalitarian societies that demonstrate stewardship of all forms of life, because we are dependent on the land, water and air, habitat for our shared survival. So-called 'development' needs to restore, regenerate and rethink the way we live so that agriculture fosters biodiversity and is introduced into vertical urban landscapes and so that fertile land and forest lungs are not lost to concrete jungles. The focus of the discussion is on urbanisation, habitation, human and animal security.

(Footnote 36 continued)
human and dog connection and found that the levels of oxytocin rise and that the raised level is higher when the dog and the human being gaze into each other's eyes (Hare 2013: 280). Dogs are not the only animals that bond. Cats and birds also bond with humans as do horses which I can assert from my own lived experience. It is possible that solidarity can be achieved across many species. Domesticated creatures have co-evolved with us and deserve our trust and protection as do the creatures that we displace as we extend our urban footprints into the wild places. This may sound unrealistic given that we do not provide protection to displace people.

The book starts with gendered, cross-cultural perspectives on what it means or could mean to be an ecological citizen and then proposes an aspirational or blue sky concept of 'planetary passport' grounded in the necessity to protect current and future generations of life through tracking and providing safe passage and the fair distribution or redistribution of resources. Water is a synecdoche for this new form of governance across boundaries. It explores the implications of urbanisation, loss of territory, loss of species and the implications for living systems of which we are a strand (Bird Rose 1996, 2005; Atkinson, 2002; Shiva, 1988, 2005, 2007).[37] The book then explores culturally diverse ways of caring and stewardship to provide practical lessons on regeneration. In line with the Paris Declaration (1997), the approach is framed together with co-researchers with local lived experience.

The focus is on how to reinvigorate democracy and governance by focusing on greater representation and accountability in terms of food, energy and water security needs by enabling diverse people living in urbanised, suburban and more regional and rural areas to voice their ideas *after* they have explored the consequences of their choices. Complex decisions need to be made by better informed people at the local level.

The focus of the book is on the need to take into account the views of those who feel left out as demonstrated by the extent to which recent polling resulted in the British Exit (Brexit) from the European Union because people feel left behind by the global economy.

The same trend has resulted in the election of Donald Trump in the United States. Some would argue that democracy has failed because it has allowed a small majority to say 'no' to business as usual. Those who are forgotten within the boundaries of the nation state have spoken out, but the problem is deeper than the lack of representation within nations.

Rationale for Planetary Passport: Knowing Our Place Through Recognising Our Hybridity and Interconnectedness

'Policy studies should seek ways of creating a climate in which people can vote in their best interests. A crucial issue for policy is removing the factors that currently 'mask the meaningful' (Kitcher 2016:1)

---

[37] It is a priority for Australia, Indonesia as regional neighbours to protect our biosphere. It is also a priority in South Africa for similar reasons. The protection of diversity is stressed in Australia, Indonesia and South Africa through Indigenous notions of custodianship of the land and their dreaming sites as well as through ecological citizenship endeavours at the local level. Protecting biodiversity in Indonesia is also priority. For example, the Annual Council of Islamic Studies Conference (Balikpapan, East Kalimantan 2014) stressed the need to protect the habitat of indigenous animals and to preserve agricultural land, because urbanisation poses a threat to food security. Food, Energy and Water security is also highlighted by the Indonesian Research Consortium, Universitas Nasional, Jakarta, 2015).

How should we live? What do we value and why? What are the implications for human capacity and human capabilities? My research questions are informed by Nussbaum's (2011: 33–34)[38] normative ideas about capabilities but I do not take an essentialist approach.

---

[38]**Life**. 'Being able to the end of a human life of normal length; not dying prematurely, or before one's life is so reduced as to be not worth living'. Nussbaum (2011) defines life in terms of being free from factors that cause immature death, or not being able live until late age. Life capability was among three basic capabilities that Human Development Index measure since it has been put in practice in 1990.

Bodily health. Being able to have good health, including reproductive health; to be adequately nourished; to have adequate shelter. **It** includes good nutrition, decent living standards. According to the World Health Organization (WHO 2016) health is 'a state of complete physical, mental and social well-being and not merely the absence of disease or infirmity'.

Bodily integrity. Being able to move freely from place to place; to be secure against violent assault, including sexual assault and domestic violence; having opportunities for sexual satisfaction and for choice in matters of reproduction.

Senses, imagination and thought. Being able to use the senses, to imagine, think and reason —and to do these things in a 'truly human' way, a way informed and cultivated by an adequate education, including, but by no means limited to, literacy and basic mathematical and scientific training. Being able to use imagination and thought in connection with experiencing and producing works and events of one's own choice, religious, literary, musical and so forth. Being able to use one's mind in ways protected by guarantees of freedom of expression with respect to both political and artistic speech, and freedom of religious exercise. Being able to have pleasurable experiences, and to avoid non-beneficial pain. **Sense,** imagination**, and thought**: Nussbaum (2011) defines this capability in terms of imagination and thought. The arts, sciences and humanities could be taught using action learning to increase the relevance of the curriculum for employment that supports and regenerates the environment

5. Emotions. Being able to have attachments to things and people outside ourselves; to love those who love and care for us, to grieve at their absence; in general, to love to grieve, to experience longing, gratitude, and justified anger. Not having one's emotional development blighted by fear and anxiety. (Supporting this capability means supporting forms of human association that can be shown to be crucial in their development). Nussbaum argues for public policy to protect the 10 capabilities as key elements for human dignity (2011, p 79).

6. Practical reason. Being able to form a conception of the good and to engage in critical reflection about the planning of one's life. (This entails protection for the liberty of conscience and religious observance).

7. Affiliation.

(A) Being able to live with and toward others, to recognise and show concern for other human beings, to engage in various forms of social interaction; to be able to imagine the situation of another. (Protecting this capability means protecting institutions that constitute and nourish such forms of affiliation, and also protecting the freedom of assembly and political speech.) **Affiliation**: the ability for people to create sense of belonging to others (including other species such as animals and plants) and to institutions.

(B) Having the social bases of self-respect and non-humiliation; being able to be treated as a dignified being whose worth is equal to that of others. This entails provisions of non-discrimination on the basis of race, sex, sexual orientation, religion, caste, ethnicity, national origin.

8. Other species. Being able to live with concern for and in relation to animals, plants and the world of nature.

9. Play. Being able to laugh, to play, to enjoy recreational activities.

10. Control over one's environment.

Instead the expanded pragmatism that informs the engagement of people in every day decision-making is also inspired by Stuart Hall who asks what is the social, cultural, economic and environmental context that shapes who gets what, when, why and to what effect?

The social, economic and environmental issues today pose a convergent challenge for representation, accountability and sustainability.

Governments these days ought to address many, interrelated variables that are viewed differently by different stakeholders within increasingly diverse nation states. The residents of nation states are not all citizens, they also include the voiceless whose interests are not represented.

Democracy represents only those within the boundaries of the nation state, but today we have more displaced people than ever before and the numbers who will be displaced are likely to increase. The well-being of many people is dependent on social, economic and environmental factors that extend beyond the boundaries of the nation state.

The issue of the extent to which the people are being represented by politicians within the boundaries of the nation state is also highly debatable with the revelation that big business and politicians seem to have allied interests, namely self-interest rather than the representation of the needs of the majority.

Thus the need to hold governments to account by people whose life chances are not being protected by governments is becoming more important.

The notion of rights and responsibilities outlined in the social contract should be extended to enable ecological stewardship and regeneration of living systems. The ability to move beyond containerist thinking is vital to address the big issues of the day that span conceptual and spatial boundaries.[39] Our dependency on the land needs to be placed at the centre of the design for a future borderless democracy. This would mean that joint responsibility for climate change, human rights could be

(Footnote 38 continued)

(A) Political. Being able to participate effectively in political choices that govern one's life; having the right of political participation, protection of free speech and association.

(B) Material. Being able to hold property (both land and movable goods); having the right to seek employment on an equal basis with others; having freedom from unwanted search and seizure. In work, being able to work as a human being, exercising practical reason and entering into meaningful relationships of mutual recognition with other workers.

[39]Voting no longer needs to be the only way to shape the political agenda. It needs to be co-determined with locally based people who have lived experience of their specific needs. In my previous books I have outlined the notion of nested governance systems that enable representation and accountability at the local, national, post-national regional level of biospheres, in order to privileged the voices of the marginalised. Similarly in this book the idea is that people should be able to move across boundaries and that the notion of planetary passport should replace the notion of citizenship with the notion of global citizenship that protects the social, economic and environmental rights of human beings. Instead of merely critiquing current challenges this book builds on previous research (McIntyre-Mills (2014a, b) to design an alternative non-anthropocentric approach that protects both people and the planet based on an ethic of care.

addressed in the shared goal to protect food, energy and water security through protecting habitat for current and future generations of life.

The new political landscape requires a paradigm shift from management to governance. The starting point is to address the five areas of priority highlighted by Archibugi as:

- Control over the use of force;
- Acceptance of cultural diversity by ensuring internal sovereignty based on constitutional rules that enable freedom and diversity to the extent that the freedom and diversity of others is not undermined. External sovereignty is 'replaced by global constitutionalism'(Archibugi, in Wallace and Brown 2010:321);
- Strengthening the self-determination of people based on participatory democracy and the 'absence of domination' over other states;
- Monitoring internal affairs, based on a form of monitory democracy and governance;
- Participatory management of the global commons as the fabric of life.

Consciousness is a continuum across all life

The book is structured to develop an alternative view of democracy and governance based on working across conceptual and spatial boundaries.[40] It proposes a non-anthropocentric approach to valuing habitat to protect living systems and to rethink the way in which global organisations such as the UN could protect the basis of life through promoting new forms of engagement (in line with Held's 2005 concept of a Global Covenant to protect the Global Commons). The following chapters explore the ethical implications for public administration and public policy of:

- Dualism and the ethics of human/nature divide and speciesism.
- Hybrid relationships across humans and other species and the rise in species extinction.
- Human–machine identities in which we are becoming more like machines than sentient beings who show compassion to others.

It examines the impact of participation at the local level (Evans 2014) through awareness raising about the implications of consumption choices on well-being and 'wellbeing stocks' (Stiglitz et al. 2010) for current and future generations.[41] It also examines the ethical values of participants in terms of idealism, pragmatism and the

---

[40]The new trends (based on a critique of Zeger, 2016, Flinders Symposium) are as follows: Diverse Stakeholders and intensive communication, Diverse authority, New skills and new ways of working, Many new ideas, Conflicting values and debate about what constitutes ethical decisions, debate about spatial and conceptual containerism versus extension.

[41]Ethical decision-making about consumption, in order to enhance well-being stocks through sustainable living and consumption has been addressed in detail in two Tier 1 companion volumes in the Contemporary Systems Series entitled: '*From Wall Street to Wellbeing: Joining up the dots through Participatory Democracy and Governance to Mitigate the causes and Adapt to the effects of climate change'*, ISBN 978-1-4899-7465-5 and McIntyre-Mills, J 2014 '*Systemic ethics and non-anthropocentric stewardship: Implications for trans disciplinarily and cosmopolitan politics*', ISBN 978-3-519-07655-3.

virtue-based approach (Preston and Sampford 2002; Taylor 2009; McIntyre-Mills 2014b) and the implications for ecosystems (Fisher et al. 2009) and how this impacts on beliefs and intentions towards multiple species within shared, liminal and separate habitats.

I hypothesise in the Journal of Consciousness Studies (McIntyre-Mills 2010) that participation through awareness and consciousness-raising will influence the way in which people value the environment and engage in healthy relationships with the land. This hypothesis is based on the notion of neural plasticity in that the brain shapes the environment and, in turn, is shaped by social and environmental factors (Bateson; 1972; Beer 1994; Greenfield 2000).

Climate change is a significant problem in Australia and elsewhere. There is little doubt that accelerated climate change will adversely affect well-being and sustainability in Australia (Flannery 2005; Pretty 2013; Stiglitz et al. 2010)—particularly if we continue to consume at current rates (Davies &World Institute 2008)—resulting in significant devastation and a compromised quality of life in terms of habitat, food, energy and water security.

The impact of climate change globally has been underestimated (Lovelock 2009; Rockström et al. 2009) and if the 'UN Future Earth' initiative and the 'Paris Agreement on Climate Change' is to achieve a lowering of emissions that meet the so-called Tuvalu Test that measures consumption choices (Murray 2007)—then Australians (and other Developed Nations) will need to limit the size of ecological and carbon footprints even further so that rising sea levels will not impact the life chances of humans and other species on islands such as Tuvalu and Kiribati, for example. Aboriginal cultures teach us about stewardship and relationships with the land, but these relationships are being lost in non-Aboriginal cultures. As Major Sumner, an Ngarrindjeri Aboriginal elder from the periodically drought-ravaged lower Murray River in South Australia and custodian of the river stresses, *we are the land and the land is us.*

Re-establishing relationships with the land is at the heart of effective cultural ecosystem management (see http://www.mdba.gov.au/what-we-do/working-with-others/aboriginal-communities/ringbalin).

As a non-Aboriginal researcher, I build on 18 years of research with Aboriginal and non-Aboriginal Australians.[42] There is evidence that many non-Aboriginal

[42]Cross-cultural engagement on well-being and climate change has shaped the research (see '*Identity, Democracy and Sustainability*', McIntyre-Mills and De Vries, 2011, '*From Wall Street to* Wellbeing' (McIntyre et al. 2014), '*Systemic Ethics*' (McIntyre-Mills 2014) and 'Thinking Matters' (McIntyre 2015) pertaining to consumption, consumerism and well-being in Australia. The process of participation is the subject of these recent publications and other monographs , entitled '*User-centric policy design*' (McIntyre-Mills 2008), based on research conducted in South Australia and funded by an Australian Research Council Linkage Grant and '*Critical Systemic Praxis for Social and Environmental Justice*' based on research conducted in the Northern Territory and funded by Local Government (McIntyre-Mills 2003), respectively made the case that participation enhances attachment to policy ideas and narrows the gap between perceived needs and service outcomes. The latter book develops an argument based on considering the consequences of social, economic and environmental policy decisions for current and future generations

people desire more environmentally sustainable lives, but little is known about the influences on choices around the management of land, water and food that affect the environment. Government response to human well-being is often based on economic development, which inadvertently increases consumerism, resulting in greater environmental degradation. We know, therefore, that *environmental sustainability, consumption choices and human well-being* are intimately linked, but there is little knowledge about how this linkage can be addressed in policy and governance terms through limiting consumerism and appreciating our place amongst other species. Attempts to address climate change are often based on policy information that is not grounded in citizen experiences and fails to address what we do know about human behaviour or choices.

How can we increase public education and engagement in the environment that balances individual and collective well-being and simultaneously protects the environment when we know little about this nexus? Consumption and anthropocentric consumerism are of central concern given the need to limit the size of our ecological footprints through better governance and enhancing capabilities (Nussbaum 2006, 2011) to live in non-anthropocentric ways that protect the planet.

New local forms of engagement and governance (Held 2004; McIntyre-Mills 2004) are needed to protect residents against environmental hazards and the subsequent economic and social consequences. Major challenges in local communities include the need to achieve or maintain: (a) Access to safe domestic housing (including energy and water), appropriate education and employment; within (b) livable cities with liminal biodiverse spaces; that are in turn supported by (c) sustainable regions. Some scholars (Christakis and Flanagan 2010; Erlich and Erlich (2010, 2012); Faist 2009; Dryzek 2010) argue that a lack of cultural understanding about our interlinked and entangled lives affects human rights in so far as it shapes access to water, energy and food security.

There is a groundswell of democracy that seeks to save the planet from our consumption choices, but little is known about those consumption choices and what motivates people to consume excessively. It has been argued that after a certain point, increased consumption does not lead to more happiness (Pretty 2013). Thus, increased consumption, driven by increased economic growth, does not make people happier. Indeed, in more egalitarian societies, all people consume less and

(Footnote 42 continued)
of life by engaging in conversations with informants about how we ought to live. These books were presented in an invited seminar in August 2014 at the Schumacher Institute in Bristol as part of the United Kingdom Systems Sciences conference program at the suggestion of the Research Director of the Schumacher Institute, Dr. Jenneth Parker. In '*Wall Street to Wellbeing*' (based on the Australian Local Government funded research) I discuss in detail public engagement to enable people to make better social, economic and environmental consumption choices and thus to think about well-being and sustainability. Specifically, this research builds on the findings of two of my previous projects (1) a Participatory Action Research (PAR) project with Aboriginal Australians focused on well-being (ARC linkage LP0560406) and (2) a PAR project on decision-making to promote adaptation to climate change (Local Government Grant https://wellbeing.csem.flinders.edu.au/) and (3) Ethical non-anthropocentric approaches to decision-making.

are less status conscious. Thus, the link between greater equality or well-being and the prevention of global warming appears to involve 'limiting consumerism' (Wilkinson and Pickett 2009: 221) and appreciating our dependency on other species in a shared and separate protected habitats. It is assumed that greater social and economic equality will provide the key to reducing the cultural pressure to consume at the expense of the majority in this generation and the next. These are policy challenges locally and regionally that need to be addressed if this agenda is to be advanced. Specifically, it will require balancing individual and community needs to achieve living simply, well and sustainably to protect human, animal and environmental rights (Layard 2006; Nussbaum 2006; Faist 2009; Shiva 2012).

We face the prospect of food deserts and increasingly impoverished communities in cities without adequate resources to maintain a decent quality of life. The following challenges (that will need to be urgently addressed by current and future generations) include food, energy and water insecurity in an increasingly urbanised world,[43] resulting in increased hunger, unemployment and crime.

In 'Wall Street to Wellbeing' and 'Non anthropocentric approaches to ethics' I try to address the notion that 'we are the land'. Rose (1996) describes it as follows:

> "The land grew the people and the people grew their country. Human spiritual and cultural processes within this landscape and with other life forms were, and continue to be, dynamic, proceeding, expanding, changing, growing, flowing, being mediated according to the movements and interactions that are natural to human, and non-human, associations. Rose names, in essence, 'being human', as transforming processes, continuous movements of activity and energy, as people are involved together in business of making sense of their actuality as they make relationships with both the corporeal and the non-corporeal world (1996: 36)"

We evolved through the ability to show empathy to others and to reciprocate and not only through our ability to compete. The good global neigbourhood program could be a way forward to evolve in ways that respect others (including the voiceless) and the environment on which we depend. This book addresses ways to address the challenges and opportunities through public education that responds to current challenges, namely the need to live with diversity in increasingly diverse cities. It stresses the need for public education beyond the boundaries formal educational institutions to protect the environment and to develop our capabilities to think across conceptual and spatial boundaries. This means learning to work across disciplines and cultures. In a globalising world, we are faced with diversity on a daily basis. Populations are growing in increasingly urbanised areas where the divides between rich and poor are widening. According to the United Nations High Commissioner for Refugees, Antonio Guterres (2014):[44]

[43]Whilst undertaking a 3-day workshop with the MOSA one of the key areas of concern was food security.

[44]http://www.unhcr.org/52d692429.html UNHCR Remarks by António Guterres, United Nations High Commissioner for Refugees, Second International Humanitarian Pledging Conference for Syria, Statements by High Commissioner, 15 January 2014.

> Within a few years, Syria has gone from being the world's second largest refugee-hosting to becoming its fastest refugee producing country. … UNHCR has registered far over 2.3 million Syrians as refugees in the region, and governments estimate the total number of those who fled at over 3 million. … Few refugee influxes have generated this profound an impact on their host countries, with dramatic demographic, economic and social consequences. The pressure is felt heavily in all areas of daily life, as budget deficits are increasing, growth suffers and jobs, salaries and price levels are affected across the region, leaving local families struggling to make ends meet.

More than 50 million displaced people who are non-citizens and thus considered to be outside the frame of reference of state protection. The enclosure acts based on Hardin's paper and Locke's argument based on the notion of the Tragedy of the Commons is part of the problem. As detailed below there is no research to support this argument. Current challenges today such as social, economic and environmental issues are associated with the convergent challenges such as the following: a growing numbers of displaced people, poverty, conflict, food insecurity as a result of shrinking agricultural land, water shortages, growing urbanisation, energy shortages associated with using non-renewables and disasters such as the nuclear disaster on the scale of Fukushima. These are some of the examples raised by the United Nations and Vandana Shiva and Maude Barlow. It is time to connect science to humanity.[45]

Mindfulness and Transformation

The emphasis of this book is to explore the role of consciousness not merely on mental and physical health but on how our thinking shapes politics and the landscape.[46] Can we train the mind to become more flexible and adaptable through public education on understanding our interconnectedness with other forms of life? Just as we aid thinking through mental exercise or we aid physical agility through Pilate's exercises perhaps we can work on the mind and body and increase our plasticity and flexibility across the life span? Surely if all that separates us from other primates is a small percentage of the genome—then we need to respect our connections with other creatures and we need to take seriously the contribution we

[45]Service for protecting—through recognising our place in the universe—more senses not just visual. Trying out and being open to being proved wrong!!! Ideas must stand up to testing. The way the 1% lives causes unhappiness for the majority.

[46]This book develops an argument based on considering the empirical consequences of social, economic and environmental decisions on the quality of life of current and future generations. It draws on 30 years of by teaching critical heuristics and my research exploring: how should we live, in order to respond to the growing challenge of meeting the needs of people living in cities.

should be making as stewards of the environment and other creatures by recognising and appreciating the biodiversity of which we are a strand?[47]

The work of Snyder et al (see Snyder, 2011) and Pert (1999) on the 'Molecules of Emotion: Emotional Intelligence and the way our thinking shapes our body and how emotional memories are stored in our cells' (Grodzki, 1999) is now widely accepted in mainstream biology and neuroscience. Lipton, author of 'Biology of Belief' (2007) develops an argument that genes only provide the potential or the pattern for proteins and that the trigger for activating the production of certain proteins always comes from the environment and perception of the environment. Although he has been criticised for overstating his new age argument, neurogenesis makes the point that the environment and our sense of perception can help us to overcome the notion of prescription. So the notion that genes *necessarily* determine all behaviour and life chances is too simplistic. Environment and *perception* of the environmental signals also play a role. Where Lipton's work becomes problematic is where it overgeneralises and provides solutions. But in some ways his argument supports the arguments made by Neil Turok[48] in '*The universe within*' (2012) that human perception and a sense of the spiritual remain important and ought not to be dismissed in the way Dawkins (1976, 2006) suggests. The mind of the human being remains a vital aspect in triggering genes and also making sense of the environment.

Lipton argues that we are human receptors[49] who make sense of the world around us. We take in signals and we send off signals. Thus the material body and the mind play an equally important role. The work of Candace Pert (1999) on 'The molecules of emotion' makes a similar argument about how our thinking and our emotions

---

[47]The most intriguing pieces of research I have read recently lead me to question the place of human beings in the universe and the implications for the way we construct our worlds. Clearly we are arrogant in our assumptions about our place in the universe. The fact that chimpanzees may lose territory as a result of our growing anthropocentric footprint will impact on our biodiversity. What will it mean if our closest human relatives are no longer able to survive? We will destroy some of our biological heritage: 'Goodall Witnessed chimps performing a specific swaying dance around large waterfalls, in thunderstorms, and heavy rains. This dance suggests a sense of ceremony and appreciation of the natural world, which Goodall speculates, might be related to awe and wonder, that could lead to one of those early animistic religions where people worship water and sun and elements they can't understand' (Kehoe 2016a; Kehoe et al. 2016).History allows us to learn from our mistakes. Our closest human relatives include chimps who may be closer to us than previously anticipated. They have demonstrated behaviour that could be interpreted as ritual. What does this mean for us as members of the human species who have lost a sense of our dependency on nature? The Eurocentric origins of many of our approaches is problematic Like Australian academic Rose Bird (1996, 2004) I work on identity, research and alienation. I am trying to develop a response to this alienation by applying an approach to research that decentres the researcher and empowers participants to address well-being and the implications it has for our identity and 'being in the world' and the way we are consuming resources.

[48]Turok's parents were anti-apartheid activists who believed in taking on worthwhile challenges and making a difference. This is the approach that Turok has taken. Commencing with research with Steven Hawking and now at the Perimeter research institute. He wanted to understand the nature of the big bang and poses the idea that the universe is expanding.

[49]http://m.youtube.com/watch?v=iI4WoeP8d_4.

shape our human bodies. This rather mechanistic explanation or the previous lock and key notion of receptors and molecules of emotion explained by Pert (1999) has been updated in 2013 in one of her last public lectures as a form of pulsing organic matter[50] and she stresses that what separates humanity from chimpanzees—with whom we share more than 99% of our DNA is our ability to make sense of categories or patterns with the help of our more developed frontal cortex.

The book addresses ways to address the challenges and opportunities through multicultural education that responds to current challenges, namely the need to live with diversity in increasingly diverse cities. It stresses the need for public education within and beyond the boundaries formal educational institutions to protect the environment and to develop our capabilities to think across conceptual and spatial boundaries. This means learning to work across disciplines and cultures. In a globalising word we are faced with diversity on a daily basis. Populations are growing in increasingly urbanised areas where the divides between rich and poor are widening. We face the prospect of food deserts and increasingly impoverished communities in cities without adequate resources to maintain a decent quality of life. The following challenges that will need to be urgently addressed by current and future generations include:

- Climate change—food, energy and water insecurity in an increasingly urbanised world[51] in which poverty (hunger, unemployment and crime) are a way of life for the vulnerable.
- Education to address human capacity and capability to regenerate local environments where necessary and to live sustainably.

With greater mindfulness of our shared fate, could we learn to understand that the divisive ('us /them') politics that plays out across nation states needs to be redressed through co-determining new relationships based on new forms of republicanism and regionalism? Instead of profit and productivity, we need to recognise well-being as the bottom line (Stiglitz et al. 2010) defined in broad terms that support diverse dimensions. But this cannot be achieved within the bounded system of the nation state or through narrow disciplinary approaches. The values that shape our current social and economic choices are the root cause of a way of life that benefits a minority at the expense of the majority in this generation and the next.

As part of the journeying approach during my sabbatical I was tried to apply the Indigenous notion of learning from 'the ground beneath my feet' and connecting to place. I attended the 'Mindfulness Conference' in Stellenbosch, South Africa and

[50]www.youtube.com/watch?v=8CFjt4qXE-Y, May 8, 2013—Uploaded by Science & Spirituality Part 1 April 28, 2013 Candace Pert, Ph.D. https://www.youtube.com/watch?v=5TeWvf-nfpA Mindfulness and neuroplasticity—throughout life 'Neuroscientists Richard Davidson and Amishi Jha join clinical mindfulness expert Jon Kabat-Zinn in, and what cutting-edge neuroscience is revealing about the transformation of consciousness through mindfulness and contemplative practice'.

[51]Whilst undertaking a 3-day workshop with the MOSA one of the key areas of concern was food security

also explored the notion of consciousness and what it means to think differently about ourselves and our relationships with others. The mindfulness approach can be summed up in an exchange between one of its leading practitioners Kabat Zinn and a neo-Marxist, Davis (2012) asks what difference a change in consciousness can make without applying it in practice:[52]

> "While the brief discussion between Davis and Kabat-Zinn remained abstract, actually existing experiments at the intersection of mindfulness and social change are blossoming.[53]...Mindfulness alone will not spark a political revolution, but when joined by actual revolutionaries, it might expand all of our possibilities for freedom."

Wadsworth (2010: 28) Wadsworth contributes to this area of concern by citing Winston Churchill's phrase ' We shape our houses, and after that they shape us' and draws attention to the way in which sociologists such as Giddens (1986) understood the way in which societies are shaped by people and then they shape us. Similarly she cites the work of neuroscientists (Doidge 2007) and makes the connection with the way in which society is made up of a series of nested systems writ large. They are a reflection of who we are. The boundaries we have drawn between us and them and the poverty some experience so others can live lives of luxury that could undermine the rights of future generations. The crime that is perpetrated by the excluded and deluded also needs to be examined. The case is made in this book for making connections with many ways of knowing and with many cultures. It is vital to develop these connections, rather than to emphasise only one approach to understanding the world. Limiting hospitality to strangers is the core issue. Climate change will lead to migrations of people and this will require rethinking the way in which the contained nation state is replaced with porous states that are part of regional, republican federations. This requires moving from weak cosmopolitanism to strong cosmopolitanism. Even if the nation state continues to exist it will require new forms of regional governance. This is the context for developing post-national monitory democracy and governance. How should we live?

> ... [C]ontentment and experienced wellbeing, in a Benthamite hedonic utility sense? Or do we care about Eudaimonia or life purpose, in an Aristotelian sense? (Graham 2011: 121)

---

[52]Jon Kabat-Zinn, 2012: 249:".....And what is growth metaphorically if not an expanded awareness of oneself in relationship to the larger world and one's place in it, a deeper understanding of the interconnectedness. What is growth metaphorically if not an expanded empathy for others and for the world, a reaching out to suffering by one who already knows suffering intimately, or who could, and knows it? There is a requirement for humility here (Jon Kabat-Zinn, 2012)..."In a racially unjust world...what good is mindfulness?" asked, a Neo-Marxist. Davis offered an answer to her own question. She noted how mindfulness might become a revolutionary force if embedded in social movements that target oppressive systems. ...Rowe (2015, op cit.) reports on a dialogue between a Marxist activist and Kabat-Zinn: https://www.opendemocracy.net/transformation/james-k-rowe/zen-and-art-of-social-movement-maintenance.

[53]http://transformativechange.org/.

Remembering and Reconnecting with Country

Today the understanding of carbon-based cultures and economies is relevant to climate the starting point is a recognition that economically well-being needs to be the outcome of economic choices—not merely productivity.[54] The rise of increasingly xenophobic, nationalist responses it indicative that more and more of the marginalised including the working classes and middle classes feel 'unprotected' by policy makers and politicians.[55] Kenny (2016)[56] cites Clinton:

> To just be grossly generalistic, you can put half of Trump's supporters into what I call the basket of deplorables", Clinton told an audience of donors. "Racist, sexist, homophobic, xenophobic, Islamophobic, you name it.

The derision of the quote only served to underline the gap between policy makers and groups of people who feel that the stereotypical policy groups do not identify with the needs of the marginalised 'other'. According to Shanahan, 2016[57]:

> To Clinton supporters, the four biggest problems facing the US, in descending order, were gun violence, the gap between rich and poor, climate change and college affordability.
>
> For Trump supporters the four biggest problems were illegal immigration, terrorism, job opportunities for working-class Americans and drug addiction…

So how can we address the issue of rights and responsibilities at a local level? Could local engagement provide the basis for scaling up management of our Social, Economic and Environmental choices at a regional level?

Florini (2003) suggests the potential of the Aarhus convention. I have combined this with the policy potential of the UN Local Agenda 21 and other policies detailed below, such as Paris Development Agenda and the UN Development Goals. Thus the participatory action research aims to:

> "[A]ddress the challenge posed by the Earth Charter: "Everyone shares responsibility for the present and future wellbeing of the human family and the larger living world" (cited by Hayden in Wallace and Brown, 2010: 368)".

[54]These public interests include environmental considerations. Today the decision to for example to use energy selfishly and excessively has implications for other communities who may be more affected by extreme weather events as a result of our choices. For example, we may choose to drive an SUV when it is unnecessary or when it is possible to walk to meet our friends. We may use water excessively for forms of agriculture that will have a downstream risk to other communities needs to be governed by laws that protect the common good. Furthermore, complex decisions need to be informed by those who will be affected by the decisions Ashby's rule of Requisite Variety (1956) to establish ways to enhance representation, accountability and engagement.

[55]Noonan, P. 2016. 'Trump and the rise of the unprotected'. *Wall Street Journal* Feb 25th.

[56]Kenny, C. 2016. 'Politicians are best playing the policy and not the voter: Disparaging a rival candidates' supporters is a sure way of limiting one's own appeal' Weekend Australian, Nov 12–13, p. 22.

[57]Shanahan, D 2016. The people have spoken: fix these problems now *Weekend Australian* Nov 12–13 2016, p 19.

The challenge is to move beyond the rhetoric of cosmopolitan citizenship and to address both justice and sovereignty. The nation state needs to be held to account by an Earth Charter that is supported by overlapping regional institutions (supported by inclusive regional research institutions, policies, the rule of law including parliaments and courts).

Co-determination in regions needs to be based on a new architecture for governance, democracy and ethics.[58]

The book discusses research aimed at piloting and testing out new architectures for democracy and better governance through:

- Addressing the issue of a priori norms and a posteriori measures for transformation towards regenerative living.
- Finding ways to match social, cultural, economic and environmental decisions to perceived needs with a focus on food, energy and water security.
- Narrowing the gap between perceived needs and the way resources are distributed and the way it impacts on service outcomes.

Previously oral histories connected people to their place and the wisdom of the elders was handed down to the next generation through remembering the vital social, economic and environmental information needed to survive. Cultural memory was aided by association with myths and landmarks. Ritual, song and artefacts reminded people of knowledge they needed to survive (Kelly 2016).

In pre-literate society, knowledge was coded in many forms and could be recalled in more integrated ways albeit at varying levels of detail for those who were initiated.

This book makes a plea for reconnection and remembering our hybrid relationships with others and the land on which we depend. Developing a greater number of connections enhances consciousness (Greenfield 2000, 2003, 2008). My research to date tests the extent to which mindfulness (based on enhancing the capability of people to think about our thinking) could help us to create closer bonds with others based on the realisation that we are interdependent and thus that we need to bridge differences. The notion of the Mobius Band from the string theory workshop hosted by Louis Kauffman[59] (2002) provided a metaphor for understanding our interconnections and interdependence on living systems

[58]The design is sketched in 'Wall Street to Wellbeing' and 'Systemic Ethics' and is extended in the forthcoming volume. West Churchman's critical and systemic Meta approach to working with, rather than within the boundaries of a single paradigm inspires my work. I draw on several approaches that locate Nussbaum's Ten Central Capabilities—relevant to all sentient beings (Nussbaum 2011, 33–34). Nussbaum includes the rights of the most powerless and voiceless. They are young people, children, asylum seekers, the dis Abled and sentient beings who are commodified and traded. The need to address capabilities to protect the planet is extended in this approach by addressing planetary rights and the inadequacies in current architectures of governance and democracy. A case is made for the social contract to be extended beyond the nation state to represent social, economic and environmental justice concerns.

[59]46th Annual Meeting of the International Society for the System Sciences at Shanghai, People's Republic of China, August 2–6, 2002.

(see also McIntyre-Mills 2008). Pragmatic recognition of this fact could be the basis for working co-cooperatively to draw on diverse practical cultural wisdoms based on experiential knowledge. A recognition of the practical value of many knowledge systems allows space for diversity and space for cooperation. Thus the 'clash of cultures' is not inevitable as suggested by Huntington's (1996) thesis.

The Mobius Band (to demonstrate our interconnectedness) has been used as a metaphor for my research and supports the notion of a continuous flow of consciousness as being as relevant to education as it is to community development. Thinking matters when it manifests in our relationships with others and the environment. Inclusive stewardship or exclusive profiteering helps to enhance representation and accountability through exploring ideas and engaging in dialogue.

The challenge of balancing individual and collective needs has never been more complex. Hence the need for a new approach to democracy and governance that takes into account the views of people in local communities.

Research as Both Resistance and Re-generation

This book considers the underlying causes of the sense of alienation expressed by voters who are cynical about the extent to which the majority parties listen to their needs and it makes a case for a new form of democracy and governance that takes into account the linked causes for poverty and climate change, namely an economy that places growth before people and the planet. It proposes an alternative form of democracy, governance and ethics that values the 'web of life' of which we are a strand. In so doing, it strives to respond to the 2030 Development Agenda and suggests a way to enhance representation and accountability by extending the Millennium Goals and UN Sustainable Development Agenda. It outlines pilot studies of alternative architectures for democracy and governance and suggests a way to extend local engagement in social, economic and environmental decision-making. The potential for new architectures for democracy and governance addresses the:

- *A priori* norms and *a posteriori* measures for transformation towards regenerative living.
- Matching social, cultural, economic and environmental decisions to perceived needs with a focus on food, energy and water security.
- Narrowing the gap between perceived needs and the way resources are distributed.

Policy implications

New local forms of engagement and governance (Held 2004) are needed to protect residents against environmental hazards and the subsequent economic and social consequences. Major challenges in local communities include the need to achieve or maintain: (a) access to safe housing (including energy and water), appropriate education and employment; within (b) livable cities;[60] that are in turn supported by (c) sustainable regions.

The book makes a plea for rebuilding relationships and the need to foster collective determinism through public engagement processes. The accountability framework for the 2030 Development Agenda could be supported by the policy framework of UN Local Agenda 21 and the Aarhus Convention which suggests the need for engagement in setting policy indicators, the right to access information and to participate. Social engagement is important for public education, representation and accountability within and across national boundaries, despite the increasing trend towards inward looking policies that do not foster post-nationalist agendas, as demonstrated by the recent Brexit decision. Governance needs to respect the needs of the marginalised through rethinking the implications of narrow forms of evidenced based policy (McIntyre-Mills et al. 2014; Hesse-Biber 2014). Instead, user-centric policy design needs to foster capabilities and introduce measures of 'what works' to protect social and environmental justice—from the bottom up point of view of the service users and not just from the top-down view of service providers—with a view to narrowing the gap between needs and outcomes so as to redress austerity and regenerate opportunities for current and future generations of life on the planet. Individual and collective needs, power distance, gender rights and species rights are all important aspects as is the management of consumption choices, because the sort of consumption choices that are being made are robbing future generations of life of a chance to survive.

The idea of solving all problems with better designs and better technology is fraught with problems, because all designs begin with values. So it is vital to design with the rights and capabilities of human and non-human animals and their habitat in mind.

The increasing loss of land to commercial groups and the moneyed classes is making the rural poor in Indonesia, the Philippines and elsewhere increasingly vulnerable. Once they have sold their land, they often buy TVs or other consumer items and when the money has run out they are more vulnerable than before they lost their land, which provides a means of production. At this point, they become not just members of the poor landed class, but instead members of the lumpen proletariat. The financialisation of the economy has resulted in young people becoming vulnerable to trafficking. Although young people do not starve they desire an income to purchase consumer goods or to pay for the education of another family member.

The ability to earn 150 dollars a day selling food as a hawker in the Bandung was for example contrasted with the earning of about 5 dollars a day from agriculture in Cibodas.

[60]Opportunities exist to include creating a better balance between rural and urban areas and greening cities (Dryzek 1999, 2000, 2010) through the creation of urban agriculture in open spaces, on top of, attached to and within buildings.

The challenge is to find ways to develop opportunities to make a living by adding value to agricultural products or creating something out of nothing as a way to create a viable income. This provides independence and it can be done through thinking differently about the local environment.

In the rural area of Ciangur[61] I was told by the leader of the PKK,[62] a women's group that trafficking was becoming more of an issue in the region.

The tourist areas up in the hills are more vulnerable she said, but the traffickers are extending their networks even here.

The book focuses on ways to de-colonialise knowledge formation in public policy and makes the case for an alternative approach to governance and democracy that takes into account a range of local people's perspectives. Drawing on Hofstede et al. (2010), this research extends the approach to consider the consequences of different cultural values for governing the Anthropocene. The approach is to work with diverse people who have experience of living in urban, suburban and regional areas and to explore the way in which they think about the way in which they think about the production, storage and consumption of food. Unlike Hofstede's research, a qualitative approach is used to explore perceptions. This research is about the culture, context and ethical decision-making. Two basic approaches to ethics are defined. The first is an idealist, normative approach which is guided by values that are a priori and taken for granted. The second is a pragmatist approach, based on considering the consequences of choices that are *a posteriori* and are not taken for granted.

But we need both *a priori norms* to guide us and *a posteriori considerations* to measure the consequences of the UN Sustainable Development Goals to preserve both people and the planet. The notion of so-called 'virtuous living' needs to be explored in terms of:

- A priori norms are guided by accepted rules of behaviour whilst.
- A posteriori indicators and measures of performance. Both are needed in order to guide practices and to hold people and their elected government representatives to account.

The UN Sustainable Development Goals[63] provide the basis against which to measure achievements. But the members of a Focus Group on Food Labelling (and ways to engage the public in addressing food security) stressed that the process needs to be much more inclusive.

---

[61]https://www.statsmonkey.com/sunburst/23644-cianjur-regency-population-statistics-by-gender-jawa-barat-indonesia-stats.php.

[62]As the wife of a Local Government leader she was concerned about the extent to which parents were encouraging daughters to take on domestic jobs offered by prospectors (pimps) acting for traffickers. The number of young women returning from Malaysia and previously from Saudi Arabia (pregnant) and then having unwanted babies has resulted in midwives facing the prospect of finding homes for unwanted babies.

[63]In the capacity of an Adjunct Professor at the University of Indonesia I attended the workshop.

The participants at the event comprised public officials, business men and women and representatives of NGOs and a few universities including the University of Indonesia. The participants at the workshop stressed the need for participation. An elderly spokesperson stressed:

> "The Goals keep changing and we are told about the Millennium Goals then the Development Goals then Development 2020 and what does this have to do with how people see things in Australia, South Africa and Indonesia?"

The options are:

- 'Business as usual'—with rapid descent into climate change and systemic problems
- Making changes too slowly
- Regeneration of habitat for human and non-human animals.

Goal 1. End poverty in all its forms everywhere

Goal 2. End hunger, achieve food security and improved nutrition and promote sustainable agriculture

Goal 3. Ensure healthy lives and promote well-being for all at all ages

Goal 4. Ensure inclusive and equitable quality education and promote lifelong learning opportunities for all

Goal 5. Achieve gender equality and empower all women and girls

Goal 6. Ensure availability and sustainable management of water and sanitation for all

Goal 7. Ensure access to affordable, reliable, sustainable and modern energy for all

Goal 8. Promote sustained, inclusive and sustainable economic growth, full and productive employment and decent work for all

Goal 9. Build resilient infrastructure, promote inclusive and sustainable industrialisation and foster innovation

Goal 10. Reduce inequality within and among countries.

Chapter 1
Beyond Anthropocentricism—Why 'Taming' or 'Tackling' Wicked Problems' is Problematic

1.1 Introduction: How Can We Achieve the Values, Will and Conditions to Govern the Anthropocene?

This chapter makes a case for a paradigm shift away from business as usual and the pursuit of profit at the expense of the social and environmental fabric of life.

'Us versus them' competition is counterproductive and unsustainable. What we do to others and the environment we do to ourselves. Unfortunately, current education remains rooted in categories rather than understanding that categories are merely constructs that can be helpful or unhelpful depending on the way in which the lines of inclusion or exclusion are drawn. If we understand that we stand and fall together then perhaps what we value (and why) will change before—rather than after a collapse of environmental systems that are over exploited by an unsustainable way of life.

The will to power is held by stakeholders in the private and public sectors, but it is possible to transform current forms of democracy and governance. The history of democracy has had many rises and falls as a result of changing circumstances—some of which can be addressed by taking responsibility for transforming the way we live and the way we are governed.

My starting point is to think about the 'taken for granted'. As a social anthropologist and sociologist I have drawn inspiration from people with whom I have learned whilst undertaking fieldwork in a range of cultural contexts. I am concerned about social justice and policy concerns such as vulnerable and displaced people (migration, mental health, TB and HIV, water and sanitation-related disease and homelessness). I draw on experience in South Africa, Indonesia, and Australia and with students across post-national regions.

The issues facing Australia and the region are of trying to pursue 'business as usual' against the tide of social, and environmental crisis. Ironically, in policy terms the Australian government is working with the business as usual 'team Australia' approach to 'win the day'. But the competitive approach to growing business,

J. McIntyre-Mills, *Planetary Passport*, Contemporary Systems Thinking,
DOI 10.1007/978-3-319-58011-1_1

stopping the boats[1] raises many questions about identity, ethics and a denial of our place in the world.

The aim is to explore alternatives and to discuss some ways of achieving well-being whilst the focus is on human rights, discrimination and outlining the notion of a planetary passport. Some key concepts are perceptions, the way in which issues are perceived and the way in which they are framed.

The environment of the problem is often limited so that terms of reference deny that we are living systems (Wadsworth 2010). According to 'the business as usual' paradigm:

> By 2020, Australia's population will increase by up to 14 million. Population growth is a vital ingredient of business and economic growth. Increasing the number of people in the working age population will reduce the burden on each tax payer cause by an ageing population's potential investment opportunities in infrastructure such as water, gas, electricity, roads, housing and related industries[2]

1.2 Accountability for the Loss of Human Security Ought to be the Next Step for Social Justice: The Environment is Eroded to Prop Up the Failing Economy

Biodiversity is threatened by markets that control life, for example the Trade-Related Aspects of Intellectual Property Rights Agreement of WTO (TRIPS) enables the corporatisation and commodification of life:

> Parties may exclude from patentability plants and animals other than micro-organisms, and essentially biological processes for the production of plants and animals other than non-biological and micro-biological processes. However, parties shall provide for the protection of plant varieties wither by patents or by an effective sui generis system or by any combination thereof. This provision shall be reviewed four years after the entry into force of the Agreement establishing the WTO. (Article 27.5.3 (b) cited by Shiva 2010: xviii).

Policy decisions flow beyond spatial boundaries and affect future generations. The refugees from Tunisia and Libya are given the boot on Lampedusa as Italy

---

[1]"Indonesia's previous foreign minister, Marty Natalegawa, was blunt in his criticism of Australia's policy of turning back asylum seekers attempting to leave Indonesia for Australian waters. He was especially forthright about the issue when it included Australian ships entering Indonesian territorial waters…" Kingsbury, D, 2014 'Don't expect Indonesian support for tow-backs 'The drum' http://www.abc.net.au/news/2014-10-27/kingsbury-turning-back-the-boats/5844926 Professor Damien Kingsbury holds a Personal Chair in the School of International and Political Studies at Deakin University.

[2]Is bigger really better? Count, Summer 2014 Issue no 118.

strives to keep control of the refugees. The attitude towards borders appears to be hardening as the economic climate worsens.[3]

By treating water as a commodity, it is traded and people have to pay for an essential resource. We regard air as free to breathe, water is another basic right.[4]

The argument developed in this book is that a new architecture for governance and law needs to find a way to arbitrate across conceptual and spatial boundaries in order to protect the global commons and to ensure not merely the fair distribution and re-distribution of resources, but a different way of viewing relationships.[5]

The current structures are unsustainable. Vast numbers move across borders as refugees from war, natural and social disaster. The risks of leaving remain lower than the risks of remaining even if they face years imprisoned in camps.

How will the law protect those who lose their territories and become stateless? Can they continue to be punished for disasters that could in fact be linked with the toxic life styles and the extreme levels of consumption by the 1%?

Currently there seems to be a strong return to sovereignty and a return to competition for scarce resources at the national level. The systemic effects of carbon emissions and living choices are addressed in this book by means of a revised Tuvalu tests and other forms of management and governance.

But according to Saul et al. (2012: 166): states can only be held to account (at present) if they have signed up to the Kyoto Treaty in 1990 or the United Nations framework convention on Climate Change (UNFCCC):

> ..[I]t is very difficult to establish legal liability for historical emissions (say before the United Nations Framework Convention on Climate Change (UNFCCC) or Kyoto Protocol in the 1990) which have predominantly contributed to climate change, because such emissions by companies or governments at that time were not unlawful. The case for liability only becomes stronger following the adoption of emissions reductions targets since the 1990s, from which point excessive emissions

The need for fine-grained local and post-national regional accountability will become increasingly important. As does the need to continuously re-frame international law to address current policy concerns. At a recent conference on creating the future in Melbourne, November 2012, Professor Susan Greenfield summed up the issue of the plasticity of the brain, meaning that the brain makes neural connections, the more connections, the more mindful we become. It is worrying to find out that research shows that voters tend to vote on the basis of how physically

---

[3]http://www.watoday.com.au/opinion/politics/we-can-do-better-than-italys-warehousing-on-lampedusa-20091014-gwh7.html.

[4]The terrifying notion of having to work to pay for air was explored in a recent Hollywood film.

[5]The past election in Australia which resulted in the Abbott government was run along the lines of 'whatever it takes'. Both parties seemed to try to outdo the other on border protection. But the focus on climate change was almost non-existent. The current election has narrowly returned the incumbent Malcolm Turnbull who replaced Abbott. The regional tensions around food and energy resources play out in the competition in the South China seas which has received a ruling by the UN, namely that China should respect the rights of access to the waters in the South China Seas of all nations and that the rights of the Philippines and their territorial waters ought to be upheld.

attractive the candidates appear, rather than on their ability to join up the policy dots to make competent policy decisions to address the social and environmental challenges (Ariel 2008; Rodrigeaux 2015, pers. comm.).[6]

The making of connections is a result of our own experiences. So even two identical twins will have the potential to develop different connections as result of their experiences. To what extent can the use of conceptual tools enable us to extend our sense of the world and to enable us to see ourselves as members of a wider group to whom we are responsible and with whom we have solidarity?

Can leadership that fosters 'neurogenesis' or making of connections make a difference—or is it all about power and the will to power?

Although activities like playing the piano will help to develop new neural connections, even thinking about playing can develop these connections. So the idea behind this research was that perhaps by getting people to think differently about the way in which they live, and to consider, what if I were to do things differently. The goal was to find out if it could enable the participants who do these mental exercises, to transform the way in which they live? I did this research on mindfulness first with people with complex health, housing and social inclusion needs. Many were addressing unemployment and had a range of associated needs and health related problems—including recovering from drug and other alcohol-related illnesses. Addiction to a way of life can be both a cause and the effect of many other problems, as Beer (1974) stressed in his research. This is why the approach to research needs to be one of engagement and being the change.[7]

The second stage is to address ways to enhance caring for others and the environment through positive and negative sanctions detailed in Chap. 4.

Florini (2003) suggests the potential of the Aarhus convention could be scaled up to enhance accountability. This book argues that social, economic and environmental accounting[8] could be assisted by means of participation to make this so-called triple bottom line accountability viable in ways that meet the Tuvalu test and thus could provide a way to address the rhetoric of the Paris Agenda on Climate Change (2015) and enable voluntary ecological citizenship to take hold as a new culture, backed by some pragmatic indicators of the consequences of our choices

[6]What is worrying is that students asked to rank candidates according to whether they appeared attractive or not actually did so when requested without refusing to do so. This is most likely the result of pressure to comply by a person in authority and is thus an artefact of research design. I hope that the rational students were careful not to annoy their teachers, but not so irrational that they would actually select candidates on the basis of their attractive appearance.

[7]'Systemicracism': discriminatory patterns and practices which involve more than the action of a few individual attitudes, but rather the systemic practices of racism built into society's major institutions (Feagin 2006).The pain of rejection based on categories such as age, gender, culture including religion, ideology or physical or mental dis-ability also trigger anger. So this appears to be a long leap from neural research. But it isn't.

[8]The GFC and the environmental melt down seem to be seen as two problems—not one. Human beings cannot achieve new growth through increased extraction of profit. Instead living elegantly and well through re-distribution and simpler life styles supported by green economies, reciprocal sharing of resources and skills in green communities, supported by energy-efficient systems.

and a way to hold organisations to account. Thus, the participatory action research described in 'Transformation from Wall Street to Wellbeing' (McIntyre-Mills et al. 2014) grasps the nettle to: 'address the challenge posed by the Earth Charter: 'Everyone shares responsibility for the present and future wellbeing of the human family and the larger living world' (cited by Hayden, 368).[9]

Dualist[10] thinking pervades our consciousness and is reflected in socially unjust and environmentally unsustainable designs for society. Designs need to be supported by constitutions, based on a priori *norms*, and consequentialist or a posteriori *approaches*, based on testing out ideas within context and with future generations in mind.

Global axioms to protect future generations and the poor of this generation need to guide legal constructs and legal decisions at the local level.

The principle of subsidiarity could rest comfortably with global axioms—provided that the caveat is made that—*people at the local level can be free and diverse to the extent that their freedom and diversity does not undermine the rights of others or future generations of life.*

This challenge of balancing individual and collective social, cultural, political and economic concerns needs to be buttressed by regional parliaments and courts that focus on social and environmental justice at a post-national level.

How can we increase engagement in the environment that balances individual and collective well-being and simultaneously protect the environment when we know little about this nexus? Stewardship for future generations underpins the philosophy of Aboriginal custodians of the land and their dreaming sites. However, there is evidence that many non-Aboriginal urban citizens wish to spend more time

[9]The challenge to move beyond the rhetoric of cosmopolitan citizenship and to address both justice and sovereignty. The nation state needs to be held to account by an Earth Charter that is supported by overlapping regional institutions (supported by inclusive regional research institutions, policies, the rule of law including parliaments and courts). This book also builds on this research and reflects on how public education could be enhanced through 'if then policy scenarios to help engage people in thinking about the ethical implications of their choices. Ideally, a universal respect for social and environmental justice could enable subsidiarity as a means to support freedom to the extent that it does not undermine the quality of life of this generation or the next.

[10]In many ways, this book is a reflection of lifelong learning. But recently, I also had some time during a short sabbatical to reflect on what it means to know our place and to understand our hybridity and our interconnectedness. The book begins with a critical consideration of the capabilities approach of Martha Nussbaum who is a social democrat whose understanding of property in some ways undermines the capabilities approach that she so carefully proposes, because she does not explore the implications for 'a fairness' underpinning current frames of what property entails. In this regard, the work of Butler comes closer to engaging with the economic system that is the root of the problem (along with the notion of the nation state and the limited social contract). The work of Judith Butler who holds the Hannah Arendt chair, stresses the right to voice our needs and to be listened to and heard. Her starting point is a focus on the other, vulnerability and a recognition that we are vulnerable and should demand the right to be heard and cared for. She starts from the position of vulnerability, not from strength or resilience, because her focus is on interdependency. A rights focus can be very individualistic and this can lead to dualism and this is the problem with the capabilities approach of Martha Nussbaum. However, it does provide the liberative potential as a guideline for human rights. But it does not go far enough.

living slower lives, walking wherever possible, riding bikes, growing local food, recycling and re-using and consuming less, instead of living stressful, competitive lives that save time, but waste resources as they are reliant on fast food in 'throw away' containers, fast travel and a 'time is money attitude' (McIntyre-Mills et al. 2012, 2013, 2014). These findings clearly link environmental health and human well-being and raise the question of what can we learn from mobile and egalitarian, place–attached people (Vaske and Kobrin 2001) whose history is recorded in the landscape? (Guddemi 2006; Rose 2004).

My personal learning is informed by African and Aboriginal mentors who explained the implications of 'colonisation of the mind' (Rose 1992) and how it leads to dualistic thinking that disconnects from the land on which we depend.

First nations remember their stories by referring to landmarks, rather than ink marks on a page. The plants, animals, rocks and rivers hold stories and act as memos for our thinking. Unlike societies who think only in abstract and record their ideas on paper in books, the Aboriginal sacred text is the land. They are its caretakers and in ideal terms according to Rose (1992)[11] they see some life forms 'as family' which makes all the difference as to how they relate to living systems.

Aboriginal people have totems that they do not eat. These totems are their animal and plant family. It provides a way of protecting the land as certain forms of life are not regarded as food by these groups. And this supports the maintenance of biodiversity. Also it makes people identify with these other forms of life in terms of a priori *norms expressed* as spiritual taboos to protect some species at the expense of others. For example, the life chances of a threatened pelican have an emotive impact on those who also wish to protect it against the fur seals who regarded the pelicans as a food source. Another example is the mass death of fish in the River Murray as a result of rising salinity levels which was a great concern to the Indigenous Ngarrindjeri—who stressed the need to stop building weirs that impacted water flow and hence on the well-being of the fish. This is perhaps why Major Darrell Sumner[12] has taken the stance that he has recently taken on the need to cull fur seals who are attacking pelicans—seen as 'totems' and 'family'. Protecting 'us against them' needs to be viewed within the context of past social and environmental injustice.

In Alice Springs, I learned the creation stories of the McDonald Ranges by finding the caterpillar (Mpantwe) dreaming site and appreciating the symbolism of birth,

[11]https://www.youtube.com/watch?v=qp3Ktlfy0Hw&app=desktop.

[12]In advocating for Indigenizing approaches to the environment a critical mindset is needed to ensure that a single paradigm is not considered to have all the answers. The case of the fur seals in South Australia is just one example of this clash of views. Whereas Darrel Sumner advocates for the river Murray, some of his advocacy for the clan totem, the pelican has resulted in his requesting a planned cull of fur seals. Elder-says-he-will-cull-seals-if-sa-government-doesn't/6674590. This has been rejected by the Greens who advocate for humane approaches. http://www.abc.net.au/news/2015-08-05/aboriginal—Aboriginal elder says he will cull more problem seals if SA's Department of Environment won't—ABC News (Australian Broadcasting Corporation) http://www.abc.net.au/news/2015-07-02/plan-to-cull-mongrel-fur-seals-in-south-australia-criticised/6590166 Proposal to cull fur seals in South Australia slammed by Greens, Humane Society—ABC News (Australian Broadcasting Corporation).

death and transformation and butterfly (Yeperenye) dreaming are profound as they trace the cycle of life, death and re-birth. The land is shaped by human beings and natural forces. The latter have lapsed into exploitation rather than protecting the land. Deborah Bird Rose discusses 'colonisation of the land and the mind' as one of the reasons we no longer care for country in the same way as we did when we knew our place and maintained our connections. Her anthropological approach is *multispecies ethnography*. She addresses the human/natural divide by showing how for example the extinction of species tells us something about our own lives. The heat stress of fruit bats and mass extinctions warn quite graphically of the impact of rising temperatures.

The attitudes to plants animals and human beings has been shaped by the 'us/them' competition based on the notion of winning at the expense of others or the so-called 'zero sum approach' in capitalist society. It is applied to plants, animals and human beings who are unwanted and treated as weeds, pests or unwanted asylum seekers. This is a policy choice that could be changed.

The exploitation of sentient beings (workers and animals) and the environment for profit is no longer sustainable. The notion of planetary passport developed in this book is a heuristic device to argue for protecting this generation of life and the next through re-thinking rights and responsibilities in terms of governing our human footprints and protecting habitat.

The starting point for this book (which begins where Systemic Ethics (McIntyre-Mills 2014) ends is that the addiction[13] of the powerful to consume luxury goods is at the expense of future generations of life.

Donaldson and Kymlicka (2011: 253) argue that our exploitation of animals without respect for their rights is at our own expense. It could be added that our instrumental approach to science and industry has exploited the voiceless through endless painful testing and harmful farming methods and abandoning the knowledge of traditional farmers (the majority of whom are women) concerning crop diversity, for example. Modernist science is not the only answer. We need to restore, regenerate and retain spaces for biodiversity if we are to have a more sustainable way of life below 1.5 °C so that territory will not be lost to rising sea levels. Ideally we should do the right thing because we care about people and place, but even if we do not care we need to do the right thing because it is in our own interests to do so because we cannot survive in isolation or desolation.

---

[13]The desire of the rich to consume and the desire of the majority of the poor to leave the ghetto are expressed through emulating the rich. Socio-demographic research shows that debt and bankruptcy in developed nations—such as the USA and nations within the EU—are driven by advertising, easy credit and the desire to 'keep up appearances' by emulating the standards of the very rich (Frank 2007; Wilkinson and Pickett 2009). Making and invoking treaties and conventions that control commodification needs to be a priority and needs to be ongoing if justice is to be maintained. The approach could be buttressed through federations supported by, for example the Lisbon Treaty (Horvath and Odor 2010) which requires that social, economic and environmental legal considerations be met. Unfortunately the structural and process mechanisms of the EU are not able to manage the distribution of power or funding (Rhodes 1997). Clearly if the EU is to survive it needs to be supported by a means to balance individual and national interests with the collective good of the union. More and more members of the union are disenchanted for a number of reasons.

Human beings who are migrants, seasonal workers, homeless, asylum seekers or opting out of mainstream society inhabit a so-called 'liminal zones' (Donaldson and Kymlicka 2011), because they are not recognised members of a sovereign state.

Non-human sentient beings also occupy these liminal zones and need to be accorded rights in line with the capabilities approach. Liminal creatures need protection by human beings and their introduced pets.[14] The xenophobia playing out as border protection is repeated in spaces and places such as ports, train stations, detention centres.

The hunted human fleeing war in Syria and the hunted animal fleeing the wild as their habitat is destroyed in the name of progress (for urbanisation) needs to be protected as so-called 'denizens' with rights (Donaldson and Kymlicka 2011) simply because they exist, not because they have instrumental use value.

Human, animal and plant biodiversity is essential for life as we know it and the notion of governance should be extended to protecting the Anthropocene. Without viable soil, water, air and safe energy supply the food supplies *for all* will be compromised, unless the rhetoric and general agreement at the Paris Climate Change Conference is implemented.

Recognising the 'other' through appreciating that fellow creatures are sentient and that relationships of trust are based on more than voicing feelings. Donaldson and Kymlicka (2011: 25) cite the words of Barbara Smuts who writes of the:

> Presence we recognise in another when we meet in mutuality is something that we feel more than something we know…in mutuality, we sense that inside this other body, there is someone home (Smuts 2001: 308).

This book builds on the arguments developed in Zoopolis (Kymlicka and Donaldson 2011) and makes a case for recognising the rights of animals in a priori terms spanning domestic household animals with whom we have personal relationships, the rights of wild animals to their own territories and the rights of animals with whom we share territory. Encroachment is a matter of perspective and we have couched it entirely in anthropocentric terms. Drawing the line requires working with stakeholders to consider who should be included, what should be included, why and how. For example, Michael spends many days each year in the Channel Country of Queensland working in the gas fields. When he travels across the gas fields he rescues frilled lizards from gas wells by creating ladders (with the right gradient) so that they can ascend to safety. Paradoxically his empathy for lizards is not extended into a deep sense of the implications of non-renewable energy sources.

[14]The relationships between people and places are changed when living with dogs and cats, for example. We feel safer with a dog at our side and we have a responsibility to ensure that the dog does not act as a predator see Philo and Wolch (1998). Cats have protected human beings from plague and disease and they are also beloved members of families. But they can also destroy the vulnerable liminal creatures such as marsupial mice and they make short work of lizards. The management of pets is our responsibility. Dogs and cats are less likely to act as predators if they are given collars and attention and safe environments where they can engage in recreation without wreaking havoc on the local creatures. It has been found that cats do not wander and stray away from home environments if they are well integrated into family life.

He supplies trapped lizards with water and fruit to tide them over whilst they work out that they can indeed escape from the wells. He said the gaze of a frilled lizard down a well connects with the creatures.[15]

[15]He rang me today to say that he found a dead pelican under the car port at the Cosmo Hut in the Channel Country. It is a known spot for water when the lakes are dry. Unfortunately there was no one there to give it water. It died of thirst and exhaustion during the extreme heat in early December. The crows Heckle and Jekyll wait for his vehicle so that they can share his lunch. In this dry part of the Cooper basin, water is at a premium at certain times of the year and their favourite treats are the whole tomatoes that they spear with their beaks. Today he rang to tell me that he gave them mini pizzas which they had not eaten before. They first observed these new offerings from a nearby tree, Then Heckle flew down and folded the pizza in half and then in quarters and flew away. He was followed by Jeckyl who only folded the pizza in half before hurriedly departing. The previous day they had demonstrated their agility with bananas, quickly peeling the banana before flying away with the banana speared on their beaks!

Photo credits

Dragon fly, spider web and lizard in the dry and wet season in the Channel Country near Innamincka by Michael Trevor Rook.

The dry desert months are relieved by rain. The following pictures are also taken by Michael during the rainy months in the Channel Country.

Alternative ways of framing human–nature relationships in legal, ethical and spiritual terms are illustrated by the Bolivian[16] and Ecuadorean constitutions in line with their belief in Mother Earth or Pachamana.

Despite the fact that the rhetoric is undermined by the political reality of business as usual, it provides a first step. The next step is providing the necessary 'planetary passport' that inspired loyalty to the planet as well as a monitoring system to protect living systems.

Cosmopolitan engagement needs to help participants to consider the potential consequences of their choices. These approaches span narrative (to explore complexity and to enable a creative flow of ideas), dialogue and questioning to re-formulate boundaries. Scenarios were also explored spanning 'business as usual, 'small steps towards a sustainable future' and achieving a sustainable future through building up 'wellbeing stocks' (Stiglitz et al. 2010).[17] According to realists the problem with cosmopolitan politics is where do governments of nation states draw the line as to how they support those who seek refuge and asylum? The norms of hospitality are unrealistic they say. Our fates are interlinked. The onus is upon all nations to re-think the notion of democracy and governance. The economy cannot continue to extract profit at the expense of the well-being of current and future generations.[18]

---

[16]The Guardian, Sunday 10 April 2011 18.17 BST http://www.theguardian.com/environment/2011/apr/10/bolivia-enshrines-natural-worlds-rights "Bolivia enshrines natural world's rights with equal status for Mother Earth: Law of Mother Earth expected to prompt radical new conservation and social measures in South American nation ... The country, which has been pilloried by the US and Britain in the UN climate talks for demanding steep carbon emission cuts, will establish 11 new rights for nature. They include: the right to life and to exist; the right to continue vital cycles and processes free from human alteration; the right to pure water and clean air; the right to balance; the right not to be polluted; and the right to not have cellular structure modified or genetically altered. ...The law, which is part of a complete restructuring of the Bolivian legal system following a change of constitution in 2009, has been heavily influenced by a resurgent indigenous Andean spiritual world view which places the environment and the earth deity known as the Pachamama at the centre of all life. Humans are considered equal to all other entities".

[17]Stiglitz (2010) the ex-head of the World Bank along with his colleagues has adopted similar ideas based on his experiences in response to an invitation from Sarkozy to address the poverty, and global financial issues facing Europe. It recognises the systemic social, economic and environmental challenges. Stiglitz et al. (2011: 15) use a multidimensional measure of well-being. According to Stiglitz et al., the essence of the commission's findings is that wealth needs to include 'stocks for the future'. Arguments for and against the national, transnational and supra national organisations have been raised by idealists, pragmatists and realists. A way to bridge the divides is through expanding pragmatism to consider the consequences for current and future generations through addressing well-being and developing governance systems to ensure that stocks for the future are developed through limiting the way in which production processes and distribution practices benefit the minority at the expense of the majority.

[18]Where to form here? How do we work with complexity? Or perhaps we merely need a simpler approach, namely a recognition that what we do to others and to the environment, we do to ourselves. The strong cosmopolitan stresses the importance of recognising the value of the environment on which we depend and thus recognises that we are either part of the change—or complicit in contributing to the problem of 'consuming the planet'. Thus Elliot and Urry stress the

A sustainable local community is determined by a sustainable region in which food, energy and water supplies are considered as major determinants for well-being. No community can be expected to transform from a high carbon life style (or aspiring to this life style) without feeling part of the design process and owning the decisions as to how resources should be used.

The ability to connect with others and country can be learned.[19] Since April 2010 Ngarrindjeri Elder Major Sumner has brought people together annually to dance and heal the spirit of the river for the Ringbalin River Country Spirit

(Footnote 18 continued)

implications of developing a non-carbon economy for cultural transformation. Transformation will require changing the way in which we care for others (Mukta 2010, 2005). This will imply re-conceptualising how we engage with one another and our attitude towards commodifying people, animals and the environment. The neo-conservative market uses people and the planet without care of the consequences. It extends the argument by Kant on hospitality to strangers by stressing the importance of the recognition of transnational agreements and law on environmental concerns and Human rights (Nussbaum 2006). The strong cosmopolitan will define human security (Kaldor 2003, 2004) in terms of meeting human rights for all and avoiding a Eurocentric approach when addressing human rights. Cosmopolitanism as a concept has developed through responding to the critics such as Babar and Derrida who emphasised the difficulties associated with universalism and the colonial and imperialist overtones. The market failures Held (2005: 15) anticipates fall short of what has actually occurred. The problem is not only concerning externalities that are not factored into calculations of the degradation to the environment, it is a way of thinking and 'being in the world' that shifts the extraction of profit to where labour is cheaper and where governments and citizens are less likely to complain about degradation of environment. Short-term profits are made at the expense of future generations. This undermines the conditions of employment so that full time employment becomes less available. The globalised market needs post-national controls to ensure control of the movement funds (see McIntyre-Mills 2011). Enabling a transformation of identity through consciousness raising, humility, and governance is the challenge for ethical systemic governance which has become increasingly difficult in the wake of the systemic crises. Morality becomes politically disposable, because of a lack of understanding of our interconnected existence. Strong cosmopolitans recognise that the containerist vision of the world is over. Our so-called containment anxiety, linked with our identity and need to define ourselves, cannot be addressed by living a schizophrenic existence in which we see ourselves as benefitting at the expense of other nation states. We already have regional conflicts fueled by energy shortages and competition for the last of the nonrenewables. Satellite monitoring from the above seems to be more of a priority at the moment by the USA than space travel. Digital communications are already widely used, but their potential is under tapped as stressed in this chapter and other research on well-being, representation, accountability and sustainability (McIntyre-Mills and de Vries 2011). The greatest challenges are the consequences of inaction that will potentially pose an existential risk to humanity. These challenges include representation of the increasingly diverse populations within nation states along with accountability to ensure that resources (e.g. water, food, and energy) are used fairly, equitably and sustainably in local and regional biospheres. Better forms of engagement are needed to enhance the capability of people to understand that attitudes towards consumption have profound implications for social and environmental justice.

[19]http://www.australiangeographic.com.au/topics/history-culture/2010/09/gallery-the-river-country-spirit-ceremony/aboriginal-river-dance_image1

https://www.youtube.com/watch?x-yt-ts=1421914688&v=v98tPtsIjSM&feature=player_detailpage&x-yt-cl=84503534#t=49.

Ceremony.[20] But the differing world views within and across groups needs to be addressed through enabling space for diversity to the extent that the rights of all sentient beings are taken into account.

Views differ and diversity needs to be accommodated to the extent that the rights of others are not undermined. Protecting 'family' and land is emotive, particularly in the context of colonial and post-colonial governments that have placed profit before the protection of country, as defined as follows by Tom Trevorrow, Ngarrrindjeri Elder, 2002 (cited in Ramsar Planning and Management—some recommendations from Ngarrindjeri experience, Partnerships and Stewardship Group 2014):

> The land and waters is a living body.
> We the Ngarrindjeri people are part of its existence.
> The land and waters must be healthy for the Ngarrinderis to be healthy.

For the stewards of country land/water/sky includes spirits and all living things. Past, present and future are connected through the everyday choices that are made to honour country. My understanding that thinking matters began when I was asked as an honours student to undertake research on why there was a spike in schizophrenia cases in Cape Town, South Africa. Suffice it to say, the so-called 'spike' was a result of the systemic discrimination in Apartheid South Africa by studying the life chances of a network of people who claimed illness and misfortune related to their position in the Apartheid state and sought support within and beyond the healthcare system. I did research on behalf of the head psychiatrist at Valkenburg hospital and in the process re-framed the research brief to work in Guguletu Sect. 1.3 from where many of the patients came. One of the patients introduced me to Adelaide Dlamini, who became a mentor and in turn introduced me to a supportive healing network of diviner healers who demonstrated the power of altered states of thinking and the power of community care. They demonstrated that emotions and thinking shape not just our political live chances but also our physical well-being. So one project flows from the next and is informed by what has been learned from co-researchers. Now the work is around how to translate this understanding of the plasticity of the mind (Greenfield 2000) and nature into better forms of policy and governance as detailed in 'Systemic Ethics' and 'Planetary Passport'.

The need to stress ways to explore knowledge systems together in order to enable non-anthopocentric stewardship is vital (Rigney et al. 2015; Bignall et al. 2016; Hemming, Flinders, Ngarrindjeri Regional Authority, Trevorrow, 2016).

The following quotations illustrate the political context of environmental protection in just one part of South Australia. It serves to illustrate the importance of avoiding binary oppositions or us/them schizophrenic thinking when working

[20]This re-establishes a traditional ceremony that hadn't been performed since white settlement. It means the coming together of the tribes. http://www.mdba.gov.au/what-we-do/working-with-others/aboriginal-communities/ringbalin.

respectfully within and across cultures and disciplines to co-create ethical responses and the different ways in which rights and responsibilities to sentient beings are viewed within the notion of a stewardship discourse. South Australia has a history of colonisation, development and dispossession chequered by belated acknowledgement of Australian citizenship, an Australian constitution that continues to discriminate against First Nations, rejection of sacred sites and a lack of understanding of the relationships across self, other and the environment.

One of the low points is the Hindmarsh Bridge affair in which the sacred knowledge of Aboriginal women elders was disregarded, because it was misunderstood and contested by rival narratives within the Aboriginal community who were not aware of spiritual dreaming stories (Mead 1995 cited in McIntyre-Mills 1996: 75).

The research with Local government came about after the launch of co-designed participatory democracy and governance software to weave together stands of experience (like strands of river grass) and to explore decision making to enhance well-being by with and for Aboriginal service users and providers. It was funded by the Australian Research Council, Neporendi Council and the (then) South Australian Department of Health as a way to enable Aboriginal stakeholders to teach non-Aboriginal stakeholders by drawing on their lived experiences. It drew on the Indigenous philosophy that thinking matters and that many, interrelated factors together shape well-being. It is not something that can be achieved through a health department, because it is about how we relate to others and the environment. The software (designed to map many pathways was launched the day before the Rudd Apology to the Stolen Generation) together with my co-researchers Assoc Prof. Douglas Morgan, Senior Lecturer, Bevin Wilson, informatics designer Denise De Vries and several mentors including senior researcher Kim O'Donnel. Doug was at the time also chair of Neproendi, an advocacy organisation by with and for Aboriginal people living in the Southern Region. Bevin was a senior mentor, because he drew upon his own experience as a previous chair of Neporendi. We said we would like to extend the research to local government as a way of enabling participatory democracy and governance to support social, economic and environmental well-being that included non-Aboriginal Australians. The reason why the middle class area was selected was because the intention of the design was to look at excessive consumers with a view to narrowing the gap between those with heavy footprints and those with light footprints. At this next stage of the research I worked with some other researchers but Bevin remained mentor. The scenarios presented as a basis for the Local government research flowed from what I had learned from these mentors and the many informants in Alice Springs, where I had previously worked and where my mentors stressed the importance of well-being as the outcome of many interrelated aspects that were explained in dreaming stories, such as Mpantwe (Caterpillar Dreaming) and hinted at through place names, such as 'Broken Promise Drive' for the road which cut through a sacred space near Cromwell Drive'. This is the official place name on the Golf Course Estate. The way we think shapes the land and the land shapes us. The irony of conquering nature along with many other narratives by leaders helped inform the design of scenarios which were then explored in the next project.

The notion of spirituality and a sense of the sacred in everyday life was stressed, because the landscape itself carries and reminds people of who they are. I learned firsthand that thinking matters. My reading of Deleuze and Guattari as part of the project to extend the work of the open systems thinker C.West Churchman, came about through meeting Jae Yu, a South Korean, devoted to the study of the importance of translating thinking into practice through being and becoming the change. The connections we make through owning a problem and finding ways to connect with others can be explained as so-called emotional connections or 'assemblages'. We called the collection: 'Rescuing the enlightenment from itself' to highlight the problems associated with linear, closed systems and the harm that it does to others in so many direct and indirect ways.

The implication of the differing ways of seeing the re-growth in seal numbers versus seal population recovery is illustrated below. It has implications for ethics and public administration in that environmental protection agencies need to find ways to work respectfully with interest groups to resolve differences (which may have resulted through histories of oppression and disregard) whilst also ensuring that sentient beings live lives that are worth living. We need to take this connection between people and creatures and the land as central to our survival and it has been understood as core to the First Nations philosophy:

> An Aboriginal elder from the south-east of South Australia has taken matters into his own hands to try to manage the growing long-nosed fur seal population, saying he has already killed several of the animals. Ngarrindjeri elder Darrell Sumner, from the Coorong at the mouth of the Murray River, said the seals have never lived in the area before and are killing the main local totem, the pelican, along with other native birds.
>
> "Just by the jetty there, you'll find four or five dead pelicans just sitting there," Mr. Sumner said.
>
> Down along the Coorong on the barrages, there's dead pelicans everywhere.
>
> The fishermen can also vouch for that when they're pulling the nets in.
>
> The long-nosed fur seals are a protected species and have been spotted 100 kilometres up-river from the river mouth. Their usual domain is the ocean.
>
> There are now about 200 long-nosed fur seals in the lower lakes and the Coorong, which is known for its vast native bird population and fishing industry.
>
> Mr Sumner said he had already killed some of the animals, even though killing one of the endangered seals can attract a $100,000 fine or a two-year jail term.
>
> "I don't care what the Department of Environment says," he said.

The fisherman agree but the Greens are concerned that the species is threatened and is only now rebounding and that any management needs to be done humanely.[21]

[21]ABC News (Australian Broadcasting Corporation) http://www.abc.net.au/news/2015-07-02/plan-to-cull-mongrel-fur-seals-in-south-australia-criticised/6590166 Proposal to cull fur seals in

> Adrian Pederick is putting a private member's bill to State Parliament seeking a program to manage seals because of concerns from the fishing industry and others in his electorate south of Adelaide, a region which includes the Coorong, the Murray and the river's lower lakes…The animals' numbers have soared to about 100,000 in recent years.
>
> Mr Pederick suggested the population was rising annually by about 5 per cent and he favoured a "humane" and stainable shooting program over other tactics to scare seals, which had been used in other states. Conservation groups have said seal numbers are just rebounding after they were almost wiped out by the sealing trade in the 18th and 19th centuries. But local professional fishing crews said the booming long-nosed, or New Zealand, fur seals were attacking their nets to eat the catch and preying on native birdlife, including pelicans.[22]

1.3 Phronesis, Ethics and Designing a Response

Three approaches to decision-making to guide ethical choices are provided and each of these will be discussed in relation to decentering dualistic/colonial mindsets.

- **A priori guidelines (from prior or before you act)** are normative guidelines[23] about what we ought to do based on rules to guide us in a changing world. The first issue is that some of the guidelines may not take into account the complexity of a situation.
- **A posteriori (post or afterwards)**—questioning what would be the consequences of actions[24] if we decide to do this action versus another action. The considerations are based on 'if then scenarios' or hypothetical notions that try to guide our choices. But the first issue is that scenarios are not the same as actual conditions, which could be much more complex. The second issue is who decides and what are the implications in the short, medium and long term for living systems, i.e. people and the planet?
- **Virtue-based ethics** emphasises being virtuous and cultivating inner virtue so as to draw on these characteristics when making decisions. Honour, justice and benevolence are three core virtues (Denhardt 1991 in Preston et al. 2002: 30) but blind application of virtues can lead to problems. We need to foster internal virtues and not only external goods, power and status. The former are the basis

(Footnote 21 continued)

South Australia slammed by Greens, Humane Society—ABC News (Australian Broadcasting Corporation).

[22]http://www.abc.net.au/news/2015-07-01/shooting-fur-seals-coorong-lower-lakes-private-members-bill/6585088.

[23]These include deontological approaches, idealism/Kantian/social justice approaches. Martha Nussbaum, for example draws on Aristotle in order to develop the capabilities approach.

[24]These include pragmatic approaches. For example Peter Singer discusses the consequences of our choices in his book 'One World'.

for achieving the public good. The danger is that as we move towards more privatisation and more 'new management' style approaches that emulate the private sector, the move can be towards indicators of performance rather than fostering the inner goods that are necessary for public service and for achieving the common good in the global commons. The good citizen, as described by Dobson (2003) needs to strive for inner virtue so that decisions can be made that strive towards ideals but also bear in mind that sometimes we need to think through the implications of different personal policy choices by thinking through the consequences in the short, medium and long term and for all living systems. The Indigenous lens is that if governance is based on a recognition that 'people are country' then there would be a different approach to the way we live. Bignall et al. (2016) and there would be no need to put it all together again. Active listening is perhaps the most important form of engagement in order to heal the divides.

1.3.1 Putting It All Together Using Critical Systemic Heuristics

I am tempted to cite Dobson's (2012) article on 'Listening: the new democratic deficit':

> Given that pleasure and pain can be spoken of in the categories of right and wrong then at the very least we should perhaps regard the pain and pleasure noises of non-human animals as political noises.

My research questions were inspired in part by Stuart Hall who asks what is the social, cultural, economic and environmental context that shapes who gets what, when, why and to what effect? As time passed and I was confronted by more and more social, economic and environmental issues I was increasingly inspired by West Churchman's Design of Inquiring Systems Approach to unfolding values and sweeping in social, economic and environmental considerations.[25] CSH is adapted from West Churchman's approach based on 'if then scenarios' and many ways of knowing to consider the anthropocentric and non-anthropocentric consequences in the short, medium and long term. Churchman (1979) developed a Design of Inquiry

[25]The body of work inspired by this approach and aspects of social cybernetics (Bausch, Christakis, Flood, Haraway, Jackson, Romm, Stafford Beer, Van Gigch, Ulrich, Midgley are also helpful in formulating more systemic research on living systems and our place in the bundle of life Most importantly the organic praxis of Shiva, Deborah Bird Rose, Max Neef and Wadsworth on Living systems is increasingly relevant to my current work on ecological footprints and social justice. Ulrich, who distills the work of West Churchman's DIS approach into 12 is/ought questions that help to shape boundary judgments when working with stakeholders and different kinds of knowledge (logic, empiricism, idealism, dialectic and pragmatism) but it does not go far enough.

Approach to enable critical heuristics or considering what if we were to make this decision or that decision. The approach helps to enable ethical literacy spanning the three main approaches to ethics whilst bearing in mind the importance of thinking about many different ways of knowing and the importance of drawing on evidence to inform rational decisions. But Churchman was also aware that emotions, values and perceptions filter the way in which we see the world and the decisions that we take. Thus, he stressed the importance of self-reflection to examine our values and to consider the social, cultural, political, economic and environmental context in which we are making these decisions.

I have also been inspired by Albert Hirschman's three options for action, namely: 'loyalty, voice and exit', but these are no longer sufficient because the scope of current problems is not merely at the organisational or national level. They are at the planetary level. My praxis strives to reveal 'in the small new ways of seeing the whole' (Adelman 2013: 9). This has implications for ethics, democracy and governance. My praxis aims to:

- **Build** the capacity of people undertaking policy research or implementing governance to think about the consequences of their choices for self, other and the environment.
- **Co-create** policy and governance literature based on testing the principle of subsidiarity and Ashby's rule of Requisite Variety and explores the policy implications for complex decision-making.
- **Extend social** theory through re-framing and re-considering boundaries (conceptual and spatial) in relation to social and environmental justice.
- **Contribute** to systemic ethics by: (a) expanding pragmatism through addressing 'what if' heuristics and 'if then scenarios' to enable individual self-reflection, group considerations and pilots of participatory democracy and governance. (b) Considering (i) identity and relationships, (ii) boundaries and flows and (iii) Policy decisions based on drawing the ethical line through questioning taken for granted ideas about the state, market and society together with those who are to be affected by the decisions that respect our relationships with others and the environment of which we are part.

Addressing diversity in perceptions is a central concern of my praxis. My role is one of building the capacity of graduate students in a range of cultural contexts including South East Asia, Canada, New Zealand, for example[26] and undertaking

[26]To date, I have supervised 26 Ph.D. students to completion and several MA theses. Current Ph.D. supervision continues to address the common theme of complex wicked problems. So addressing health, housing and social inclusion in Australia was a logical continuation. Clearly well-being requires thinking about the environment and when the project for ARC linkage grant was completed we decided to extend our inquiry into well-being and the need for protecting the environment on which we depend. Emphasis on principle of subsidiarity; Emphasis on testing out ideas with those with lived experience.

my own program of research[27] by working with diverse stakeholders in a range of policy contexts to enhance social, economic and environmental well-being through addressing, in particular:

- Complex health housing and social inclusion needs and complex social, economic and
- Environmental decisions to mitigate the causes and respond to the effects of climate change.

Wicked problems comprise many diverse variables that are interred related and that are perceived differently by different stakeholders (Rittel and Webber 1984; Flood and Carson 1993).

As the nation state increases in size it is increasingly difficult to respond to the diverse needs of citizens and non-citizens. The nation state only protects those who are citizens through the social contract. But non-citizens are not protected and the environment is not protected by nation states that complete for resources.

My program of research on complex health housing and social inclusion was funded by the Australian Research Council and UHURI and Aboriginal Health. It tested the hypothesis that the greater the level of participation the better the match between service users and service providers.

We tested the principle of subsidiarity (namely that decisions need to be made at the lowest level possible and that the complexity of the decision needs to be matched by the complexity of the decision makers (Ashby's Rule of Requisite Variety).

The research found that participation is an end in itself as it enabled service users and providers to develop rapport around shared activities at Neporendi and when that organisation closed down when it lost its funding the activities moved with the key project leaders to an Aboriginal community and job support project in the same area.

The next step in the program of research was to extend the approach to health and well-being to consider the environmental context of health by adapting and

---

[27]People with whom I have worked: Staff and students at University of Indonesia, National State Islamic University, Padadjaran, Ministries of Social Affairs, Religion and Finances, for example. Emeritus Professor Alexander Christakis, Dr. Ken Bausch, Dr. Tom Flanagan and members of Global Agora team, Professor John Roddick (Informatics, Flinders University).

Professor Ann Roche (NCETA, Public Health, Flinders), Dr. Doug Morgan (previously at Flinders), Kim O'Donnell, mentor at CRCAH, Flinders University and Bevin Wilson, mentor at Yunggorendi, Dr Denise de Vries (Informatics, Flinders University).

Dr Jon Deakin elected Local Government member and APAI for ARC Linkage, Sun Binchai (Hons graduate, Informatics, Flinders University), Ivantia Mokoginta (Ph.D. student, Economics and Public Policy), David Hope (Accountant and risk management consultant for local government), Adib Mohammed Abdu Shomad (Ph.D. student, Public policy and management), Barbara Dickson (MA student, Public Policy and engagement specialist for an urban renewal department, public sector, Adelaide)

Dr. John Mugabushaka (Ph.D. graduate, public policy and youth justice officer, public sector, Adelaide).

scaling up the process with Aboriginal Australians and extending it to include middle- and high-income groups to explore their attitude to decision making on complex needs associated with rising temperatures, rising costs of food and fuel impacting on air conditioning and rising costs of water.

The prototype software (in constant renovation thanks to the work of the programmer Rudolf Wirawan who leads the Indonesian diaspora in Australia) and process involves service users and providers exploring their perceived needs in terms of what they have, what they need, what they are prepared to add or discard and turning points for the better or worse.

The findings stressed that complex health, housing and social inclusion requires a systemic response to support a bundle of social, cultural, political, economic and environmental factors. Our findings resonate with those of Max-Neef (1991).

The problem is the social contract is extended only to citizens within a nation state. It does not protect non-citizens, young people, the majority in this generation or future generations of life or the planet on which we depend.

The limited nationalist social contract is reframed through suggesting what if we could develop a planetary passport to protect the environment of which we are a strand?

What if we could become less tied to only limited ways of knowing?[28] Turok (2012) stressed that the 26 letters of the alphabet and text has made us literate. Our senses have emphasised visual and textual dimensions as a way of knowing. The next iteration of print and digital has altered our entire outlook. The visual made more important than other senses. One could argue that the over emphasis on the rational and the way it has extended the development of our nervous system is a result of our inventions of technology.

Electronic media has indeed changed us. The next steps for science, democracy and ethics could be an attempt to recognise and appreciate the importance of extending consciousness to appreciate other forms of life and to recognise the importance of the environment. The non-anthropocentric appreciation of the land and the non-human animal could be first steps towards achieving legal rights based on an expanded sense of justice.

Turok stresses that the universe and the cell have similar physical properties and the cell is perhaps the most complex unit to understand because through understanding its interaction with the environment and the way in which cells respond to signals gives us clues about the nature of consciousness. This is the starting point of 'Systemic Ethics: a non-anthropocentric approach to cosmopolitanism' (McIntyre-Mills 2014). He stresses that once we begin to understand the universe in quantum terms we will begin to move away from either or/zero/1 approaches and realise that nature is a continuum.

Quantum communicators are beginning to help research the nature of expansion and ongoing emergence and its implications for the way we understand ourselves,

---

[28]According to Turok, quantum technology could bring us closer to nature as it operates in terms of continuums, electrons in motion.

others the planet and the universe. In the digital paradigm, information is displayed as bits. They can be zero or one—the basis for computer language. But nature does not work that way in the quantum world—it is in motion so that continuous numbers can be zero and one and more information than any classical bit can be referenced. This could handle greater capabilities that could better enable the appreciation of complexity. He argues that how we make sense of information is important. More and more information alone is not enough to make a difference. For example, he stressed that Darwin had access to about 10 m of books and he was able to make sense of their contents by developing theory which informed his empirical research. The Quantum computer will be able to carry all potential books, but sense will still need to be made out of the data based on critical thinking. Turok (2012) cites the work of McLuhan and stresses that technology changes evolution: communication technology changes us and he coined the aphorism—'the medium is the message'. He suggested that we have the potential to become more connected and that we could create a global village. But this has not been realised. The social contract is only extended to citizens within the boundaries of the nation state.

The digital communication had led to our 'Eyes being in the grip of our gadgets' at the expense of being in the moment and being connected with others in real time or with the environment. But we also need to realise that technology is merely the design that we have constructed.[29] I agree with the socialist feminist, Donna Haraway who argues that we are the designers and that we need to take responsibility for what we design.

If we design Frankenstein's then we will need to take responsibility for them and we will need to try to redeem our inventions through intervention to rectify or dismantle the inventions that are problematic.

Indeed technology shapes human beings but human beings are the designers of technology and we need to be aware or conscious of our potential to be stewards or destroyers of the environment on which we depend. The conscious decision to make atom bombs or to use missiles that are controlled by drones are cases in point. The control of the technology from a distance by the designers and the minders does not reduce the moral responsibility.

The digital technology has indeed enhanced some aspects of our lives and made others obsolescent as McLuhan stressed and we also need to restore our pre-literate senses.

Can we achieve this transformation through conscious evolution? Humans are indeed the most aware of all sentient beings—human beings are nothing else than evolution become conscious of itself. But we are not becoming more aware—we are actually in denial that we are capable of making choices—including the wrong ones. We are emphasising competition rather that co-operation. Both dimensions are important for evolution.

[29]Mary Shelly—father was a radical political philosopher wrote Frankenstein, the modern Prometheus as an early warning.

This is stressed by the primatologist, De Waal (2009) who stressed that primates evolved through both the ability to compete and to co-operate. He stressed the importance of emotion and empathy for evolution in 'The Age of Empathy'. He explains that the pillars of morality are empathy and reciprocity.

Animals such as primates and elephants (and other sentients) are capable of making decisions based on a sense of fairness. Unfortunately Huxley, Darwin's colleague emphasised competition not co-operation when he discussed Darwin's research.

But technology is not alone the answer—our values are potential enemies. This is why design not technology needs to be the starting point for thinking about how to live differently and how to achieve 'earth democracy' and appropriate non-anthropocentric approaches to governance that protects food, energy and water resources.

The argument about animal capabilities by Nussbaum (2006) in 'Frontiers of Justice' is developed by disputing the notion that the social contract is an adequate basis for ethics, because it does not protect sentient beings who also have rights *simply by virtue that they are sentient*—not because they have a *use-value* or any a posteriori *consequences* for human well-being (such as lack of safe food resulting from the intensive farming of animals in unhealthy confined spaces, leading to their health being compromised and thus compromising the health of human beings). The idealist argument by Nussbaum (2006) argues for their right to a quality of life worth living based on their intrinsic rights to exist. Recent research on the capabilities of primates and other animals demonstrates that Aristotle's work needs to be expanded to show more solidarity with other creatures, some of whom can indeed understand the difference between fairness and unfairness (De Waal 2009).

If animals can understand fairness and unfairness and are capable of empathy, then surely it is time to re-think the social contract which is far too narrowly defined. The social contract extends rights and expects responsibilities to be fulfilled in return. But what about those who are voiceless, disabled, too young or without citizenship rights? What about prisoners? Rights are given only as a basis of reciprocity to those who are capable of giving something. We need to re-think human rights and responsibilities as we evolved from animals and we need to re-learn empathy and appreciate that reciprocity is not the only basis for empathy. In some instances animals will help even if they are going to receive nothing in return.[30]

Dawkins (1976) and De Waal (2009) raise the issue of empathy. Why should we be kind to each other? Is reciprocity only based on the hope that we will get something in return? But co-operation was good for evolution—the uncontrollable urge to co-operate within a group was as important for evolution as the ability to compete with outsiders.

[30] http://www.dailymotion.com/video/xd35bs_frans-de-waal_tech.

The 'will to power' (Nietzsche 1886)[31] and the will to control or survive through dominance is not the only driving force in evolution. In the animal kingdom the power of the Alpha male to compete successfully with other males is often cited as the dominant value in evolution. De Waal stresses that animals have also evolved through their ability to co-operate with one another in small groups. But now human beings are part of a bigger group and we need to continue to evolve through co-operation and to understand the evolutionary importance of co-operation.

De Waal stresses that primates make up after a fight within their group or comfort those who have been defeated. Primates have the capacity to fight and to demonstrate an empathic response to others. He stresses that empathy sits behind altruism.

Dawkins (1976, 1996) stresses that he did not wish to emphasise that individuals are programmed to be selfish only genes are programmed. He stresses that altruism in individuals plays out, because of the way that genes favour the survival of those who are able to co-operate and to compete. This is the best formulation of his argument as it clearly emphasises that people can think about their thinking and their behavior and make choices. The danger is that the so-called 'pillars of morality' (De Waal 2009), namely: 'empathy, fairness and reciprocity' are being lost. How can we justify eating food and not helping others who in need because they are fleeing natural or social disasters? How can we emphasise terms of reference that are exclusive and deny the needs of those who are outsiders?

Nussbaum (2006) in her book on 'frontiers of justice' stresses that the social contract is inadequate because it excludes voiceless beings, children, asylum seekers and animals. But Nussbaum does not develop an argument about the role of capitalism and property as a major barrier to achieving the transformation that is urgently required. In her discussion (Nussbaum 2011) on capabilities she stresses that primarily her work is focused on human development and that she is not necessarily making broad cosmopolitan claims. This book makes the case that enhancing capabilities is vital for cosmopolitan social and environmental justice

Although Peter Singer emphasises ethical consideration for animals he stresses the consequences of not thinking about sentient beings. His emphasis is not on the essential capabilities of animals.

He stresses the impact on carbon emissions as well as the dangers of intensive farming for human and animal health. What could mindfulness and decision making for stewardship entail? It needs to draw on the wisdoms of many ways of knowing, including animal wisdom and learning from the lessons of nature.

In 'Systemic Ethics' (McIntyre-Mills 2014) I suggested that the concurrent challenges could be addressed through loving nature and animals and accepting that consciousness is a continuum that needs to be respected.

[31]Beyond Good and Evil, by Friedrich Nietzsche

www.gutenberg.org/files/4363/4363-h/4363-h.htm The Project Gutenberg EBook of Beyond Good and Evil, by Friedrich Nietzsche This eBook is for the use of anyone anywhere at no cost and with almost no … http://philosophy.lander.edu/intro/articles/nietzsche-a.pdf.

1.4 Design for Meaningful Research

'Framing' research is perhaps the starting point for losing the plot. So, it is important to remain open to an area of concern by considering who thinks it is a concern and why and considering the implications of exclusion/inclusion based on critical systemic thinking. This seems such an obvious issue but as we see from the early discussion on conservative and progressive thinking we can limit it to biological determinism or we can take responsibility for the issue by looking at it critically and systemically. The first step is to consider:

- The environment of the area of concern.
- The lenses through which an issue is 'seen' as filters.[32]

1.4.1 Community of Practice Approach

My role is one of facilitating engagement for social and environmental justice through enabling policy researchers and service users to voice their needs and preferences and to address the policy responsibilities of the various players so that they narrow the gap between perceived needs and policy or service responses. For example, in April 2015 I spent a few days at Universitas Nasional setting up a research consortium on: 'Representation, Accountability and Sustainability'. One of the key themes was the importance of implementing the Paris Agenda on Development by ensuring that participants who are recipients of Aid are able to shape the agenda.[33]

The symposium and consortium meeting was held at the same time as the Anniversary of the Bandung Asia/Africa Conference in 1955. It was supposed to be attended by South African President Zuma. He cancelled his attendance and was represented instead by his deputy Ramaphosa, because of the xenophobic attacks on African migrant workers who are perceived to be 'stealing jobs for locals' or setting up successful shops/businesses that are seen to lessen opportunities for locals. The situation was worsened by comments that 'foreigners could/should go home'. Offers to send migrants home in the wake of the violence in Durban (with the support of government) were interpreted as tacit support by some. This was hastily corrected in public assurances by government.

---

[32]Unfolding values and sweeping in social, cultural, economic and environmental factors is a first step.

[33]In line with the Paris Declaration, the research will be framed together with key stakeholders in public and private secular and non-secular secondary and tertiary institutions. The Paris Agenda of 1997 stresses the importance of involving all those party to research and development to be part of the design process and evaluation of the initiative. The approach will be to engage with students and staff over a three month period through on site observation, participation, focus groups and interviews.

Democracy needs to be worked on each day to ensure that people engage in protecting their rights and asserting their responsibility. The need to demonstrate leadership for democracy is not just the role of the elected members. It is the responsibility of all global citizens. Today more and more people are mobile as a result of natural, social disasters and personal choice. Public education is vital to support the rights of those who are currently unprotected by the social contract because they are non-citizens of nation states. A cosmopolitan support such as education for justice is important within formal educational organisations and through public education opportunities including public arts in open societies and a range of media.

1.4.1.1 Praxis: Believing in Students and Empowering Them to Become Leaders Through Enabling Them to Apply Critical Systemic Thinking and Practice to Diverse, Complex Trans-Disciplinary Issues

This section addresses ways in which my own program of research has been used to facilitate graduate researchers, who are working on their own areas of concern. My role as A/Prof and higher degrees co-ordinator in the School of Social and Policy Studies at Flinders University and as adjunct professor in Indonesia and (recently in South Africa) is as a facilitator to

> influence, motivate and inspire...which 'requires a 'different kind of and level of commitment' and care ...keeping current in one's field, teaching, advising students, overseeing dissertation research and intra institutional governance responsibilities ... can prevent the kinds of long term dedication to a community necessary to effect change. One option, of course is to involve one's students... This has the immediate advantage of training a new generation of researchers ... (Lincoln 2001: 130).

In this way, I echo the notion of 'engaging sympathies 'expressed by Lincoln (2001, p. 130 of Action Research Handbook) by creating an ongoing community of practice with my students, the organisations and learning communities they serve:

> Action research is defined in terms of effecting change together with people who are co-facilitators. Humility and hopeful engagement are the hallmarks of change. As is the hallmark of collegiality rather than academic 'power over'....
>
> this communitarianism, in part an outgrowth of feminist theories of caring and connection, expresses itself as a form of ethics, the first principle of which is the interconnectedness of human life, respect for others, dignity, concern for the welfare of others and solidarity, as well as an abiding concern for ecological matters. The commitment to community- communitas and caritas, or caring concern – has begun to dissolve the old borders between knowledge producing and knowledge consuming elites, and the communities in which they study...the new knowledge is for understanding how to enable democratic action and greater social equality.

My teaching and supervision is designed to help students address complicated policy and management challenges in complicated environments. I facilitate leadership in teaching and learning through developing opportunities to learn what works, why and how and conversely what does not work, why and how in fostering the learning outcomes of international, local and non-traditional graduate students including those with disabilities or from disadvantaged backgrounds. To sum up, my role is one of engaging students, believing in their capabilities and enabling them to build their skills step by step so that they achieve their goals. I draw on the Harmon Doctrine as an example of water policy that ensures that resources are shared across boundaries. Invitations from graduates and colleagues to collaborate are evidenced by workshops for University of Indonesia, University of Padjadjaran, Institute of Technology in Bandung, Ministries of Religious Affairs, Finance and Social Affairs testify. These invited workshops and plenaries and capability to promote critically reflective systemic capacity in graduate students but also provide resources to support decision-making and leadership in relation to complex issues around teaching and learning within universities. Other indications include research with students and colleagues—through government and non-government organisations or providing workshops, for example to West Java Provincial Government, Bappenas and consultants for UNICEF or providing advice/mentoring to Living Hope (linked to the University of Cape Town's Knowledge Management network).

The theoretical perspective developed by Nussbaum (2006) in 'Frontiers of social justice' is applied to address complex challenges within and across boundaries in order to address quality of life and capabilities for all.

My own reflection spanned social, economic and environmental considerations. Climate change is a significant problem in Australia. There is little doubt that accelerated climate change will adversely affect well-being and sustainability in Australia (Flannery, 2005, 2010, 2015; Pretty 2013; Stiglitz et al. 2010)—particularly if we continue to consume at current rates (Davies and World Institute 2008)—resulting in significant devastation and a compromised quality of life. The impact of climate change has been underestimated (Lovelock 2009; Rockström et al. 2009) and local solutions have been overlooked. Aboriginal cultures teach us about stewardship and relationships with the land, but these relationships have been lost in non-Aboriginal cultures.[34] There is evidence that many desire more environmentally sustainable lives, but little is known about the influences on choices around the management of land, water, food and energy supplies that affect the environment. Government responses to human well-being are often based on economic development, which inadvertently increases consumerism, resulting in greater environmental degradation and a heavier carbon footprint. The implementation of coal plants and the lack of local government support for green energy sources are a case in point for both Australia and South Africa.

In workshops in South Africa and Indonesia, I have explored specific questions that flowed from their engagement with our shared area of concern, namely the

[34]As Major Sumner, an Ngarrindjeri Aboriginal elder from the periodically drought-ravaged lower Murray River in South Australia and custodian of the river stresses, we are the land and the land is us. Re-establishing relationships with the land is at the heart of effective cultural ecosystem management (see http://www.mdba.gov.au/what-we-do/working-with-others/aboriginal-communities/ringbalin).

need to respond to urbanisation and ways to live sustainably. Often the challenges are seen in terms of unemployment or food insecurity. Public education needs to ensure that policy makers join up the dots so that a balance between individual and collective needs can be achieved through enabling learners to achieve pathways to sustainable living. Many of the issues faced by South Africans, Indonesians and Australians are issues faced by those in other parts of Africa, Asia and Europe. The droughts in Tunisia, for example has led to protests and the flow of asylum seekers to Europe. Many also seek asylum in South Africa from other parts of Africa.

Thus, the issue of employment and unemployment is one that needs to be placed within the context of the environment of the problem. Namely a sustainable future in which learning supports social, economic and environmental well-being. After sharing my own concerns about the implications for climate change locally and regionally in workshops and showing the interconnectedness of social, economic and environmental considerations by means of a soft systems map, each participant is asked to raise their own areas of concern using critical systemic thinking tools ranging from the simple to the more complex (see McIntyre-Mills 2006, Chap. 4)

The work of Hall (1992) on cultural identity and representation discussed the diversity within the modern state and stressed the need to consider the way resources are distributed and consumed within nation states by asking: What is the social, cultural, economic and environmental context? He has stressed in 'our mongrel selves' (1992) that we need to think critically about who, gets what when why and to what effect. Stuart Hall on critical heuristics stresses that human identity today is not tied to the nation state. As a British citizen who hailed from Jamaican, a child of parents with diverse ancestry who stressed the need to strive for social recognition he won a Rhodes scholarship. The statue of Rhodes[35] has been removed as a symbol of colonial mindsets and the need to consider the direction of education. It is worth noting that in Bristol the statue of Colston (the slaver who traded in Africa) remains despite many protests.

Despite the sense of loss expressed by some that the statue of Rhodes was defaced and de-centred—it is indicative of confidence in the new South Africa where people can speak out and re-frame what is valued, what constitutes knowledge and who should be shaping what is valued. It is also understandable as an expression of frustration. Nevertheless history needs to be preserved so that future generations will learn about the past when the colonial mindset and colonial symbols are decentred. History tends to repeat itself.

> For 21 years democratic South Africa has been struggling to transform an education sector which had an inbuilt bias for over 300 years against the poor but mainly against those who were not white.

---

[35]http://www.theguardian.com/world/2015/mar/20/students-attack-cecil-john-rhodes-statue-south-africa-university-cape-down-questions-race accessed 1/4/2015

http://www.theguardian.com/world/2015/mar/20/students-attack-cecil-john-rhodes-statue-south-africa-university-cape-down-questions-race http://www.theguardian.com/commentisfree/2015/apr/13/cape-town-remove-cecil-rhodes-statue.

> Protests in South Africa cannot quite lose the long shadow of those barbaric days, and the history of the nation has been so imbued with the actions of students - from the Soweto Uprising of 1976 to the beating anger of the 1980s.
>
> Their defiance gets noticed and, depending on which side of the historical fence you currently occupy, these protests can be strangely effective.
>
> Having felled the statue of British imperialist Cecil John Rhodes back in April, emboldened students decided university fees should fall too.
>
> This past week, they took to the streets to complain at plans to raise university fees by between 10.5% and 12% in 2016......If such a united front persists, President Zuma needs to find the means or the money to keep disaffected youths on side.[36]

Curriculum development and learning processes need to be based on testing out ideas in conversation and dialogue with others through creating rapport based on many ways of knowing and by learning from all aspects of nature.[37]

1.5 Policy Opportunity

- What does this mean for constructing new forms of civil engagement to address the question: how should we live?
- To what extent do secular and non-secular curricula[38] prepare students to become interested in local government elections and democratic engagement possibilities?
- To what extent do secular and non-secular curricula support gender mainstreaming and strategic rights of young people?

Capacity building to appreciate a revised form of capabilities is the focus of this book. I advocate public education involves developing people's capability to address not merely their basic needs, but also their strategic needs, namely their ability to address rights and responsibilities. These include addressing complex, wicked challenges of poverty and climate change based on critical systemic praxis and leadership skills to work across cultures, genders and disciplines. This requires knowledge management, communication skills and research skills.

Human rights underpin all relationships within the nation state and within the broader region. This has implications for social and environmental justice. This has implications for social and environmental justice. Brundtland (1989) and Kaldor (in Held 2005: 177) have stressed that human security rests on creating a peaceful

---

[36]Farai Sevenzo 2015, letter from Africa http://www.bbc.com/news/world-africa-34656160.

[37]Reciprocity and empathy can be extended through encompassing more ways of knowing to achieve global enlightenment—that respects diversity to the extent that freedom and diversity does not undermine the rights of others.

[38]The research design was discussed with Dr. Adib Mohammad Shomad and Professor Dede, (Director of Higher Education) at the Ministry of Religious Affairs.

world, enshrined in rights expressed in the United Nations Charter (Article 55, 56) and in the constitution of the European Union. As stressed below (see McIntyre-Mills 2011) rising living costs led to food riots and the so-called Arab Spring, culminating in the Occupy Wall Street Movement.

The move towards re-membering and re-connecting with the land echoes the voices of Indigenous First national movements and the implications for the way we live, our relationships, our sense of identity and for new forms of governance and democracy.

Churchman's (1979) five domains span: logic, empiricism, dialectic, idealism and pragmatism and need to be extended to include:

- Non-anthropocentric knowing drawing on the environment as well as knowing through all our senses—including empathy and intuition along with:
- Respect for 'know how' developed through empirical trial and error
- Creativity in creating sustainable relationships with the land and
- Caring for future generations of life.
- 'Phronesis' or Aristotelian wisdom to match the right knowledge in context so as to empower learners through providing them with appropriate pathways to learning.

In the social context learning to use a design of inquiring system based on basic questions can be helpful as a way to consider where and when to draw the policy line. Testing out ideas requires preserving freedom, space for doubt, diversity and disagreement to the extent that diversity does not undermine the right to freedoms of others. An open approach to design and to research could be extended through acknowledging the praxis knowledge associated with reading the environment and living systems of which we are part. Research needs to span consciousness, cultural studies and systemic praxis to link the notion of relationships between humans and the land as a source of well-being and the broader societal need for environmental protection and effective ecosystem management. The challenges are to address planetary issues. Thus when framing research it is important to realise that the well-being of individual citizens cannot be protected unless the global commons is protected. An axiom to guide secular and faith-based education that fosters social and environmental justice for all is based on considering the consequences of our choices in the short, medium and long term and that we can be free and diverse in our praxis (thinking and practice) to the extent that it does not undermine the rights of others.

Research needs to address a different approach to the way in which we live our lives in terms of our consumption of energy resources, fresh water, transport and meat consumption and the implications for well-being and sustainable living (Pretty 2013). New local forms of engagement and governance (Held 2004) are needed to protect residents against environmental hazards and the subsequent economic and social consequences. Major challenges in local communities include the need to achieve or maintain: (a) access to safe housing (including energy and water), appropriate education and employment; within (b) liveable cities; that are in turn supported by (c) sustainable regions. Some scholars (Christakis and Flanagan 2010; Dryzek 2010) argue that a lack of cultural understanding about our interlinked and entangled lives affects human rights insofar as it shapes access to water, energy and food security.

According to Turok (2014) we need to understand that boundaries in nature can be understood by reflecting on the smallest and largest scales. These provide lessons for policy makers. Just as a cell has pores that open or close to allow in nutrients and to expel pollutants, we need to enable people to understand how to make decisions on what and who to include in the processes of decision-making. The decision as to what to include or exclude in order to make policy that is socially and environmentally just needs to be based on careful consideration, based on questions that address who, what, why, who, how and so what are the consequences.

Capacity building though public education involves developing people's capability to address not merely their basic needs, but also their strategic needs, namely their ability to address rights and responsibilities. These include addressing complex, wicked challenges of poverty and climate change based on critical systemic praxis and leadership skills to work across cultures, genders and disciplines. This requires knowledge management, communication skills and research skills.

There is little doubt that accelerated climate change will adversely affect food security and sustainability. This study looks at rapid levels of urbanisation and the implications for living in cities (Rees and Wakernagel 2008) by comparing and contrasting attitudes in Australia, South Africa and Indonesia. The threats to human security will increase (Flannery 2005; Pretty 2013; Stern 2006; Stiglitz et al. 2010) —particularly if we continue to consume in rapidly urbanising cities at current rates (Davies and World Institute 2008)—resulting in significant impacts on the size of our ecological footprints defined as:

The impact of climate change has been underestimated (Lovelock 2009; Rockström et al. 2009) and local solutions have been overlooked. The aim of the public policy and administration research is to explore the following hypothesis: The greater the level of Public Participation (a) the greater the understanding of UN development goals, (b) the greater the personal application of the goals.

The research aims to enhance strategic decision-making and to develop inclusive development strategies in line with the Paris Development Agenda (1997) which stresses that those on the receiving end of development aid need to be shapers of policy and the UN Sustainability Development Goals.

The threefold aims of this social policy problem are to:

- Analyze perceptions on sustainability with reference to the UN Sustainability goals.
- Deepen an understanding of how people perceive local challenges and experiences (Hulme 2009, 1976; Vaske 2001) to extend social theory on social change and resilience.

- Discover and evaluate a way to engage people through learning organisations to think about their consumption choices as a way to manage the size of local ecological footprints that support the UN Sustainability Development Goals?
- A number of objectives need to be achieved, namely:
- Assessing statistics on consumption of food, energy and water across the public, private and volunteer sectors.
- Addressing the basic framework of UN Sustainable Development goals through PE to create an interactive digital well-being and resilience score card by, with and for local people.
- Setting up a participatory education and capacity building through enabling people to think through the implications of their choices by using 'if then' scenarios that help people to make better public policy decisions.
- Encouraging discursive dialogue and community involvement on human and ecological security pertaining to food, energy and water.

This research explores perceptions on ecological citizenship and rights and responsibilities to manage the size of our ecological footprint on ways to manage the size of our ecological footprint and governing the Anthropocene to assess the impact of participation at the local level (Evans 2014) through awareness-raising about well-being and 'wellbeing stocks', sustainability (Stiglitz et al. 2010), resilience and biodiversity. It examines the value placed on ecosystems (Fisher et al. 2009) and how this impacts on beliefs and intentions. This proposed research will explore the perceived implications of the stewardship of biodiversity for well-being, resilience. Well-being according to Stiglitz et al. (2010: 15) who use a multidimensional measure of well-being spans (1) material living standards (income, consumption and wealth), (2) health, (3) education, (4) personal activities including work, (5) political voice[39] and governance, (6) social connections and relationships, (7) environment (present and future conditions) and (8) insecurity, of an economy as well as a physical nature. Leisure should also be valued. According to Stiglitz et al., the essence of the commissioned work's findings is that wealth needs to include stocks for the future—social, economic and environmental. The way to achieve this is in part through public education.

> Resilience is defined as the adaptive capacity of the physical environment, of an individual or of a group. It concerns factors such as the capacity of members of a community to act

[39]It addresses the UN Sustainable Development Goals and targets and in particular it addresses social inclusion through building the strategic capacity of people to participate to address food, energy and water security. The emphasis (in line with the UN) is to a) address healthy lives and promote well-being for all at all ages through inclusive and equitable public education through promoting lifelong learning opportunities for all to address gender equality and to empower all women and girls. It strives to address ways to ensure access to affordable, reliable, sustainable and modern energy for all in environments that are inclusive, safe, resilient and sustainable. The emphasis is on the most marginalised and focusing on ways to protect the poor and the affected communities, focusing on women, youth, local and marginalised communities to understand mitigation and adaptation processes.

together and to be able to modify or even transform existing ways of life (Rose 2005; Hulme 2009). Biodiversity means:

> Current rates of species loss exceed those of the historical past by several orders of magnitude and show no indication of slowing. Major drivers of biodiversity loss on a global scale are land-use changes and agricultural intensification. These processes are threatening ecosystem functioning and services on which humans depend.... In consequence, there is a deep concern that a loss of biodiversity and deteriorating ecosystem services contribute to worsening human health, higher food insecurity, increasing vulnerability of ecosystems to natural disasters, lower material wealth; worsening social relations by damage to ecosystems highly valued for their aesthetic, recreational or spiritual values; and less freedom ....Growing awareness of the importance of biodiversity for human wellbeing has led governments and civil society to set targets to reduce ...loss (see Convention on Biological Diversity; www.cbdint) (Lindemann-Matthies 2014: 195–6).

The research draws on and extends Hulme's research on why we disagree about climate change (2009) and Cornelius' research on systemic approaches to understanding emotions (1996). Values, perceptions and emotions could be better understood from a systemic viewpoint that draws on interdisciplinary theories, in this case such as to 'why more equal societies almost always do better' (Wilkinson and Pickett 2009). The 'best-worst' option, namely democracy is in need of improvement (Hulme 2009; Giddens 2009). The specific challenge that this research addresses is to match forms of appropriate participation that are accessible to diverse groups and not viewed with cynicism (Cooke 2004). Significantly these questions assess whether participatory democracy and governance enhance sustainable living and well-being. It involves public education through local governments, schools, businesses and NGO organisations. It will identify: the decision-making context; constraints to achieving outcomes; elements of three scenarios (denial of the need to change, too little action too late, sustainable long-term adjustments); and key factors (variables) in facing up to the risks. It could contribute to the SD21 policy initiative 'Sustainable development in the 21st century'. The ability of governments to secure the environmental regions—on which the cities depend—to deliver services and resources will impact on the livability of our communities and human security within our region.

1.6 The Horizon: Trans-Disciplinarity and Cross-Cultural Studies Matter

Stewardship for future generations underpins the philosophy of Aboriginal custodians of the land and their dreaming sites. However, there is evidence that many non-Aboriginal urban citizens wish to spend more time living slower lives, walking wherever possible, riding bikes, growing local food, recycling and reusing and consuming less, instead of living stressful, competitive lives that save time, but waste resources as they are reliant on fast food in 'throw away' containers, fast travel and a 'time is money attitude' (McIntyre-Mills, 2012a, b, c; McIntyre-Mills and De Varies 2012; McIntyre-Mills et al. 2014). These findings clearly link environmental health and human well-being and raise the question of what can we learn from mobile and egalitarian, place-attached people (Vaske and Kobrin 2001) whose history is recorded in the landscape (Guddemi, 2006; Rose 2004). Education needs to rediscover stewardship rights and responsibilities (Flannery 2012) and to become accountable. These goals are considered essential for the governance of sustainable resources by Florini (2003) and Graham (2011) of the Brookings Institute. More applied research is needed at the interface of capacity building, knowledge and biodiversity management, consciousness studies and systemic governance (McIntyre-Mills 2006, 2010, 2014), in order to contribute to new ways to address the systemic social, economic and environmental challenges we face today.

1.6.1 *Learning Communities Contribute to Developing Insight and Foresight to Narrow the Gap Between Service Users and Providers*

In this section, I will share a reflection on an approach to re-framing policy research that has made some contribution to social and environmental justice directly or indirectly. The reflection will stress that what makes the difference is creating opportunities for ongoing learning within the frame of the research project and through being prepared to consider the end point as the beginning of the next phases of facilitation. Some of the more traditional research projects by MA and Ph.D. students as well as my own research and activities have become seeds in the broader policy domain. I strive to share what I have learned with Graduate PhDs to enable them to become leaders and facilitators on their own research projects. In some instances I become a co-facilitator to help students network through sharing contacts to enable the research to take place or by lobbying. Researchers and policy makers need to address ways to learn from a changing environment and how to translate decision-making into praxis at multiple levels applied to multiple forms of intervention. I strive to enable policy researchers to address ways to engage participation in order to address diverse perceptions in order to enhance decision-making to address social and environmental justice. Below I offer some examples of this intention.

1.6.2 *Participatory Decision-Making on Well-being and Climate Change to Enhance Representation, Accountability and Sustainability*

My own research with Aboriginal Australians in Alice Springs (McIntyre-Mills 2003) entitled: 'Systemic Practice for social and environmental justice' resulted in my becoming more engaged in sustainability issues, because mentors such as Olive Veverbrandts and Kathy Abbot taught me the importance of linking physical and mental health with a sense of place and a sense of being caretakers of the land. Policy complexity needs to be understood in terms of many, diverse, interrelated variables understood differently by diverse stakeholders who hold strong values and perhaps see the world differently.

These new insights became the basis for further research on ways to address the concern that the current way of life is unsustainable. My research funded by an Australian Research Council Linkage grant supported by a transdisciplinary team were scaled up with colleagues in Neporendi (a systemic NGO) and the Department of Health in South Australia (McIntyre-Mills with De Vries 2008).

We adopted a PAR case working approach to address complex needs that made linkages to support social, economic and environmental well-being. The usercentric policy design approach addresses complex needs by working systemically and using software to aid the process. When the NGO lost its funding the facilitator and I worked through two others—the domestic violence association and employment agency.

The axial themes, namely: Home safety, Health, Sense of purpose, Connection/belonging (people and place) and Self-respect and confidence are detailed in 'Participatory democracy based on user-centric policy design to address complex needs' (see McIntyre-Mills et al. 2008). The next phase of the research on complex needs (was to pilot the software with local government. This was at the suggestion of a senior policy researcher at the Australian Federal government's Chief Minister's Department after listening to a presentation and participating in a workshop on the research funded by the Australian Research Council to address complex health, housing and social inclusion.

These axial themes in turn became the starting point for exploring ways to enhance governance and exploring ways to enhance the life chances.

This developed into the research with a colleague from Informatics and Engineering at Flinders University on alternative architectures for democracy and governance to promote social and environmental justice, entitled 'Identity, Democracy and Sustainability (McIntyre-Mills and De Vries 2011) to address the life chances of both citizens and non-citizens who are not protected by the social contract as it is currently framed in terms of nation states. Simultaneously an application for funding from the SA local government in 2011 was successful and led to testing the process with service users living in a high-income group with heavy carbon footprints to enable people to think about their lives and relationships with others and the environment.

The assessment overall was that although people were prepared to think about changing the way they live; their shift was to installing solar panels, grey water systems and rain tanks further steps to re-assessing the way they lived would require more support from the public sector and the private sector. They stressed that living slower lives with more time to be human and to spend time with family and friends would be preferable. But the rat race or tread mill was seen as the way to pay bills. In these high-income areas, two person incomes were regarded as necessary to maintain their current lifestyle. But at the same time it was considered to be a problematic way of life. Participants in the research were encouraged to say what they value and why and to use this as a basis for reframing what we value and making a contribution to re-conceptualising social, economic and environmental accountability as we know it today.

Whilst undertaking the research in South Australia, I was simultaneously linked with international Ph.D. students and their host institutions which resulted in invitations to host workshops on representation, accountability and sustainability and offers to form a research consortium to scale up the research in Indonesia. This is currently in progress along with other collaborative endeavours to pilot new architectures with Global Agoras since 2004 Author et al. (2013).

Policy engagement has also been achieved through working with the West Java Provincial Government as a result of post thesis engagement with Dr. Ida Widianingsih to ensure that the principles of the Paris Development Agenda are indeed implemented. The principles stress that development needs to be conducted by with and for people and in terms of their own (not imposed agendas) and in ways that put into practice the Gender Mainstreaming law that is enshrined in the Indonesian constitution.

The program of research provided the basis for Agus Bambang Irawan (McIntyre-Mills and Irawan 2014) to consider suggesting policy change to the curriculum of Indonesian performance auditing in Indonesia and to discuss ways to enhance performance accountability in Indonesia through applying the logic developed with Denise de Vries and Natasun Binchai on ways to develop participatory indicators of social, economic and environmental well-being.[40]

The learnings on the need to take into account social, cultural, political and economic dimensions from my own research on addressing the complex needs of Aboriginal Australians (McIntyre-Mills 2003, 2008) also helped to shape the research approach and design by Ph.D. student Louise Carroll, CEO of The National Foundation for the Deaf in New Zealand on the need to address the hearing needs of prisoners in New Zealand who are not accorded full citizenship rights as prisoners if they have long-term sentences. And thus L.C has advocated for the application of the Convention on the Rights of Persons with Disabilities (CRPD) to the hard of hearing sector globally in her capacity as Board Member with Human Rights responsibilities for the International Federation of Hard of Hearing People (IFHOH). She has been lead author on the CRPD Toolkit for the hearing loss sector which is being implemented globally and is an Observer for the IFHOH at meetings of the UN Committee on the CRPD.

Several applications or 'spin offs' from the research on narrowing the gap between the perceived needs of service users and providers (see McIntyre-Mills and

[40]A participatory engagement process was applied more widely with funding from the South Australian Local Government Association to enable people to think about decision making to address well-being at the local government level through addressing complex social, economic and environmental decisions to mitigate the causes and respond to the effects of climate change. This was also extended to the making policy suggestions to address performance audits and performance accounting in Indonesia with Agus Bambang, funded by Ministry of Finance in Indonesia.

De Vries 2008) entitled 'funded by ARC and SA Dept. of Health and Aboriginal NGO). These include the following, for example:

- Supervision of Dr. John Mugabushaka who acted as a Research Assistant for me during 2012 whilst undertaking his Ph.D. research to address complex challenges in DRC faced by women and children within the context of political corruption and exploitation. He made the case in his Ph.D. that the ongoing legacy of colonialism and the economic path dependency of exploitation that needed to be addressed when striving to support the life chances of women and children. Exploitation at the local level and household level was conceptualised as a synecdoche of exploitation at the national and international level. The rape of resources and the violent extraction of wealth through the military and the mining companies.
- Supervision of the Ph.D. by Riswanda (2015) on the rights of women, children and Warias (transgender groups) in Indonesia has highlighted the need to extend the understanding of rights. Instead of merely considering gender in terms of the relationships between women and men in a patriarchal context, democratic engagement also needs to take into account gender sensitivity towards those who do not fit the binary oppositional male/female category. The other aspect of empowerment to address capabilities is to consider the needs of sentient beings who are voiceless. Advocacy for the rights of creatures was explored with Dr. Adib Muhammad Shomad (Ministry of Religion) and with staff at the Universitas Nasional, in Indonesia, for example.
- Supervision of a Ph.D. currently in progress by Eshanta Loku Ariyadasa on the governance of children's homes for young people in Sri Lanka[41] aims to achieve user centric policy design.

The symbolism of weaving together strands of experience based on the weaving of baskets from River Murray grasses as a metaphor for living with nature and learning to understand that we need to preserve the landscape and to work with it was discussed in a workshop with the West Java Provincial Council. The symbolism stresses the need to take into account many ways of knowing. Knowing based on sharing ideas, (intersubjective), knowing based on technical expertise and knowing based on intuition and respect for nature. This non-anthropocentric knowing is a focus of stewardship. The emphasis is on the need to move from 'Business as Usual' to a new way of doing things. Prior to the publication of the article, 'Reframing performance auditing' with Ph.D. student (Irawan and McIntyre 2014) I gave a series of workshops in 2013 at the Ministry of Social Affairs (3 day workshop) in Indonesia, the Ministry of Finance (half day) and was then asked to

---

[41]Applied to support the design and approach adopted by Eshanta Loku Ariyadasa in Sri Lanka who holds an Australian Government Endeavour award for setting up the best children's home to respond to the needs of orphans. His PhD research is on ways to improve the governance of these homes so that they improve their capacity to govern and so that they enable children to achieve their capabilities as well rounded human beings who are aware of their rights and responsibilities and capable of striving to achieve their goals.

form an MOU with Ministry of Religious Affairs.[42] This was at the invitation of Professor Dede Rosyada and my Ph.D. graduate Adib Muhammad Shomad who compared secular and non-secular education and the need to facilitate critical, systemic thinking and practice. This led to his shaping the agenda of a conference hosted by Ministry of Religion last year at which more than 600 people gathered in East Kalimantan to discuss multicultural education and how Islamic studies can be supportive of Pancasila (a diverse Indonesia). As the only non-Islamic woman who attended the conference as a plenary speaker I raised the theme of urbanisation and food security and the need for not merely protecting well-being stocks but through creating cultural flows that enable us to share and learn from one another. I drew on the work of Shiva (2002) who discusses seed security and water security how these basics of life need to be protected by recognising that they are sacred.

As stressed elsewhere (McIntyre-Mills 2014) the water management approach is a synecdoche for the challenges for managing the global commons. The laws for water management highlight the challenges, but also the way forward for water management. The four theories of water rights were raised by Shiva (2002: 77): The first law is the Harmon Doctrine which only protects the sovereignty of states and is derived from British private property law. Sharing of water for downstream users is only moderated by the suggestion that good neighbourliness is preferable. The second approach is the natural water flow theory based on protecting the rights of up and down stream riparian users. The third is the Helsinki principle[43] based on the equitable apportionment of water, but as Shiva (2002:78) stresses: 'equitable utility is defined as the maximum benefit accruing to all the riparian states, in light of their differing economic and social needs'. The fourth is the community of interest is closely related to the Helsinki Rules on the Uses of the Waters of International Rivers, 'Each basin state is entitled, within its territory, to a reasonable and equitable share in the beneficial uses of the waters of an international drainage basin (Shiva 2002: 79).'

---

[42]This led to his shaping the agenda of a conference hosted by Ministry of Religion in December 2014 at which more than 600 people gathered in East Kalimantan to discuss multicultural education and how Islamic studies can be supportive of Pancasila (a diverse Indonesia). This is important as it us an annual event at which Islamic scholars from leading universities such as Al-Azhar University attend and share policy concerns. As mentioned earlier, I talked about the work of Major Sumner, Aboriginal leader and his focus on allowing the river Murray to flow and stressing the dangers of damming the river through building weirs or walls to enable some communities and private sector companies and farms to profit at the expense of others. This is a metaphor for containerism and greed. The other natural metaphor is the need to protect river grasses that grow in the river Murray and that helps to control levels of salinity as the water levels ebb and flow. Unsustainable development practices have led to removing the grasses. As we talked about the issues at Neporendi, many of the senior woman who are also stewards did weaving. Their telling and sharing stories of their life, their losses and how they had coped with being Aboriginal in a society that did not value their values provided a basis for enabling people to reflect on their lives and a way to address grief and to strive towards well-being. The weaving together of stories and also services to address needs became the core metaphor for user-centric policy design.

[43]http://w0ww.mpil.de/shared/data/pdf/pdf/8helsinki_rules_on_the_waters_of_international_rivers_ila.pdf downloaded 19 May.

The complexity of the use of flows, not stocks also needs to take into account community interests or values that are central to a sense of well-being The notion of the water flows is important in this regard as an example of policy architecture that ensures that resources are understood to span boundaries and that they are not contained within organisations or nation states. Shiva stresses that 'Seed Life' cannot be commodified, nor can 'Water Life'. I gave examples drawing on stories told for example by MA students from Laos who stress that fish stocks are threatened by damming the river for hydroelectric power up stream in China.

At the conference I talked about the work of Major Sumner, Aboriginal leader echoes this learning applied to allowing the river Murray to flow and stressing the dangers of damming the river through building weirs or walls to enable some communities and private sector companies and farms to profit at the expense of others. This is a metaphor for containerism and greed. The other natural metaphor is the need to protect river grasses that grow in the river Murray and that help to control levels of salinity as the water levels ebb and flow. Unsustainable development practices have led to removing the grasses which have a negative impact on the habitat of many forms of life. As we talked about the issues at Neporendi, many of the senior women who are also stewards sat and talked whilst weaving baskets from the river grasses.

Their telling and sharing stories of their life, their losses and how they had coped with being Aboriginal in a society that did not value their values was considered to be the basis for enabling people to reflect on their lives and a way to address grief and to strive towards well-being. The weaving together of stories and also services to address needs became the core metaphor for user-centric policy design.

The participatory design process and the use of metaphors and scenarios that were used in the local government research in South Australia was inspired and adapted from the Mont Fleur Scenario Approach (Kahane 1992) used to inspire policy design for transformation in South Africa.

The scenarios included (a) policy based on denial or 'the ostrich approach', (b) policy based on doing 'too little too late' or the 'lame duck approach' or policy based on making small adjustments for the long haul or the 'flamingo approach'.

Participants in this research were asked to consider the implications of policies along a continuum of denial, too little too late or making small ongoing adjustments towards using resources sustainably and living differently. In the Mont Fleur engagement process policy that pushes for more immediate and extreme changes is symbolized as the flight of Icarus too close to the sun. When addressing the scenario approach with Local Government in Adelaide South Australia it was adapted to enable people to think about 'business as usual' and not making any changes, making 'small adjustments' and living in ways that support well-being. The idea was to establish to what extent the engagement in thinking about thinking made any difference. The approach was piloted with a middle and higher income community with a heavy carbon footprint to assess the extent to which thinking about life style and climate change made a difference to their choices. This symbolism was discussed in focus groups with participants and the need to move from 'Business as Usual' to a new way of doing things.

The process did not lead to substantial change as most of the residents discussed making material changes such as installing rain tanks or grey water systems of solar panels, because these helped to reduce the costs of climate change. But they stressed the challenge was to find time to live differently and have time for slow living, leisure and friends and family. This research in turn provided the basis for Agus Bambang Irawan, a Ph.D. student to consider suggesting policy change to the curriculum of Indonesian Performance Auditing in Indonesia and to discuss ways to enhance performance accountability in Indonesia through applying the logic developed with Denise de Vries and Natasun Binchai on ways to develop participatory indicators of social, economic and environmental well-being through weaving together strands of experience on what works, why and how.

1.7 Conclusion

The areas of concern that I work on span governance, democracy, ways of knowing and ethics, including: (a) Mindfulness and decision-making for stewardship, (b) Design and stewardship for current and future generations of life, (c) Gender-sensitive decision-making that appreciates the spectrum of identities, (d) Well-being and living flowing systems in terms of many ways of knowing (science, arts, cultures, religions and non-anthropocentric wisdom), (f) Conversations for decision making across cultures to enhance rapport and to foster connections and to foster awareness. Through working with many Ph.D. students and my own research teams which cross fertilize and overlap, it is possible to extend PAR to a range of contexts and to foster a groundswell of projects that help to make a difference to social and environmental justice.

Chapter 2
People and the Planet: Implications of Hybridity for Ethics and Consumption Choices

2.1 Introduction

This chapter reflects on the area of concern: How should we live? I start by considering whether well-being needs to be explored in terms of a sense of purpose or in terms of a sense of perceived pleasure, what is valued more:

> contentment and experienced wellbeing, in a Benthamite hedonic utility sense? Or do we care about Eudaimonia or life purpose, in an Aristotelian sense? … While scholars surely can measure both, from a societal and policy perspective, which dimension of happiness should policy aim to maximise? (Graham 2011: 121)

My ongoing research respond to these questions by exploring the ethics of living in diverse environments ranging from the city to the suburb to the rural and undeveloped, wild habitat. Do our daily choices make a difference to our happiness and do people value purposeful living versus pleasure? How do diverse cultural lenses shape our perspectives on purposeful living versus pleasure?

> There is now a world culture but we had better understand what this means. It is marked by an organization of diversity rather than a replication of uniformity (Sahlins 1990: 237).

It explores the way in which local cultural experience frames the way in which the links between people and the planet are understood. Perspectives on hybridity and interconnectedness and the implications for ethics shape what globalisation means locally. The global homogeneity does not only shape human beings 'from above' through powerful agents, it can also be reshaped from below.

The central question posed is what constitutes an ethical choice for living? What should be our guideline? I start by reconsidering Boulding's (1956) General Systems Theory—the skeleton of science which maps the different ways of knowing:

J. McIntyre-Mills, *Planetary Passport*, Contemporary Systems Thinking,
DOI 10.1007/978-3-319-58011-1_2

1. Static structure
2. Dynamic system
3. Cybernetic system
4. Open system
5. Genetic level
6. Animal level
7. Human level[1]
8. Social organisational level
9. Transcendental level
10. Human survival and transformation of values.

The challenge is to find a way to move from level 8 through to level 10. People now respond daily to global events that are informed by exposure to the media and to people who choose to move across nation states as cosmopolitan mobile knowledge workers, tourists, or as dislocated people forced to move from place to place.

Whilst the work of anthropologists today is based on multisite ethnography and the cultural flows that inform daily life, previously it was a much more static experience. Hannerz (2003) for example stresses that 'being there' used to be a way in which anthropology was undertaken over years in one or perhaps two places. Today it is possible to draw on everyday life that spans multiple sites in comparatively short spaces of time aided by multimedia as well as fieldwork telescoped into a few weeks or months that is then continued through Skype or email. This helps to gain a sense of how everyday life is both local and global and infused with its own contextual interpretations. The sense of connection with a wider word is now commonplace amongst most people (even those who not particularly mobile) but who have everyday access to TV images or the Internet. People make sense of their exposure to outside world through the lenses of their own experiences and their own culture. Thus, local cultures can be said to indigenize global culture to a greater or lesser extent.

For example, an Indigenous healer—Adelaide Dlamini— who mentored me as a key informant whilst I did research as an MA student at the University of Cape

[1]Harari (2015) argues that: "We humans control the world because we live in a dual reality. All other animals live in an objective reality. Their reality consists of objective entities, like rivers and trees and lions and elephants. We humans, we also live in an objective reality. In our world, too, there are rivers and trees and lions and elephants. But over the centuries, we have constructed on top of this objective reality a second layer of fictional reality, a reality made of fictional entities, like nations, like gods, like money, like corporations. And what is amazing is that as history unfolded, this fictional reality became more and more powerful so that today, the most powerful forces in the world are these fictional entities. Today, the very survival of rivers and trees and lions and elephants depends on the decisions and wishes of fictional entities, like the United States, like Google, like the World Bank – entities that exist only in our own imagination" July 24, 2015—Uploaded by TED https://www.ted.com/talks/yuval_noah_harari_what_explains_the_rise_of_humans/transcript?language=en#t-68345.

Town commented on TV footage as we shared a cup of tea and watched the marriage of Lady Diana Spencer.

Her home in Guguletu Section 3 was located in an Apartheid ghetto, not far from a large green painted Mosque on the Cape Flats close to Cape Town's airport. This area was home for local residents living in so-called 'matchbox houses' and the migrant hostels for workers permitted into the Cape in compliance with the infamous 'pass laws'. Once their employment contract was completed they were expected to return to their so-called 'homelands'. The pass laws were central to the maintenance of Apartheid.

When she viewed the crowds lining the streets, she commented wryly that she was glad that she did not need to feed all the people attending that wedding! Elements of the procession were particularly meaningful to her, such as the plumed feathers with which she identified as similar to the feathers on her 'xhentsa' or dancing stick to symbolically brush aside negative illness of the spirit.[2]

Her comments reflected her everyday experience as a resident of Guguletu and a healer who led a Diviner School. She regularly organised healing dance ceremonies for those in liminal or transition states. These healing dance séances or 'inthlombe' were hosted for those who felt called to heal themselves and to become healers of others (see Van Gennep 1960; Turner 1987; Mills 1983, 1985a, b; McIntyre-Mills 2000).[3] This liminality was expressed as a sense that life needed to change and symptoms were presented to her as a physical or mental disease sent by the ancestors. The healing journey is called the process of 'ukutwasa'. The ancestral calling is one of cultural transformations from one state to another. Each stage of the transformation was marked by a 'rite of passage' (Van Gennep 1960) and celebrated by sharing food, brewed maize beer and protein from a ceremonially sacrificed animal. A feast marked each stage of an initiates learning process. Prior to urban living this included learning to become herbalist and diviner of problems by appreciating the potential and pitfalls in a shared habitat including people within the living systems. She explained that being a diviner of the causes of ailments of the human spirit also required understanding our sense of connection with others and our interdependency.

In the above quotation, Sahlins (1990, cited in Moore 2009) explains that although people are exposed to the so-called 'global homogeneity' they do so very differently in each local context and they draw on their own values and experiences to make sense of what they see and what they experience. Although a culture of globalisation is shaped through global media and mobile elites, it is also shaped by the everyday decisions taken by local people and local communities about their

---

[2]Her connections to the air, the land and the sea were symbolically expressed through a feathered head dress, skin from a sacred cow on her ceremonial drum and a shawl made of sealskin.

[3]The events involved dance accompanied by polyrhythmic drumming to help induce a sense of altered consciousness and receptiveness to communion with the living as well as past generations of ancestors.

environment. In this way the local cultures change or 'indigenize' aspects of global culture. 'The new planetary organization' that he refers to be described as:

> 'a Culture of cultures,' a world cultural system made up of diverse forms of life… (1990: 237).

Culture is the way in which human beings respond to one another and their environment. It is adaptive, it shapes and is shaped by conditions. It cannot be confused with biology or be said to lead to inevitable political or cultural connections or divides unless people think about the consequences of their decisions in context.

The approach used in this book differs from the notion of culture as made up of separate dimensions (Hofstede et al. 2010) but like Triandis (1995; Triandis and Gelfand 1998) I argue that human beings have both the potential to compete and to cooperate and that the challenge is to develop the capability[4] to balance individual self-interest with collective concerns through enabling better decision making about how we choose to live. This is based on the axiom that we can be free and diverse to the extent that freedom and diversity does not undermine the rights of future generations of life (McIntyre-Mills et al. 2008, 2014).

Thus, this chapter is about the culture and ethics of place. Two basic approaches to ethics need to be defined. The first is an idealist, normative approach which is guided by values that are a priori and taken for granted. The second approach is a pragmatist approach based on considering the consequences of choices that are a posteriori and are not taken for granted. We need both a priori norms to guide us and a posteriori considerations to measure the consequences of UN sustainable development goals to preserve both people and the planet. The notion of virtuous living needs to be explored in terms of: (a) A priori norms are guided by accepted rules of behaviour and (b) A posteriori indicators and measures of performance. Both are needed in order to guide practices and to hold people and their elected government representatives to account.

Drawing on Hofstede et al. (2010) my research extends the approach to consider the consequences of different cultural values for governing the Anthropocene and the way in which power and distance are understood. Instead of understanding the concept as relationships across people, the concept is extended to consider relationships with other forms of life as well. The approach is based on reflecting on diverse life chances of people who have experience of living in urban, suburban and regional areas and to explore the way in which they think about everyday decisions, pertaining to production, storage and consumption of resources and the consequences for social, economic and environmental well-being in terms of their own household, community and habitat.

---

[4]As human beings we have the capability (Nussbaum 2011) to respond to social, cultural, political, economic or environmental challenges by cooperating, competing or understanding our interconnectedness and then designing a new culture that transforms our ability to survive as stewards of living systems.

A meeting with a Sundanese healer in Bandung (who is a school friend of an academic at Padjadjaran University) discussed the notion of what he thought being a virtuous human being entails and how this virtue relates to citizenship rights and responsibilities. Our in depth conversation is detailed elsewhere but he underlined that protecting habitat ought to be the role of the virtuous citizens and that in Indonesia it is important for the well-being of everyone.

He stressed the many layers of 'being human' that are recognised by Sundanese culture and that connecting with others includes connecting with animals, the natural world and our spiritual selves.[5] We discussed the meaning of being virtuous and living well and he said he also believed it was being true to a calling and recognising the threads of connections that bind us together.

Banner stating 'Stop Global Warming: Save the Earth' across a congested street in Bandung close to the University of Padjadjaran

Making the choice to use public transport, such as trains, could be regarded as an example of being 'a virtuous citizen', as suggested by Professor Bambang about living choices he made as a resident of a commuter area near Jakarta, because it is focused on doing the right thing by minimising the size of our carbon footprint by not 'driving a vehicle with less than 3 people' and he stressed that this experience could also be pleasant:

---

[5]So many people living in urban environments feel lost. Hence the title of my forthcoming volume, entitled: 'Getting lost in the city'.

> When I insert my ticket [in the train] I feel like a good person. Also people these days are more likely to stand up and make space for people on trains. They are more polite. This makes using trains a more pleasant activity and perhaps more people will be encouraged to use public transport as a result.

Focus group discussion at the University of Indonesia in 2016 organised by Prof Bam bang

2.1.1 Decentring Anthropocentric and Ethnocentric Mindsets and Learning from Country

Country is an Indigenous term for interconnected living. The research journey reflects on the way our lives connect with others and strives to learn about the 'memory space' (Kelly 2016) of Indigenous cultures who rely on oral histories connected with people and country. Ironically, our memory space although widened by the Internet has contracted in terms of our appreciation of geographical space and our interdependency as living systems.

Knowing our place in the universe and recognising our hybridity is a series of self-reflections and essays drawing on many diverse ways of knowing. It includes papers written during my sabbatical in Japan, South Africa, UK and Indonesia. The journey provided time during the sabbatical to spend time comparing and contrasting social, economic and environmental challenges and to reflect on what unites us.

The first part of my sabbatical enabled me to complete some reading, whilst the second part enabled me to develop and map out the next stage of my programme of research on representation, accountability and sustainability. To this end I spent time in Japan, South Africa, United Kingdom and Indonesia where the widening gap between rich and poor is evident in Yokohama, Cape Town, Gauteng, Bristol and Jakarta. Challenges for both South Africa and Indonesia include urbanisation that will result in growth of cities and 65–75% of the population living in city environments. This could result in the creation of food deserts and competition for resources. Whilst undertaking a 3-day workshop with the Ministry of Social Affairs in Jakarta (MOSA) one of the key areas of concern that was raised by the participants was food security.

2.1.2 Consciousness of Who We Are and What We Stand for

Whilst in South Africa, I explored the cultural museum in Pretoria and the Arendt nature reserve to view rock art and to explore the notion of mindfulness and consciousness. It resonated with my experiences in Kakadu visiting the rock paintings of Aboriginal Australians who as hunters connected with the land and the animals on which they depended for food.

My introduction to the Kung and trance dancing was provided by Professor Martin West at the University of Cape Town in my first general introduction to social anthropology. As an enrolled law student I decided to attend his lecture which resulted in my deciding to read social anthropology rather than law taught within Apartheid South Africa at that time.

Katz (1987) a psychologist worked together with a social anthropologist to learn more about the San/Kung approach to healing and well-being. He wrote about their communal way of life. The book 'boiling energy' by Katz (1982) does justice to the relationships fostered between the Kung and the importance of resolving differences through humour and dialogue. But the connections between the Kung and the animals they hunted and environment from which the women gathered food is not given enough attention. The relation between Gwa (drug to help train the mind to enter an altered state), num the energy that could be used for healing and Kia the state of altered consciousness is discussed convincingly. But without getting to the heart of the matter, namely that by intuiting the animal nature of the creatures they hunted they were trying to become one with their spirit. At a pragmatic level this was to ensure a better outcome for the hunt. But it also underlined their understanding that they were dependent upon the land and the animals. As their access to the land and hunting rights decreased as the surrounding farmers enclosed the land, the Kung has had to sell their labour and buy their food. They are right on the edge of the capitalist economy and in some ways, the Kung's way of life has changed dramatically. The healing ceremonies are now used to serve not only the Kung, but have become commodified by some healers who prefer to sell their services. Many serve their own community and also sell their healing services. And so the spiritual dimension of their lives becomes at risk of becoming a fee for service.

Further reading and research indicates that ceremony helps to remind hunters of animal behaviour and to re-hearse successful hunting techniques. The mindfulness technique is also used when healing. It encourages people to think differently about themselves and their relationships with others. Katz explained that energy or num wells up within the healer as he enters an alternative state of being. This sense of connection through 'awareness' (num) is to achieve an altered state of consciousness (Kia). The mindfulness approach that is currently gaining momentum resonates with this attempt to achieve an altered state of consciousness without any form of drug to aid the process. Katz stresses that a drug based on the roots of a plant (Gwa) is only used as a means to an end, namely training initiates to think differently and to 'let go' of their current state of consciousness in a journey of learning to be a healer.

The so-called 'boiling energy' of the Kung is an enhanced or trance state which brings about trembling and sweating indicative of being able to transmit energy to another through laying on of hands. But the main point of the ceremony is bonding to heal wounds within the entire community through ritual dance, clapping and singing and to reconnect with their animal selves. Men as hunters connect with their prey through closely studying them, imitating them and trying to think like them. Hunger was the result of unsuccessful hunts by the men. Their diet was based on staples—roots and fruit, small animals gathered or caught by women to supplement their diet. Understanding their links with animals and the environment was central to their daily spirituality. So connecting with the land and with animals was central to their Kung identity. These are themes that were not discussed by Katz, because as a community psychologist he was focused on anthropocentric themes. The San lived close to the land and depended on it for their survival. Through altered states of consciousness, they became one with the beings they hunted,[6] in order to survive in a dry landscape.

The notion of energy (num) was understood to heal others and to enable people to connect with the life force through entering into a trance-like state.

As West Churchman stresses we are a part of our subject matter and if we do not examine our own values, namely religion, morality, politics and aesthetics then we are likely to view the world through these lenses without realising that we are filtering the world and perhaps not understanding what we are seeing. Nevertheless he did gain many profound insights by being with the Kung and realising many of the limitations imposed by Western culture. The lenses of the healer made him focus on some aspects not others. Just as his gender limited his connections to women. Listening carefully, working with a colleague who spoke (more of) the language Katz gained insights over the few weeks that he spent with the Kung,

[6]http://www.biomedcentral.com/1472-6882/14/240 jstor rock art https://www.aluka.org/heritage/collection/ra#san, https://www.youtube.com/watch?v=et_xtm9msx0.

because he was able to build on the deeper insights of his colleagues who worked there for much longer periods of time. All researchers need to be aware of their limitations. What did resonate was the importance of simple grounded reflections on what he saw and heard. The voices of the people were central in telling stories about a way of life that was being eroded.

The past requires us to look down at our roots and ruins under our cities to which we will return. It also requires us to look up and to realise that past civilisations or past life forms exist within our solar system of which we are a small part. The past provides lessons from past civilisations and life forms. The present requires us to look to the hungry and thirsty people crossing from Africa and Syria to Europe. The future requires us to contemplate and to say: 'there but for the grace of god go I, so what can I do differently'?

A journey back in time to the belief system of the San reminds African people that from the earliest times we communicated with the land and believed in our hybrid connection with the land and the animal life on which we depended for food. The following photographs were taken at the Pretoria Cultural Museum and at the Arendt Nature Park outside Cape Town in 2015.

The archaeological digs at Schroda were emphasised in the same exhibition as examples of material culture that predates other finds demonstrating early civilisations in Africa many thousands of years before white settlement.

Other sections of the cultural museum celebrated the connections across old and new art forms. Clay pots spanning centuries of design were displayed alongside a rich diversity of artefacts from the many religious and cultural groups in the City of Tshwane (Pretoria). The mosque and Hindu temples in Marapastad were celebrated alongside the role played by early places of education as Fort Hare in raising the consciousness of human rights activist against the apartheid system.

The empowerment associated with being African and black and not a white settler/colonialist was recognised by the Black Consciousness Movement. But Mandela stressed the notion of Ubuntu as a way forward rather than conflict.[7] This truth and reconciliation approach supported by Bishop Desmond Tutu could wear thin if the gaps between rich and poor are not addressed. The statue of Mandela at the Union Buildings overlooking Pretoria has become an icon for people to visit as a place to renew their understanding of how far South Africa has progressed, but it is also a reminder of how far it needs to travel. I looked across to the hills opposite the Voortrekker monument stands and recalled the speech many years ago given by

[7]Nelson Mandela engaged in dialogue with journalists or as a negotiator, in order establish a conversation and a rapport. This was the basis for working with other people. It epitomises 'being the change' and the importance of developing relationships and maintaining relationships through treating people as ends and as whole people, not merely as a means to an end. In this way he brought about change through trying to establish or create connections with very different people from many cultures and different walks of life. He did not strive for a simplistic approach based only on competitive yes/no dialogue. Some of his principles were fixed but he was always prepared to talk and to respect aspects of people's humanity even if he did not agree with them.

Eugène Terre' Blanche, leader of the resistance movement called the 'Afrikaans Weerstand Beveging'. I had attended it together with colleagues in the University of South Africa sociology department. I worried that his eloquence may lead to a larger following of his racist supremacy ideology.

At the time, I reflected on the past as I stood with my guide Mr. Lucas who expressed his current concern about the way unemployment threatens all South Africans, despite the progress that had been made to date.

Whilst at the Cultural Museum in Pretoria, I reflected on the role of doctors who were complicit in the silencing of Steve Biko, the Black Consciousness Leader. I recalled how I used to walk past a police station in Nelson Mandela Bay[8] (then known as Port Elizabeth or I Bhayi in Xhosa)—aged about 12—en route to the Walmer Library. It was in this police station that the assaults took place.

The role of doctors in South Africa has become notorious. But we need to remember that what happened in South Africa is being rehearsed elsewhere. Personal reflections on my own experiences provide the basis for suggestions as to how we could address better ways to govern the Anthropocene through not being silent and speaking out in the name of participatory democracy to address the human rights of people displaced through natural and social disasters, described by a conservative commentator as the:

> naïve rhetoric of participatory democracy. In that deadly phrase the adjective completely annihilates the noun. (Sheridan 2015).[9]

The need to speak out in nation states that do not necessarily represent the view of the vulnerable becomes increasingly important as legislation such as that aimed to silence whistle-blowers working within Australian government or contracted to government is evident in the current legislation[10]:

> More than 250 doctors, nurses and psychologists have protested against laws that threaten jail for those who speak out about violations in detention centres. It follows a similar protest in Melbourne last week.
>
> Medics and their families packed the steps of Sydney's Town Hall and covered their mouths to send a message of opposition to the secrecy provision of the federal government's new Border Force Act. Under the legislation 'an entrusted person' must not disclose protected information, with a breach potentially resulting in two years' jail.

---

[8]Previously called Port Elizabeth before the name changes to reflect post-colonial and post-Apartheid South Africa.

[9]Left wing lunacy in Britain. *Inquirer*, Weekend Australian p. 15.

[10]http://www.theguardian.com/australia-news/2015/jul/11/australian-doctors-rally-over-threat-of-jail-for-speaking-about-asylum-seeekers.

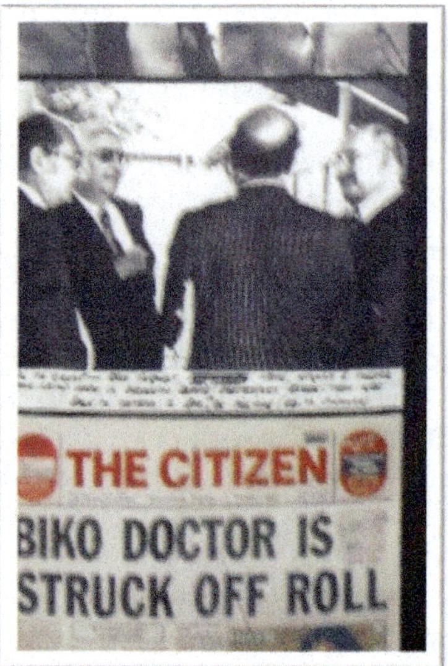

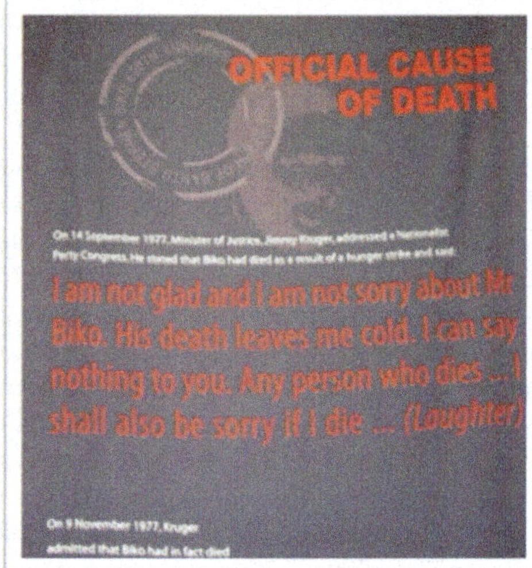

Mandela engaged in dialogue with journalists or as a negotiator in order to establish a conversation and a rapport. This was the basis for working with other people. It epitomises ‘being the change’ and the importance of developing relationships and maintaining relationships through treating people as wholes, not merely as a means to an end. In this way he brought about change through trying to establish or create connections with very different people from many cultures and different walks of life. He did not strive for a simplistic approach based only on competitive yes/no dialogue. Some of his principles were fixed but he was always prepared to talk and to respect aspects of people’s humanity even if he did not agree with them.

The ability to see the big picture in terms of what we value and why, ranging from social thinkers to physicists[11] should be

[11]Held (2005) cites data cited from the World Bank that three billion people live on less than $2.50 per day. The will to make a difference is the challenge (Held 2005: 33–34): “We may lack the will but it cannot be said that we lack the means. … What do we require to make a substantial difference to the basic wellbeing of the world’s poorest? Again the statistics are available. … Required would be $6 billion per annum on basic education, 9 billion per annum for water and sanitation, and 12 billion per annum for the reproductive health of women, 13 billion per annum for basic health nutrition. … These figures are substantial, but when judged against the major consumption expenditure in the US and EU they are not beyond our reach. Moreover if all the OECD agricultural subsidies were removed and spent on the world’s poorest peoples this would release some 300 billion per annum. … In addition a small shift between military and aid budgets – 900 billion and 50 billion a year globally would make a marked difference to the human security agenda….” But the Sydney Peace Prize winner Shiva (2002) sums up the challenge for this century as one of preventing the commodification of life and stresses the need to learn from nature and across disciplinary specialisations.

encouraged.[12] The statue near the Union Buildings symbolises a big hearted approach to life's challenges and a leader able to embrace diversity and to rise above Apartheid to reframe the South African Rainbow Nation.

[12]"... [A]all the indications are that the universe is at its simplest at the smallest and the largest scales. ... One can draw further parallels with the selfish, individualistic behaviours that are often the root cause of our environmental and financial crises. Within physics I see the idea of a 'multiverse' as a similarly fragmented perspective, representing a loss of confidence in the prospects for basic science. Yet I believe all of these crises will ultimately be helpful if they force us, like the quantum physicist, to remake our world in more basic and far sighted ways. ... If we can only link our intelligence to our hearts, the doors are wide open to a brighter future, to a more unified planet ... to quantum technologies that extend our perception". (Turok 2012: 256–7).

According to Shiva (1988, 2002, 2011) multinational companies and the World Bank and the World Trade Organization have made it possible to patent the conditions of life and to link the so-called Green Revolution in India with terminator seeds that do not self-generate because they have been genetically modified, resulting in spiralling costs associated with the purchasing of seeds and the pesticides needed (despite the claims by the manufacturers). Shiva (2011) argues that this has led to many farmers being unable to afford to buy seed and that some are driven to suicide. Shiva argues furthermore that the attempt to criminalise farmers who store old varieties of seed could lead to undermining the seed diversity and that this could lead to increased food insecurity as a result of vulnerable monocultures of foods. The idea that the very basis for life – seeds and genes –can be patented is part of the process of commodifying people, animals and the fabric of life. Palombi (2007) stresses that patents supported by international trade undermine both the developed and the developing world through eroding human and planetary health. Shiva's (1988) praxis (like Gandhi's) is to find ways to intervene where it is most needed.

The images above and below appear at the entrance of the University of South Africa as a reminder of the role of education in building the future. Recently, the fee protests in South Africa led by the so-called 'born free group' of students focuses on the need for free, Indigenous education. Some of the most strident call Mandela a so-called 'sell out', because he did not support the nationalisation of resources, such as mines in South Africa. This has been linked with other criticisms with the current ANC based on a more nuanced response by Xolela Mangcu, UCT sociology professor and winner of the Oppenheimer Award:

> Despite my criticism of Mandela, I reject the idea of him being a sell-out. Nobody spends 27 years of their life in prison and deserves to be called a sell-out. It's ridiculous.[13]

Xolela is currently writing another biography on the life of Mandela. In an article discussing his award, he stresses that he criticised Mandela for his lack of emphasis on race and that he had responded critically to an early article written by Mandela on the need to 'abandon a focus on race'. He disagreed with Mandela and stressed the need for Black Consciousness. He is also inspired by the work of Steve Biko.

[13]Omar, Y. 2016. http://www.uct.ac.za/dailynews/?id=9777.

To his credit, Xolela stresses that Mandela invited him to his home to discuss his views on race and that he listened respectfully to his views and that his own views on race had 'evolved':

> One of my first interactions with Mandela was about an article he wrote called "Wither, Black Consciousness", which he wrote in prison and was published in a 2002 book by Mac Maharaj. I wrote a strong response, criticizing his criticism of Black Consciousness. He invited me to his house and said his ideas had since evolved, that he wrote that article in 1975, but that after many years he had begun to appreciate the power of Black Consciousness in giving pride and dignity back to black people.

He then goes on to say:

> The current leaders of the ANC could learn from him. He appreciated criticism; the current leadership seems to have adopted a 'you're either with us or against us' mentality…
>
> I've always been very critical of Mandela's avoidance of race. The re-emergence of racism in this country is evidence that this approach does not work. It is no longer justifiable.
>
> Not that Mandela was perfect… But where Mandela had Mbeki, Tambo and Sisulu to hold him in check, the current leaders have nobody to counteract them in their organisation, and that's what makes them so dangerous.

The removal of the statue of Rhodes on the UCT campus was a symbolic expression by students to decolonize education and also a response to the sense of exclusion by students struggling to pay their fees, an issue which has rallied students across South Africa in the so-called 'fees must fall' campaign:

> In a way, the collective assembling of bodies is an exercise of political will, and a way of asserting, in bodily form, one of the most basic presuppositions of democracy, namely that political and public institutions are bound to represent the people, and to do so in a way that establish equality as a presupposition of social and political existence (JB in Butler and Athanasiou 2013: 196).

The emphasis on decolonisation is about the need to balance the rights and responsibilities of the individual and the collective.

> South Africa's student movement has scored a big victory after plans to raise university fees were scrapped following weeks of nationwide protests that culminated in a mass gathering outside the main government offices. The largest demonstration took place at the Union Buildings, South Africa's seat of government in Pretoria. More than 10,000 people gathered on Friday to demand that proposed fee rises be scrapped, and called for the president, Jacob Zuma, to personally address their concerns.[14]

The need to address policy concerns of young people[15] who are struggling to finance education is important as is the need to redress the current imbalance in the development of university campuses in urban environments through rewarding the best and brightest through urban-based scholarships. The move from rural to urban areas is causing increased competition for scarce university places. Vocational Education in Rural Universities and Training colleges could help to promote the value of agriculture to bright students who could be tasked with the vital issue of food and water security. The protests and closures of universities in South Africa are indicative that higher education is failing young people and future generations as it is not addressing poverty and climate change. The unemployment rate in South Africa reached 26.7% in 2016, according to the Quarterly Labour Force Survey produced by Statistics, South Africa,[16] the highest level since 2008.

The Blue Economy approach developed by Pauli (2010) outlines more than a 100 ways to create opportunities through environmental thinking that does not privilege the environment at the expense of people. Instead his approach is to find ways to enable the unemployed to benefit through working on environmental challenges, his motto is: 'There is no unemployment in eco-systems' (Pauli 2016).[17] He stresses the need to provide opportunities through design that taps into the abundant talent and environmental opportunities that can be found and to ensure that the designs protect both people[18] and habitat. This is a systemic approach that could ensure that people come up with solutions that do not create binary oppositions between people and the environment. It is unnecessary to argue that for

[14]http://www.theguardian.com/world/2015/oct/23/south-african-students-protest-pretoria-tuition-fees-rise#img-1.

[15]http://www.news24.com/SouthAfrica/News/hard-to-predict-the-next-step-for-fees-must-fall-20160814.

[16]Statistical Release PO211.

[17]https://www.speakersassociates.com/speaker/gunter-pauli.

[18]Individualism is understood to mean the human thinking and behaviour focussed on the self and collectivism is understood to refer to group concerns and relationships with others including living systems.

people to flourish the environment must suffer. Pauli begins by stressing that systemic Blue Economy options are rooted in justice for living systems. The Blue Economy is important because it differentiates itself from the Red Economy which extracts profit from people and the planet in ways that are unsustainable and lead to a debt to future generations of life. It also differentiates from the Green Economy which places the environment at the forefront and perhaps can be seen to privilege middle and elite classes of people who not only have their immediate survival needs met, but who live comparatively privileged lives away from the squalor and slums in big cities who think in terms of mobility and frequent flyer miles and the inevitability of the right to travel.

His experience of working for a company that solved biodegradability but at the expense of the habitat of orangutans shaped this way of thinking:

> When even the largest manufacturers adopted our biodegradable ingredient—the fatty acids of palm oil—as an industry standard replacement for petrochemical surfactants, it dramatically increased demand for this alternative. This spurred many harvesters, especially in Indonesia, to replace vast swathes of rainforest with palm tree farms. In destroying the rainforest, much of the habitat for the orangutan. Thus I learned to my chagrin that biodegradability and renewability do not equate with sustainability. (Pauli 2010: xxvi).

Raising awareness about climate change needs to be accompanied by a means to express agency. The photograph below shows a banner reminding people in Bandung of the need to live differently.

This is an issue in Australia where some sections of the environmental movement are clashing with Indigenous leaders such as Noel Pearson and Marcia Langton who argue that pragmatism is needed and that Aboriginal Australians have a right to benefit from the environment. Marcia Langton has stressed that Aboriginal people should not be stereotyped[19] and that it is appropriate that they should be able to benefit from the resources that are mined in Australia. As first Australians they should benefit from the land. This is a pragmatic approach to development and one that needs to be discussed with environmentalists. The land is not empty of people in the remote parts of Australia, it is populated. Just as Noel Pearson argued in the contested Wild Rivers campaign it is important for Aboriginal Australians to benefit from their own land.

Langton (2012) in the ABC Boyer Lectures stressed that it is time to for a different approach to Indigenous employment. She makes a passionate case for pragmatism based on considering the consequences of policy and political choices for all. In an article in Meanjin (2015)[20] Langton engages in a discussion with

---

[19]Langton, M and Longbottom, J 2012 Community futures, legal architecture. Oxon Routledge. She defines indigenous as: 'applied to territorially based ethnic groups that were culturally distinct from the majority population of the nation states in which they find themselves, that were politically marginalized and who identified as indigenous' (Simpson 1997). Simpson (1997) Indigenous heritage and self-determination: the cultural and intellectual property rights of Indigenous people. Copenhagen Denmark. IUCN.

[20]https://meanjin.com.au/essays/the-question-of-constitutional-recognition: "The Question of Constitutional Recognition: Marcia Langton talks to David Leyonhjelm."

David Leyonhjelmin and makes the point that what matters is not race, but recognition of being first Australians and in her Boyer lectures she makes the case (like Noel Pearson) that Aboriginal Australians have a right to benefit from the land. The denial of Aboriginal Australians the right to exploit the land is tantamount to a new form of green colonialism. According to Langton (2012)[21]:

> In these lectures, I will examine the underbelly of the resources boom and the standing of the Indigenous population in the Australian economy—especially those who live in northern Australia and the remote regions which are the geographic heart of this activity. Mining is the only significant industry in remote communities, and dependence on it may leave these communities in a precarious position when an operation closes. High levels of dependency on mining can be detrimental for Indigenous and rural and regional communities, so development aimed at increasing economic diversity is needed. Now there is talk that the resources boom has peaked. How vulnerable to the mining downturn are these Aboriginal businesses.

The same kinds of arguments were made by Prof K. at Universitas Nasional in Jakarta who stressed that Indigenous people in Kalimantan want to be able to make a viable living from the land. The challenge is to find a way to achieve this in ways that balance the needs of people, indigenous primates and their habitat. A biologist K at Universitas Nasional who researches ways to protect Orangutan stressed that unless more was done to provide viable living options for people they would not protect the primates. He also stressed that hunting is seen as enjoyable and people continue to believe that Indonesia has unlimited resources.

I raised the problem of the corrupt sale of forests to developers for palm oil and the resultant loss of habitat for the primates. People in the cities do not make the link between cheap palm oil and the loss of habitat.

Prof K. an environmentalist concerned about caring about the Orangutan said that previously bush meat was abundant and enabling the Indigenous people to understand that bush meat cannot be obtained without sustaining the animals needed to be a starting point[22] for protecting habitat along with protecting corrupt politicians and corporate leaders. The sense of abundance and right to consume is shared by people across rural, suburban and regional areas as well as in the Ciangur region of Java. Education is not the solution, according to Prof K, it is providing people with options so that they are not so desperate that they have to hunt and destroy forests and not so desperate that they are willing to sacrifice their land to greedy corporations. He then went on to reflect that:

> Hunting is a cultural experience that people like to pass on to their children. The hunt itself is enjoyable.

---

[21]http://www.abc.net.au/radionational/programs/boyerlectures/boyers-ep1/4305610#transcript.

[22]I reflected that hunting and being in the bush is regarded as training by many Indigenous people. Surveillance and laws to prevent poaching and deforestation are thus viewed by some with ambivalence, unless they are part of the process of designing and implementing them.

I agreed that in South Africa and Australia, hunting is also regarded as a cultural experience necessary for those engaging in sheep farming to ensure sufficient grazing for the sheep, whilst Aboriginal Australians talk of the joy of hunting and sharing Kangaroo as a way of passing on survival skills. Knowing the limits of environments and retaining balance was a skill that is becoming increasingly important. Kangaroo meat is a good replacement for sheep and cattle, Orangutan meat in east Kalimantan is not. An understanding of our affinity with other sentient beings and the closeness of our primate connections was underlined by a member of the audience at a public lecture I gave on the evening of the 29 Sept, 2016 in Jakarta. A man from Flores said that he believed it important when studying Indonesian perceptions that places like Flores be included, because their voices are often unheard. Similarly, another stressed that women's issues should be central and I concurred stressing that all those who feel more marginalised need to be given a voice. Individual and collective needs, power distance, gender rights and species rights are all important aspects as is the management of consumption choices, because the sort of consumption choices that are being made are robbing future generations of life of a chance to survive.

The idea of solving all problems with better designs and better technology is fraught with problems, because all designs begin with values. So it is vital to design with the rights and capabilities of human and non-human animals and their habitat in mind.

Access to software for social inclusion and decision making does not in itself entail sustainable praxis or re-generation. Some people are becoming addicted to gadgets and the same edition of Jakarta Post[23] that I studied en route to Jakarta stressed that a law was being formulated to prevent young children from becoming addicted to internet gadgets a very early age. Even children around a few months old are playing digital games and losing the capability to make connections with other children and adults, let alone with plants and animals. The notion that plants are just lifeless things is a problem according to Ibu Inez who said she was shocked at the way people throw cigarettes into the plants and shrubberies around the Universitas Nasional building:

> They are not plastic, they are living things. Also the children do not realize that dogs and cats are living creatures.

I reflected that in New Zealand in 2015, they passed legislation that recognises that animals have rights and that sentient beings should be protected. There is still a long way to go in addressing this value in many parts of the world. Ibu Inez at the Universitas Nasional stressed that she lives close to the university campus and that she walls to work and also takes public transport some of the way. When she walks through the last few streets in her neighbourhood she observes the way in which people treat the voiceless. She stressed that treatment of household pets made her wonder about what could be done to improve their living conditions.

---

[23]http://www.thejakartapost.com/news/2016/09/29/parents-support-plan-limit-gadget-use.html.

I see in a cage – just one poor pet, I see children throwing stones at a dog and pulling his tail… children these days do not relate to animals and nature. They are not automatons.

The theme raised by Inez is apt. The treatment of caged animals destined as food, civet cats[24] and pets in homes and the Bandung zoo raise many concerns as does the treatment of a skeletal, stray dog in Cibodas with a skin disease. Attitudes to cats are very much more positive that to other animals. But even so I saw many who were woefully thin, scarred and mangy at the Gede Bagel Market. How are animals perceived? My guide explained to me that dogs are regarded as unclean for many reasons in Islam. One is practical, namely the number of rabid dogs that pose a health hazard. The other is their blatant sexuality which makes them regarded as unclean domestic animals. My colleague explained that her own grandfather kept dogs and regarded them as helpful on their farm. They were treated decently, but not as pets. Many keep dogs for hunting wild boar in the Ciangur area, but they are also not pets or companion animals.

The dog which was suffering, I witnessed in Cibodas, begged for food and was chased by a vendor who was trying to sell his wares from a bicycle to passersby. I chased the dog throwing the only food that I had to him which caused interest and some concern for my safety. This contrasts with the attitude of a woman who ran a

[24]The vision of Pak Rudolf Wirawan is the need to use technology for sustainable living to foster coffee production in West Java by protecting the coffee plants and the civet cat that eats the coffee beans as they ripen. He wants to track the ripe coffee by working with the civet cats so that as it ripens the coffee beans are picked by tracking the movement of the civets and working with them. If you pick the coffee beans too soon or too late it is a problem. This could also help to protect and value the civet cat in his opinion. He also wants to encourage people to plant coffee beans and that this would help to maintain a clean and green Indonesia. His notion of using technology as a solution to tracking those who try to burn forests illegally by using surveillance drones is considered problematic as the forests are too dense.

small shop (warung) outside the Padjadjarang campus. She came from Aceh and provided food to a street cat to which she agreed to give meat on my behalf as the cat was pregnant. I left a small sum for her to feed the cat, which she assured me she fed anyway and that she visited her on a daily basis.

A contextual analysis of material from a youth leader who fosters excellent entrepreneurship (and who also has a great fondness for cats that visit the many incubator hubs that he has set up) demonstrates that like many other business people he commodifies animals that are destined for the plate, such as Muskovey Ducks, for instance, he says:

> Therefore, it is necessary to evaluate, improve, and innovate it through adding more value … to the *entog* product.

A farmer in Cibodas sees his cows as valuable as milk producers and in need of specific care. He stressed that even if they were to 'become dinner' they needed to be well groomed and well cared for. He gave them names and he worked closely with the vet while also ensuring that each cow was groomed and the store in which they lived was cleaned and maintained by sweeping and washing down the flooring. He was also concerned that their diet was affected by the lack of variety as he needed to rely on green fodder collected from the forest. He stressed that their diet needed to be supplemented by corn but that the farmers did not want to sell corn for fodder as they could get a higher price using more value-added products. He also stressed that the manure from the cows was made available to some of the poorer farmers so that they could increase the fertility of their soil. Biogas is the other byproduct from these cows which enable the members of his cooperative to live sustainably.

The notion that environment should be placed before people has been criticised as a form of Green Colonialism in Australia and in Indonesia.[25] The notion that human and other animals are codependent on habitat could be a way forward. Another way through this polarisation is enabling young people and the environment to benefit through a creative approach to systems thinking that recognises the opportunities that exist, such as rubbish upcycling to create art products from waste, compost for food production, new agricultural products that add value, research into new forms of biodegradable packaging, new forms of energy by enabling those who are most marginalised to have a voice and an opportunity. Policies also need to decolonialise knowledge formation through recognising traditional knowledge pertaining to plants and animals that take into account a range of local people's perspectives.

The UN Sustainable Development Goals provide the basis against which to measure achievements, but as members of a Focus Group on Food Labelling and

[25]The research explores the extent to which people understand the implications of their food, energy and water choices. To what extent do people understand that what they decide to eat has an impact on our ecological footprints? Some food requires more energy and water, depending on whether the food locally grown and locally consumed. The growth of cash crops for marketing at a distance can add costs to people and the planet. Eating a punnet of strawberries out of season flown in from elsewhere a makes a difference to the planet. Particularly if it is sold in a plastic container, which is often the case.

ways to engage the public in addressing food security was the topic of a UN event held in an upmarket Jakarta Hotel. The participants at the event comprised public officials, business men and women and representatives of NGOs and a few universities including the University of Indonesia. In the capacity of an Adjunct Professor at the University of Indonesia, I attended the workshop. The participants at the workshop stressed the need for participation. An elderly spokesperson stressed that the Goals keep changing:

> We are told about the Millennium Goals then the Development Goals then Development 2020 and what does this have to do with how people see things here in Indonesia?

The need for people to have a say in shaping designs is important for matching responses to need but it is also important as a means to educate people. The food labelling issues raised by the UN did not address concerns associated with the ethics of Palm oil or Palm sugar production and the need to be able to identify products that contain any of this item. Food and habitat and what constitutes an ethical diet will be perceived very differently by those with allergies or those with concerns about what constitutes an ethical diet such as vegetarians and vegans.

The call to address poverty and climate change by Desmond Tutu and the call to live in ways that reflect the philosophy of Ubuntu is part of this policy challenge. To draw on and cite Tutu: We are people through other people and we need to be keepers and not consumers who live at the expense of others (including sentient beings) and future generations of life. He goes on to say:

> It speaks of the very essence of being human. When we want to give high praise to someone we say, Yu, u nobuntu, hey so- and so- has Ubuntu,
>
> Then you are generous, you are hospitable, and you are friendly and caring and compassionate. You share what you have. It is to say, my humanity is caught up, is inextricably bound up, in yours….We belong in a bundle of life. We say a person is a person through other persons. It is not I think therefore I am. It says rather; I am human because I belong, I participate, and I share. A person with Ubuntu is open and available to others, affirming of others, does not feel threatened that others are able and good, for he or she has a proper self-assurance that comes from knowing that he or she belongs in a greater whole and is diminished when others are humiliated or diminished, when others are tortured or oppressed, or treated as if they were less than who they are…. Cited from Tutu (2013) www.thinkexist.com/quotes/bishop_desmond_tutu/ 5 June 2014

2.2 From Working Within Boundaries to Recognition of Flows

The personal and public policy choices within the boundaries of a nation state have implications for others. I hypothesise that participation through awareness and consciousness-raising (McIntyre-Mills 2010) will influence the way in which people value relationships with others and the land. This hypothesis is based on the notion of neural plasticity in that the brain shapes the environment and, in turn, is shaped by the environment (Bateson 1972; Beer 1994; Greenfield 2000).

Internationally the gap between rich and poor has widened and the price of inequality is born by people and the environment. Most importantly, the current social contract does not go far enough to protect the needs of non-citizens, namely young people, asylum seekers as well as the voiceless who are unable to express their needs. This book makes a case for or transformational education based on listening to the land and to the lessons we can learn from our human animal connections. The work of Deborah Bird Rose and Donna Haraway is central in this regard. We are the boundaries.

We know that environmental sustainability and human well-being are intimately linked, but there is little knowledge about how this linkage can be built upon to improve both areas. Attempts to address food, energy and water are often based on policy information that is not grounded in lived experiences and fails to address what we do know about human behaviour or choices. The inherent link between engagement in civil society and community well-being (rather than the economic bottom line) needs to be the focus of research through which to explore the nexus across well-being, consumption choices and the environment. We need to develop a deeper understanding of how the intangible aspects of perceived well-being can be appreciated. But we also need to appreciate them in relation to the links across well-being and sustainability.

In Japan, I attended the International Sociological Association Conference where I chaired a panel and organised the XV111 stream for RC 10 in Yokohama, 13–19 July, 2014 on 'Wellbeing, Participation and Digital Democracy'.[26]

2.2.1 *Cultural Transformation: How Democratic Is Democracy if It Does not Foster Human Security?*

Mary Douglas, a social anthropologist who has inspired systemic thinkers, distinguishes between 'the sacred' and 'the profane' and stresses that norms guide social choices and they are reinforced by authority and religion. She argues from a so-called structural functionalist perspective and according to her, religion has a purpose and is functional for social order. The 'sacred' covers all aspects within the pale (fence) or the boundaries of a culture and these norms are regarded as 'ethical' and supported by civilised society. The 'profane' covers all those aspects that are *beyond the pale* and pertain to the wild, the 'untamed' or the uncivilised. So this argument begins to challenge the independence of religious values. In some contexts, values are individualistic and this can support materialism and selfish

[26]I was nominated South East Asian rep in recognition of my contributions to transdisciplinary research and sociology. I presented a paper and organised a digital platform with Prof Vera Vratusa so that those who could not attend in person could engage with others. McIntyre-Mills (2014), Planetary passport: the challenges for social and environmental justice, The challenges posed by the global Pan Optacon, Penal states and dis appearing states in an increasingly vulnerable and unequal world. July 16, 2014-4th ISA Forum of Sociology—Facing an Unequal World: Challenges for Global Sociology. 13–19, July 2014, Yokohama, Japan. Recording at: http://vocaroo.com/i/s14IGmoPgPez see http://isarc10internetforum.wikispaces.com/Session+2+ISA+14.

decisions. In other contexts, the group's values prevail because people are communal and this can support collective concerns. But God can be invoked to support both individual and collective concerns.

The San Bushman draws on a sense of connection with the land animals and believes that egalitarian norms should prevail. They draw power from a sense of their interconnectedness with the land, animals and one another. Fiske (1992) extends the work of Mary Douglas and develops an argument that hierarchies and communal sharing characterise the two axes for guiding behaviour. These are in turn linked with different kinds of relationships associated with treating people as equals and distributing resources equally or alternatively giving a price or value based on benefits or entitlements. He sums up the relationships and ways of organising and implications for ethics and morality as: 'Community sharing', 'Authority ranking', 'Equality matching' and 'Market pricing'.

The point I want to make is that both rational and emotional dimensions are important for ethics. When we place too much emphasis on culture and religion as a basis for the sanctity of choices, we also have to deal with different viewpoints. This is problematic within Western culture and even more so across cultures. This is why I make a case not for religious cooperation or cultural cooperation or a case for the clash of cultures as being inevitable (as per Huntington 1996) but to emphasise that we are interdependent. Pragmatic recognition of this fact could be the basis for working cooperatively to draw on practical cultural knowledge for the survival of living systems.

Humans evolved from primates and we share the capacity for empathy, reciprocity and fairness. In fact, we evolved through our ability to cooperate and not only to compete (De Waal 2009). Thus, the emphasis is on developing a new basis for transforming the way in which we live. The emphasis needs to be on what we all share in common, namely the need for food, energy, water, safety and the capabilities to live a good life. This is stressed by Nussbaum (2011) who uses the concept of capabilities and wisely avoids too much emphasis on culture per se. Culture is a way to adapt to an environment. It needs to be seen as responsive, rather than as a pre-given.

2.2.2 From Ark of Covenant to Global Covenant for Space Ship Earth

The encoding of knowledge for survival in landscapes, skyscapes and seascapes is evident across Indigenous oral traditions. The memory aids were linked with the places where people lived and travelled. Stories of origin myths such as the caterpillar dreaming story in Alice Springs are memoed in the landscape of the McDonald Ranges and in the sacred sites where the caterpillar hatches (McIntyre-Mills 2003). Local custodians are reminded of the stories by looking at the landscape and the stories remind them of their role and connection with the ancestors. The creation stories are personalised and the links to the land are renewed through symbolic acts that have a practical and pragmatic goal, namely to balance individual and collective needs. The oral stories are 'owned' by the local people who keep them alive through singing, ritual, art work which are in each act of creation and co-creation aids to remembering their care taking roles as country stewards.

I make a case for renewing our respect for learning from many ways of knowing as they are vital for human survival rather than elitist forms of knowledge that grew as we became increasingly urbanised.

Today, the top-down approach to controlling knowledge could repeat the same mistakes made by previous civilizations that lost a sense of connection with the living systems on which they depend. The vast archaeological ruins of agricultural nation states that ended as a result of top-down hierarchical power, population growth, loss of habitat conflict, loss of food and water and conflict.

Building capacity and capability through leadership requires new approaches that reframe the way in which we do governance. It requires a multilevel and multiagency approach within and across overlapping regions. It needs to avoid blue print approaches and to work in a responsive and contextual manner. But at the same time avoid nationalist agendas that undermine post national regional concerns. Transdisciplinarity and cosmopolitan politics are today increasingly threatened as local people struggle to survive and to limit the worst aspects of globalisation, namely protecting the wealth of the few at the expense of the majority in this generation and the next.

Personal well-being is linked with the well-being of culturally diverse people living in a biologically diverse environments. The oral traditions of place enabled local people to develop a deep, empirical knowledge of place, based on learning what works why and how.

Source https://en.wikipedia.org/wiki/Ark_of_the_Covenant and https://en.wikipedia.org/wiki/Portal:Space credit William Anders, 'Earth rise'.[27]

[27]**Art of the Covenant**: photograph has been released into the public domain by the photographer, details at https://commons.wikimedia.org/wiki/File:Cath%C3%A9drale_d%27Auch_20.jpg. **Earth-moon**: photograph is in the public domain in the US as it is a government work, details at https://commons.wikimedia.org/wiki/File:Earth-moon.jpg. It is a grey area if US government

2.3 Protecting Spaces for Diversity and Biodiversity

Three options underpin human behaviour, namely cooperation, competition and a realisation of our interconnectedness. In the liminal spaces in which we live our shared lives, we need to find ways to live with human and biological diversity. This involves creating opportunities to connect with others and the places where we live.

Three options for living need to be preserved and protected to enable options of integrated co-located living (cooperation), living alongside or separately but respecting diversity. These three options support cooperation and interdependence based on recognising our hybridity.

2.3.1 Ethical Choices: Competition, Cooperation and Interdependence Based on Recognising Our Hybridity

Is nature 'red in tooth and claw'?

> "Is mother-love vile because a hen shows it, or fidelity base because dogs possess it?" (Collected Essays vol. 3, p. 152). Nature is not simply red in tooth and claw, and humans are not fundamentally brutish or noble. We are both – just like our primate cousins and our ancestors, and just as Huxley claimed (Lyons 2016)[28]

Is competition the main driving force in evolution as Dawkins (1976) suggests or is human nature equally rooted in our ability to cooperate with one another on the basis of 'empathy, reciprocity and fairness'? Hobbes argues in Leviathan (1651) that the social contract is a way to tame selfish passions. But those who do not believe in the inevitability of human competitiveness such as De Waal (2009) make the case that we are indeed capable of cooperation because we evolved through our ability to compete and our ability to cooperate. Today more than ever, we need to realise that the narrative of 'us versus them' expressed in narrow forms of nationalism and xenophobia will not help us to survive the common threat to humanity which requires cooperative effort to curb carbon emissions and to change the way in which profit is extracted from the majority in this generation in the interests of an elitist and unsustainable way of life.

(Footnote 27 continued)

works are also in the public domain in other countries although you could probably rely on point 14 of NASA's guidelines on using material

[28]Lyons, S. 2012 A Most Eminent Victorian: Thomas Henry Huxley 76 Automne|2012: 85–104 and another version at https://philosophynow.org/issues/71/Nature_Red_in_Tooth_and_Claw.

Arendt[29] stressed that humanity can do better if we retain our ability to think critically—and this requires on going truth telling—democracy depends on us every day. In so doing she stresses engagement every day in exploring the way we live our lives. Similarly Butler (2009) argues in 'the examined life' that humanity is vulnerable and that we need to acknowledge our 'interdependence'. Butler argues that capitalism has stressed independence, competition and being 'fit to do a job'. The whole social contract has been based on this assumption that people who are good citizens within the state should be protected. But what happens when this fit worker finds him or herself unable to work? Globalisation has indeed led to displacements where some people can choose to work as a migrant across sites with a view to returning 'home' where family rely to a greater or lesser extent on the incomes generated and repatriated. They benefit from the wider networks created by the mobile migrant workers. Unfortunately for some, there is no 'home' where they can return because of conflict or disaster. What happens when the fit worker becomes displaced through illness or disaster? Clearly, the road worker in Bandung will not receive benefits as a casual day worker if he injures himself and needs to return to his village in Ciangur or some other regional area.

The scenes of Syrian refugees fleeing from war across national boundaries before they are shut down are a case in point. The inhumanity of the war in Aleppo is summed up in the scattered toys in bloodied dust as viewed on the Australian Broadcasting News in December 2016. What can possibly justify the inhumanity of bombing civilians? The silence from United Nations and most so-called civilised nations is alarming and raises the need for all organisations (particularly at the international level) to be open to criticism, hence the need to monitor the UN from below.

As Sahlins and others have said anthropology is about revealing in the small new ways to view the world. Globalisation has also resulted in a global economy that has ramped up carbon emissions that have resulted in the era of the Anthropocene —or human created culture that has impacted on the entire fabric of life on which we depend.

The rising sea levels experience by islands in the Pacific as stressed at the Paris Climate Change Conference in Dec 2015 is a case in point. The challenge for social agents of change is to lobby for more emphasis on the common good and to move beyond the zero sum approach and to understand that individual survival and well-being is dependent on group survival and well-being, even if there are short-term gains (in some instances) to living at the expense of others.

---

[29]Hannah Arendt stresses in her work on 'the banality of evil' that evil is comprehensible. In her discussion of the Eichmann trial after the holocaust in Germany she explains that Eichmann could not be singled out as being the monster at fault for the system of discrimination or extermination of the Jews.

The role of symbolism remains important. The needs for security and belonging are universal.

Ascending towards spirituality needs to become an everyday experience once again as it was in the lives of San/Kung hunter gathers who achieved a sense of community through ritual connection with one another and with the land. But as Katz (1982: 252) stressed even a few decades ago this sense of spirituality was becoming increasingly eroded as the previous hunter gathers become absorbed into the cash economy and became prey to so-called ‘peripheral capitalism’.

Setting aside our old ways of thinking as a step towards transformation that can be symbolised through spatial rituals that symbolise conceptual change, such as of ascending and descending stairs, symbolising the setting aside our old thoughts through making material representations and discarding them before envisioning a new approach. The rituals of oral culture remain in our literate culture today as important reminders of integrated knowledge systems that were pragmatic in terms of their ability to enable people to survive.

The knowledge in oral culture needed to be recalled through memorising stories linked with structured myths and landscape prompts that helped to recall details. Survival depended on remembering and connecting up the different pieces of knowledge on what worked, why and how that had been gleaned through centuries of empirical testing out of ideas. The re-hearsing of ritual and engagement in symbolic acts that helped to recall and remind were important ways to prepare themselves for navigation, travel, fishing, hunting, healing, meeting up with neighbouring tribes or any of the other activities in pre-agricultural daily life (Kelly 2016). Once people settled in one space and started to farm and own property which was codified and protected by written laws to protect the interests of owners.

The notion of categorised and contained spaces became increasingly relevant. This segmentation of knowledge followed suit and powerful interest groups framed the disciplines of knowledge and the terms of reference according to which knowledge was restricted or shared. The powerful controlled what was deemed knowledge and this trend has accelerated as media control what constitutes knowledge. Lessons from ancient civilizations that settled in hierarchical societies that relied on extracting surplus to sustain an unsustainable way of life can be learned from the archaeological ruins. In 2005, the International Systems Sciences was held in Cancun. A visit to Chichen Itza on the Yucatan Peninsula of Mexico sparked the question why did this civilization come to an end? The first response was to reflect on the growing size of a population that must have developed in the bureaucratic, agricultural society and that it could no longer sustain itself through living in harmony with nature. The oral histories that may at first have been stored in this memory space slowly grew into a means to control the population and to exact blood sacrifice when food and water security became increasingly problematic for the growing population. Two things were evident when visiting this site. The first was the dwarfing of human beings who approached the pyramid and the second was the sense of awe, wonder and fear which the architects of the temple inspired through the design which enabled the light during the spring and autumn equinoxes (Kelly 2016: 259) to emulate a serpent crawling down the side of the pyramid.

> When societies move from predominantly hunter gather lifestyle to a more agrarian society, increasing levels of secrecy become part of the power structures.... (Kelly 2016: 143).

The oral histories which had enabled an integrated way of life were no longer shared and instead became a means of control. The lesson for today remains relevant knowledge needs to be shared and extended through testing out ideas but core knowledge based on recognition of our dependency on nature needs to be held sacred. The hierarchical control of knowledge by the elite is encoded in a 'memory space' (Kelly 2016).

The rituals derived from the days when oral culture was the only way to store knowledge remain and play an important role in reminding and reconnecting disjointed and segmented modern lives. For example, before entering into a Buddhist shrine the Japanese faithful symbolise setting aside their thoughts and striving for renewal through clearing their minds before connecting with the Buddha. By travelling to Kamakura, people are able to experience a less urbanised environment a few kilometres from Yokohama. The process of emptying the mind is assisted through walking through idealised garden versions of a controlled nature and leaving their negative thoughts behind by symbolically as pieces of discarded, shredded paper and later writing their prayers on small wooden discs that remain as prompts for others to express their thoughts. Alongside, the more spiritual aspects of renewal the more pragmatic collection of money and celebration of those who donate food and rice wine are evident as well. The donations at the feet of the Buddah were regularly collected by Monks and the wine barrels were loyaly displayed near the entrance of the sanctuary without any self conscious attempt to pretend that money was unimportant to the spiritual continuity of the place.

A space for meditation

Globalisation has led to displacements and lack of time which can erode kinship when people move to find jobs or have less time to invest in people and the place where they live. People need to be reminded of their connectedness and dependency through symbolism of humility and a sense of awe for the creative and destructive forces of nature. Facing Mecca, contemplating the face of Buddha or remembering the parable of Christ washing the feet or going to the Bush or the Desert to contemplate nature and to empty the mind become increasingly important.

This is not a naïve wish to hark back to the past but to realise that transformation is not about striving towards a new modernised future without drawing on the wisdom of nature or losing our connections with animal relatives.

The few protestors who were concerned about the prospect of job losses in the transport sector in for example Yokohama—are reminders of those who are on the edges of the capitalist society in other cities where a changing economy challenges not only blue collar workers such as ticket collectors (who will be replaced by technology) but also taxi drivers, for example, whose margins are threatened by on line applications that offer cheaper fares. The old economy is shedding many workers which makes the role of the blue economy ever more important through enabling people to identify opportunities through new designs.

The city landscape remains linked with the natural world through remembering human vulnerability to nature but also human capability to redesign and rebuild in ways that prevent and regenerate after natural disasters. The symbolism of the power of nature in the wave/waterfall imagery was a powerful reminder on our arrival in Yokohama, that the city had recently been rebuilt as a result of massive tsunami. We were met at the airport by young delegates bravely sporting the image on their shirts and kimons. The disaster and rebuilding theme was stressed by Buroway in his plenary speech at the XV11 World Congress. The great wave design was adapted and worn by the students welcoming the delegation to Yokohama conference.

The Great Wave: *Source* Wikipedia commons[30]

[30]https://en.wikipedia.org/wiki/The_Great_Wave_off_Kanagawa#/media/File:Great_Wave_off_Kanagawa2.jpg.

The symbolism of the Kanagawa wave was discussed informally by systemic thinkers Professor Paul Hays and Tom Flanagan at the Future World Symposium in Cyprus (2015) as a powerful metaphor. Flanagan explained that if you look carefully at the picture you see Mount Fuju behind the wave. The mountain can be regarded are solid reminders of our ability to weather the storms and to prevail if we work towards a culture of resilience.

People face increased risks associated with climate change such as rising food, energy and housing costs. What emotional connections do we have to particular kinds of life? As cities developed, the differentiation in roles increased as did the gap between people and the environment. Instead, people need to rely on trade exchanges or labouring for others in order to provide for their daily needs. People, animal and the environment became commodities from which value could be extracted for profit through trade or labour.

Taming the space through greening areas and introducing connections with nature and animals using symbols, cartoons and anthropomorphised pets in shopping centres that dwarf and alienate and at the same time tantalise the senses merchandise. Hunter gatherer bags are on display and the shopping mall advertises using plastic inflatable icons and cartoon animals with friendly 'child like' faces. Whilst bonzaied trees and tamed vegetation in containers remind us of our habitat and the need to connect with a liveable environment. When all else fails plastic trees are placed in dark malls to remind us of the environment we are losing.

This is why more green spaces and more habitat for diverse living systems are needed. The attempt to provide a sense of connection is attempted in this artificial environment in which people appear dwarfed and alienated. This response to commodification and alienation in big cities is inadequate. The pseudo-ritual of transpersonal connection is symbolised by taking part in shopping festivals advertised in the shops using cartoon creatures that at some instinctive level connect with our wish to connect with real animals in a real environment. Katz (1982: 307) writes of transpersonal connections with others in Kung society as detailed above. If we are to heal ourselves and society then we need to foster reconnections to the land of which we are a part and to which we inevitably return. Collecting for animal welfare and keeping pets helps urban residents to re-connect with animals in the city environment of Yokohama.

Current society is in denial expressed as a form of dualism (Katz 1982: 299). It cites the work of Churchman (1979) and stresses the need to apply healing far more expansively beyond the health system to consider the environment of the problems facing urban dwellers. Death and life need to be seen as a cycle of renewal rather than as a divide. The narcissic culture of eternal youth and striving for perfection in a luxurious city environment has become a focus for advertising. The reality is that the luxury for the majority is illusive and it comes at a cost in terms of the way cities are currently designed. In Chap. 5, the urban life of extreme poverty for some in Cape Town, South Africa is discussed.

2.3.2 *Urban Living Shapes Our View of the World*

Although many may regard living in cities as temporary and a way to enable survival in rural and regional areas, the issue is that the rural and regional areas are disappearing as discussed in later chapters on poverty in South Africa and Indonesia as the rate of urbanisation increases.

If a comparison is made across three increasingly urbanised areas in West Java, for example, it appears that the movement to the cities to find work can appear logical, because the unemployment rate is lower. But even though the poverty level according to the HDI is lower in Depok the costs of living in urban centres render the urban wage as no guarantee that one will have a much higher standard of living. We can compare quality of life across:

The highly urbanised area of Depok City at the University of Indonesia, Jakarta with a population of 1,898,537 with a high unemployment rate of 9.54 and poverty levels recorded at 2.4% possibly because of the ability to rely on the informal sector to survive.

The suburban area of Jatinangor that is becoming increasingly affected by loss of land as urban sprawl extends, with a population of 4407. The poverty rate is 11.85% with an unemployment rate of 6.5%.

The rural area Cimis in the Ciangur region, a food production area with a population of 570.991 and an unemployment rate of 11.43% and a poverty rate of 5.63% which indicates that having access to land helps to prevent the extreme poverty experienced when land is lost. In the rural area of Ciangur[31], I was told by the leader of the PKK, a women's group that trafficking was becoming more of an issue in the region.

The tourist areas up in the hills are more vulnerable she said, but the traffickers are extending their networks even here. As the wife of a Local Government leader, she was concerned about the extent to which parents were encouraging daughters to take on domestic jobs offered by prospectors (pimps) acting for traffickers.

---

[31]https://www.statsmonkey.com/sunburst/23644-cianjur-regency-population-statistics-by-gender-jawa-barat-indonesia-stats.php.

The number of young women returning from Malaysia and previously from Saudi Arabia (pregnant) and then having unwanted babies has resulted in midwives facing the prospect of finding homes for unwanted babies. My informant commented that her sister (a midwife) had fostered three abandoned children by women who had worked in Saudi Arabia.

The vulnerability of young people (male and female) has resulted in Bandung and Jakarta, for example being unsafe for children who are not accompanied on their way to and from school. According to one of my informants, Bandung is known for trafficking both women and children. The city attracts because it is possible to earn 150 dollars a day selling food as a hawker in the city in contrast to arming of about 5 dollars a day from agriculture.

The challenge is to find ways to develop opportunities to make a living by adding value to agricultural products or creating something out of nothing as a way to create a viable income. This provides independence and it can be done through thinking differently about the local environment.

Unemployed and underemployed people who have no hope are needed to be given hope. At the end of the day, people care about meeting their daily needs and giving a future to their children. What can be done to enable women and young people to have a foothold in a financialised rural economy, a voice and a future?

Current research focuses on ways to decolonialise knowledge formation in public policy and makes the case for an alternative approach to governance and democracy that takes into account a range of local people's perspectives. The aim is to look at current challenges with a focus on the most vulnerable women and young people.

For example in Indonesia, the regional area of Ciangur with a population of over 2 million of whom 48.2% are women. The region relies on out migration by women and remittances from overseas. The majority work in Malaysia, Taiwan and Singapore and previously Saudi Arabia. The financialisation of the economy has resulted in young people becoming vulnerable to trafficking. Although young people do not starve, they desire an income to purchase consumer goods or to pay for education of another family member. What is needed is more work on a design approach that identifies sources of abundance. Some recent field visits which will be detailed in a forthcoming publication explored the impact of urbanisation and ways to create job opportunities. Waste and bio digesters are an answer for producing energy,, whereas biodegradable and renewable building materials can also create jobs. Bamboo can be used for a host of building purposes and have a zero footprint. The purpose is to explore the implications of food, water and energy security for the most vulnerable.

The first project is a biodigester in Cibodas that generates cooking gas as well as fertiliser for farmers. This inexpensive, effective project contrasts with a failed project in urban Bandung run in Cibodas by a young leader of a dairy cooperative who enabled a group of entrepreneurs to work effectively together to produce biogas, manure for fertiliser as well as a range of milk products in Cibodas.

The cooperative manager and youth leader gave personalised care to his cows. He worked closely with the vet and ensured that each cow was groomed and the store in which they lived was cleaned and maintained by sweeping and washing down the flooring. He was also concerned that their diet was affected by the lack of variety as he needed to rely on green fodder collected from the forest. He stressed that they needed a diet supplemented by corn but that the farmers did not want to sell corn for fodder as they could get a higher price using more value added products. He stressed that the manure from the cows was made available to some of the poorer farmers so that they could increase the fertility of their soil. Biogas is the other byproduct which enables the members of his cooperative to live more sustainably, but they still rely on the grid as they need to extend the number of bio digesters to enable more families to reduce their living costs.

In cities for some the only opportunities can be found in working with rubbish. Unfortunately, the approach to rubbish management in cities is wasteful. The comparison of two biodigester projects provides a case study of an expensive failure and an inexpensive success. Regenerating the environment through buildings that produce energy and absorb waste and provide ecosystems for pets, farm animals and liminal creatures as well as oases of sanctuary for wild creatures and spaces for human beings to reconnect with Country needs to be addressed by decolonising the mind.

The second case study is based on corrupt implementation by overseas developers and the public sector who did not foster civil society engagement and who did not provide paid employment for those who processed the rubbish. The result is a failed project where people live in amongst rubbish behind a food market and water distribution centre which is likely to result in higher levels of morbidity and mortality for all who live there or who consume the food and water sold from the marketplace.

The use of rubbish as a metaphor for the lack of care for people with minimal income cannot be ignored. The unclean industries and unwanted people are often forced from the centre of a city. It is only when the rubbish is not only nearby but likely to cause a problem for other groups that it is given more attention. Ironically biogas has been used in Bandung since 1911, which makes the payment for the new technology to a Development Bank by bureaucrats who do not leave the office in Jakarta or Bandung even more problematic.

The central food market in Bandung is ironically called the 'Hygienic Fish Market'. It is an exemplar of the way in which mismanagement of a biodigester can cause more harm than good. The market is in front of the failed waste processor funded by an international Development Bank and implemented by an NGO. After 1 month, they left the project (costing over 1 million Australian dollars) to be managed by the local community. No one was trained in the use of the bio digester. The area has become a dumping ground for unsorted waste. Alongside, the dysfunctional biodigester is a private water distribution company. The water hawkers fill their water carts from the water and distribute it to the other traders, including an ice hawker. The volunteer workers and rubbish pickers live in shanties in amongst the putrefying waste. I also had a conversation with a student who collected data on the dump management and recorded her primary findings. The failed project was studied (soon after its implementation) by senior student at Padjadjaran University. At the time, she visited it had been in operation for 1 month, but even at that stage it was clear from her photographs that the preparation area for chopping vegetables for the bio digester was unused. The installation for gas production was also not operating and the rubbish was being piled (unprocessed and unsorted) as if it was a repository for dumping.

Clearly, only one salary was being paid to a technical assistant who did not have the means to employ other people to help manage the burgeoning amounts of waste he was supposed to sort and process. The lack of community involvement and commitment to the project has resulted in a wasteland of rubbish alongside a water distribution source and a market for vegetables, poultry and fish. The sale of chicken meat and fish alongside vegetables and waste raises questions about public health and the lack of rights for sentient beings were evident to all. The sight of live animals being kept in cages unsuited to a live worth living was confronting to say the least. But in a situation where people are so disrespected, it is hardly surprising that they have little compassion. The live chickens in cramped baskets appeared next to dead chickens in a ghastly echo of the Shambles Butcher markets in fifteenth Century Britain. The need to address animal rights was raised in conversation at one of the training sessions at Padjadjaran University. The notion of engaging with the government in Bandung was also regarded as difficult as the mayor could stand for president. The notion that blame should be attributed at low levels in the community is commonplace. Several meetings with ministry officials were held but the issue of the market was raised only by making indirect reference to the organisation and management of biodigesters and the protocols in managing markets safely.

Clearly, the Bank Project is not considered to be the preserve of anyone and that the NGO appointed through Jakarta was considered to be at fault. Research in Bandung on resilience reveals how the University of Padjadjaran twins with the West Java Provincial Government to enable rural urban development through an incubator programme that supports new business ventures such as adding value to farm products, supporting young people to learn how to manage co-ops, extending small initiatives and fostering self-belief.

The traditional economic model argues that growth in population helps to sustain the economy. But it is unsustainable and needs to be reframed (Stiglitz et al. 2010; McIntyre-Mills et al. 2014). The copy of the Jakarta Post (29th of September) that I had read on the plane en route to Indonesia was filled with reports on the forced removal of people from areas designated as unsafe. Urban villages in Jakarta were being removed in the interest of flood control[32] and urban renewal. Some claimed that their deeds to the land on which they built gave them permission from the time of Dutch colonialism. Others stressed that Jakowi had promised that they would not be moved. Critics have stressed that the governor Ahok has ignored human rights in his hurry to complete projects prior to the next round of elections in 2017. He has stressed[33]:

> The Ciliwung River restoration program is a project under the Public Works and Public Housing Ministry. It uses the 2016 State budget for the project so the normalization has to start immediately…

According to Elyda and Anya (2016), the loss of connections and a sense of place has left people dazed:

> I was born here in 1968, I have spent all my life here….
>
> Bukit Duri's young people found the right place to release their energies at Ciliwung Meredeke's Hall is one of the areas where young people met: "We frequently sang and played guitar there…"

[32]Accessed 3rd Oct http://www.worldbank.org/en/news/video/2012/01/17/rehabilitating-jakarta-waterways-to-mitigate-flood-risk.

[33]http://www.thejakartapost.com/news/2016/09/29/ahok-sticks-to-his-guns.html.

This is one of many villages that are being lost through clearance programmes as flood prevention and urban renewal programmes lead to the end of the small urban villages. The inhabitants are moved to low cost apartments.[34]

Most of the academics with whom I discussed this considered this an inevitable move given the flooding and stressed that the alternative housing could meet their needs.

Overall, the Human rights issues are being raised as a concern. In Indonesia, the United Nations Protection of rights working group has stressed that the rights of the disabled have received attention but the Wahid Institute stresses that abuses to religious freedoms could undermine the Pancasila notion of religious diversity and faith (Jong 2016).[35]

Thus, the challenges faced by the most marginalised in cities are likely to grow as urban development strives to meet the needs of the urban elites. At first, the citizens in Jakarta did not resist the recent removals.[36]

The social fabric was important for their survival. I discussed this point with colleagues and was assured that the removal was essential for the protection from flooding. Indra Budiari (2016)[37] comments:

> NGO Kota Kita has stressed that Jakarta is becoming a divided city and that the removals from Bukit Duri was indicative of this trend towards top down decision making.

I returned to Indonesia after 3 weeks of research in August and early September to attend a conference in Solo and an Alumni conference in Jogjakarta. At that stage there was rioting in Jakarta.[38] The violence in Jakarta during November against Ahok did not surprise me, because Ahok moved people as a result of recent floods. The local people resented his removals and the way they were being used to foster his political agenda, namely to clear out the slums as a precursor to his political election campaign. Politically, when the time was right, the marginalised found a way to accuse him of being 'anti Muslim and blasphemous'. This gave the necessary focus for venting their hostility towards his decisions and also an opportunity

---

[34]http://www.thejakartapost.com/news/2016/09/29/jakarta-loses-its-urban-villages-it-becomes-less-inclusive.html.

[35]Jong, H. N. 2016. UN to grill RI on rising rights abuses Jakarta Post Thursday 29th Sept.

[36]Some said they would challenge the evictions in the courts. Most have found alternative accommodation and will strive to maintain the bonds they developed in the urban villages in Jakarta.

[37]http://www.thejakartapost.com/news/2016/10/04/public-participation-needed-create-livable-city-all.html Accessed 29th October.

[38]The reason for the rioting was cited as being a comment made by the mayor who is a Chinese Christian: http://www.japantimes.co.jp/news/2016/11/05/asia-pacific/political-meddling-instigated-deadly-jakarta-riots-indonesian-president-says/#article_history … President Joko "Jokowi" Widodo said the riot showed "political actors have taken advantage of the situation." He did not identify any individual as responsible, but earlier in the week former President Susilo Bambang Yudhoyono went on national television to say he supported plans for the massive protest. http://www.thejakartapost.com/news/2016/11/03/more-muslim-groups-to-join-anti-ahok-rally.html.

to try to discredit any attempt to use his position as mayor as a stepping stone to higher office (the presidency) in the next round of elections.

Ibu Inez collected me from the airport from Terminal 1 outside the Solo Coffee shop where I recalled the work of a PhD student on the traders in Solo and wondered how he was faring. I looked forward to reconnecting with this student who earned some of the funds to support his studies by running a small store selling chickens. His work on informal traders in Solo tracked the challenges of people living on entrepreneurial earnings who faced evictions by a bureaucracy that paradoxically wanted to beautify the city but were at the same time reliant on the levies from the traders to subsidise their salaries.

The plan of moving people from the centre to the periphery was resisted in Solo because unless the traders remained near the centre they missed passing trade. Menders of vehicles were told that their spare parts and machinery were both an eyesore and caused noise. Those engaged in activities that were regarded as 'morally reprehensible' were also advised that they should move.

As an invited plenary speaker[39] at this social and political studies conference on the potential and pitfalls of E democracy and E governance, I shared the platform with Dr. Eric Loo.[40] He stressed that the Internet is only as good as the users and that the way it could be used to mobilise or polarise or as echo chambers of some at the expense of others should make us cautious of what we read and hear on the Internet. He stressed that democracy needs constant work and that he remembered a time in Indonesia when he had slept with his shoes on because he feared for his life as a small child of Chinese parents during the riots at that time in Solo.

The presenter at the same conference, Professor Bilveer Singh[41] stressed that the rise in the number of internet users would impact on the way in which the disaffected would mobilise. Whilst reflecting on the latest visit to Indonesia with a colleague from the Torrens Resilience Institute I explained that an integrated approach to urban-rural development remains vital. Abu Bakir Bashir lives in a Pesantran in Solo and continues to teach. He gave dinner to one of the presenters at a conference at Universitas Sebelas Merat in Solo at which he discussed the rising number of disaffected young people who were active internet users.

Clearly, the guest was excited by the prospect and informed us that he felt honoured to be invited for 'a lamb dish'. I listened whilst my colleague conversed with animation about how the disaffected feel that many aspects of today's economy disregard the spiritual dimension of life but they are also captivated by the

---

[39]Representation and accountability to address mitigation and adaptation to climate change: the 2030 development agenda challenge and the potential and pitfalls of e-governance and e-democracy. 3rd International Conference on Social and Political sciences:information technology impacts on social and political dynamics, (IcoSaPS 2016) Lorin Hotel Solo, November 2–3, 2016, by: Faculty of Social and Political Sciences,Universitas Sebelas Maret.

[40]School of the Arts, English, and Media—University of Wollongong, His paper: Internet's only as good as its users.

[41]Department of Political Science—NationalUniversity of Singapore, Singapore colleague, gave a Paper Title: The off line and on line Impact of ICT on ASEAN Security: A Perspective.

excitement of the internet blogs offering pathways to heroism. I restrained myself from commenting on the fact that whilst Bashir has freedom many young people have lost their lives as a result of his ideological sermons in Bali.

The capability to respect diversity extends to appreciating different ideas, but the line must to be drawn when one's own ideas about rights and responsibilities erode the rights of others. Hence the axiom: we can be free and diverse to the extent that our freedom and diversity does not undermine the rights of others, including sentient beings. The needs of the disaffected in Solo and in Indonesia are also reflected in some of the needs of the marginalised in other cities.

I also spent time at the Schumacher Institute in Bristol, a sea port town that made its fortune with through the Royal African Company.[42] The entire city is built on trade.[43] The river system was dredged to make the passage of the boats easier. The divided nature of the city continues to play out through protests against racism. A Jamaican man who held out against segregated travel on public transport which led to Bristol riots has been honoured by unveiling a plaque to commemorate his efforts.[44]

The position of a so-called 'city father' Colston is held by a statue to the slave master who funded many of the civic buildings. It seems apt that an ex choir boy led by Banksy (a pseudonym for an ex choir boy from Bristol) comments on the silenced aspects of society by creating alternative forms of public art. Protest in the form of recent graffiti[45] is his way of commenting on society. 'Mobile lovers' was created near to a club for young people who are facing unemployment in Bristol to

---

[42]http://www.flocs.com/websites/bristolslavery/people/royalafricancompany.htm.

[43]http://discoveringbristol.org.uk/slavery/after-slavery/effects-on-bristol/public-commemoration/monuments-to-individuals/ "Bristol merchants were granted the right to trade in slaves in 1698 and it did not take them long to turn the business opportunity into profit. From 1698, to the end of the Slave Trade in Britain in 1807, just over 2100 Bristol ships set sail on slaving voyages. According to Richardson (The Bristol Slave Traders: A Collective Portrait Bristol: Historical Association, Bristol Branch, 1984) this amounted to around 500,000 Africans who were carried into slavery, representing just under one fifth of the British trade in slaves of this period".

"There are many statues in Bristol. One of the most prominent, in the Centre, was erected in 1895 to the memory of Edward Colston. This statue lists the many good works done by Bristol's 'Great Benefactor', a merchant in the 17th century. But it makes no mention of his role as a highly placed officer in the Royal African Company. The Royal African Company held from 1672 to 1698, the sole British rights to trade with Africa for gold, ivory, spices and slaves. In 1998, when information about Colston's involvement in the slave trade became better known, the statue was vandalised. A furious public debate ensued in the pages of the local papers about whether or not the statue should be taken down or whether a more truthful inscription should be added, telling about his involvement in slavery. Many white Bristolians resented this questioning of the reputation of the city's most generous benefactor (he gave the equivalent today of about £10 million to schools, churches and charities for the poor in Bristol). To date (late 2003) nothing has happened to the statue. The debate still continues".

[44]http://www.bbc.com/news/uk-england-bristol-28949801.

[45]http://www.theguardian.com/artanddesign/2014/apr/15/banksy-mobile-lovers-bristol-youth-club.

enable and support the club.[46] This is because in Britain, the gap between the rich and poor has widened. The young can only vote at 18 and it has been suggested that the voting age should be lowered to 16 as they need to be able to stand up for their rights.

The merit of the art is less relevant than the protest it makes by taking and making space for alternative views that deface the veneer of respectability, because the history of slavery is not openly acknowledged in public spaces.

Outside Bristol, the Wells Cathedral choir practises in a building that has historically served many purposes. One of which was a place of worship, penance and imprisonment.

Next door is the vast tithe gate. The size of the barn gate is indicative of the size of the carts and the loads that they bore to the barn in payment to the church, a form of feudalism! The life of Thomas Bewick a naturalist and engraver who lived in the North of England developed a strong social and environmental conscience (Uglow 2006: 363) which led to his criticism of the the power of the church and that:

> it is not the bible but nature that opens …eyes. It is not prayer but the trembling of the hare that makes him renounce hunting, not 'waggon loads' of sermons' but the people of the common who alert him to social wrongs.

[46]http://en.wikipedia.org/wiki/Banksy#cite_note-bbc.co.uk-1.

My guide in the Durham cathedral made the comment that the bishops of the church Durham were amongst the most powerful in England. Clearly their role in fortifying the frontier against the invading Scots played a role in protecting the state.

Thus, religion and culture can indeed be seen through the lense of structural functionalism so that the norms for social organisation are reinforced through religion. The stained glass windows symbolise the sacred and filter the light. But whose reality prevailed? Those who were once incarcerated in Wells when it served a different purpose? The serfs who gave their tithes? Or the powerful elites? Hence, the thesis of this book is that the way forward for transformation is to focus on what we share in common as human animals: the need for food, fairness, reciprocity and empathy. The rest is likely to follow. My guide for the day made the wry comment that the poachers were kept out of the vast estates by Game keepers[47] or stewards. The note book by Jones and Woodward (1910) details the life and seasonal role of a gamekeeper, the creatures (rabbits, rooks and crows) and the environment he protected from poachers and their trained dogs. They detailed the months of the year and remind the reader that March was 'starvation month' for birds and that it

[47]Jones, O. and Woodward, M 1910 A game keeper's notebook. Edward Arnold London https://archive.org/stream/gamekeepersnoteb00jonerich#page/40/mode/2up.

could be a hard month for human beings too. My guide told me that the nickname 'nosy Parker' has its origins in the name being associated with the game keeper's role of checking up on what people were doing! Some dictionaries give some support to the comment of my guide.[48] The enclosure movements began in the seventeenth and eighteenth centuries (Fairlie 2009)[49] and the role of the game keeper was well established by the first time the nickname was cited.

The city of Bristol won the Green City award for sustainability as a result of an application developed by the Schumacher Institute. But, according to the director of the institute this is more of an encouragement award for setting appropriate benchmarks that for achieving sustainable green living. Traffic congestion in the city centre and parking congestion remain challenges. A bike project to teach young people how to fix bikes and in the process to earn them is one of the positive steps. But they do remain tokens. The cost of living in first world nations such as UK is facing the challenge of food security. For example the BBC 1 news on the 29th August reported that families are finding it difficult to make ends meet. Middle class

[48]http://beta.merriam-webster.com/dictionary/nosey%20parker which claims the first reference was in 1907 and that is was linked with the name 'Parker'.

[49]Fairlie, 2009 http://www.thelandmagazine.org.uk/articles/short-history-enclosure-britain

This article originally appeared as 'A Short History of Enclosure in Britain' in The Land Issue 7 Summer 2009.

families are shopping more carefully whilst those with low incomes are skipping meals so they can afford to pay the rent. The lack of basics such as three meals a day, affordable heating and safe housing, affordable public transport remain issues for this green city. In conversations with the staff at the Schumacher Institute it was clear that it is challenging and that they have to do a great deal through innovative approaches to accessing funding. At a conversation on 'Systemic Ethics' at Schumacher Institute on 26th and 28th August, 2014 we discussed the potential for doing things differently. The city of Bristol won the Green City award for sustainability as a result of an application developed by the Schumacher institute for setting appropriate benchmarks for achieving sustainable green living. Traffic congestion in the city centre and parking congestion remain challenges, however. A bike project to teach young people how to fix bikes and in the process to earn them is one of the positive steps. But they do remain tokens.

At the Schuacher Institute in Bristol we also discussed systemic ethics and the way in which hybridity and our connection with the land are understood by First Nations, such as Aboriginal Australians:

> We read our past in the landscape and we create its future through our choices (McIntyre-Mills 2014: 10).

The issues facing Australia and the region are of trying to pursue 'business as usual' against the tide of social and environmental crisis.

> …[C]ountries must focus on increasing the ambition of their intended reductions, and show these are credible by setting out how they will be achieved through domestic policies and legislation (Stern 2014: 26)

We discussed applying the UN Sustainability Agenda to more profound shifts such as the declaration of the Earth Charter by Bolivia:

> Bolivia enshrines natural world's rights with equal status for Mother Earth: Law of Mother Earth expected to prompt radical new conservation and social measures in South American nation… "The law, which is part of a complete restructuring of the Bolivian legal system following a change of constitution in 2009, has been heavily influenced by a resurgent indigenous Andean spiritual world view which places the environment and the earth deity known as the Pachamama at the centre of all life. Humans are considered equal to all other entities (Vidal 2012).

But testing out ideas in conversation and dialogue with others has a prerequisite, namely a willingness to engage in sincere dialogue. But power always needs to be considered in this equation and is always beneath the surface if not openly acknowledged it plays out in many unacknowledged ways that can be surfaced through discourse analysis. It is necessary to create rapport based on many ways of knowing and by learning from all aspects of nature needs to also know where to draw the line. But boundaries need to be drawn and decisions need to be made in order to 'make cuts'.

The line provides the border of what is acceptable and unacceptable. The work of Mary Douglas on pollution and danger remains relevant to critical systemic thinking. The necessity of making cuts or decisions about what is acceptable and what is not remains a challenge for current generations who face the challenges posed by Islamic State (Dean 2015) and the need to respond without lapsing to the same level as stressed by Kilcullen, 2015 who cites Nietzsche, Beyond Good and Evil, 1886:

> Whoever fights monsters should see to it that in the process he does not become a monster. And if you gaze for long into an abyss, the abyss also gazes into you.

Nussbaum (2006, 2011) stresses the need to address human capabilities within and beyond nation states. She stresses that the needs of the voiceless including asylum seekers, young people and the disabled as well as sentient non-human animals need to be respected. She stresses the need to extend the social contract to support the capabilities approach. In discussing human capabilities she stresses health, safety, bodily integrity, education, standard of living, quality of social interaction, productive valued activities, environment, play, basic needs (Nussbaum 2011: 33–34).

The Sydney Peace Prize Winner, Julian Burnside (2014) highlights that 'without justice there will be no peace'. People who arrive by boat are not necessarily illegal nor are they criminals. Members of the public are not aware that all asylum seekers are stripped of the medication and any medical supports from glasses to false limbs. Burnside stresses that they are classified as part of the problem of border control because they are without visas, but they make up less than 'one half of 1% of the 5 million who arrive each year.

This raises many questions about the inadequacies with current forms of governance to support justice.

We need to ask to what extent the current laws protect the most vulnerable? Barbara Molinario, a spokeswoman for UNHCR in Rome, said:

> "… at least 1,500 migrants have died so far in 2015 while on route to Europe – at least 30 times higher than last year's equivalent figure, which was itself a record. It comes just days after 400 others drowned last week in a similar incident".[50]

This morning (8/06/2015) the ABC announced that 6000 people had been rescued crossing to Italy this weekend in small, unseaworthy boats. The Sydney Peace prize winner Shiva (1998, 2002, 2011) sums up the challenge for this century as one of preventing the commodification of life and the importance of thinking about the interconnections across a number of issues. Human rights underpin all relationships within the nation state and within the broader region. This has implications for social and environmental justice. This has implications for social and environmental justice.

Brundtland (1989) and Kaldor (in Held 2005: 177) have stressed that human security rests on creating a peaceful world, enshrined in rights expressed in the United Nations Charter (Article 55, 56) and in the constitution of the European Union. As stressed below (see McIntyre-Mills 2011) rising living costs led to food

---

[50]http://www.theguardian.com/world/2015/apr/19/700-migrants-feared-dead-mediterranean-shipwreck-worst-yet.

riots and the so-called Arab Spring, culminating in the Occupy Wall Street Movement. The move towards remembering and reconnecting with the land echoes the voices of Indigenous First national movements and the implications for the way we live, our relationships, our sense of identity and for new forms of governance and democracy.

A sustainable local community is determined by a sustainable region in which food, energy and water supplies are considered as major determinants for well-being. No community can be expected to transform from a high carbon lifestyle (or aspiring to this lifestyle) without feeling part of the design process and owning the decisions as to how resources should be used. Young people (Osler and Starkey 2005); the disabled, asylum seekers and sentient beings (Nussbaum 2006) along with future generations live 'precarious lives' (Butler 2005). Those perceived as different are not protected (Young 2011). The ability to show compassion underpins cosmopolitanism (Butler 2011). Butler's work stresses 'the need to rethink the human as a site of interdependency'. She emphasises that humanity needs to be able to ask for assistance and we need to be able to anticipate that we will be heard and that people will respond with compassion.

Do we wish to live in a world where we do not want to help one another and in which we deny the pain of sentient beings? (Butler 2011). If we are prepared to recognise not our resilience, but our mutual vulnerability, it provides a basis for stewardship. We are all reliant on others and need to be able to depend on our connections with others.[51]

Huntington's (1996) so-called 'Clash of Cultures/Clash of Civilisations' Clash thesis has become a self-fulfilling prophecy. The realism of a clash is indisputable, but it is a clash of an ideology that strives to alienate other ways of thinking and living. Internationally, to date about 500 or more young[52] women have travelled to join ISIS. Young people of many cultural backgrounds and from a range of socioeconomic groups are travelling to Syria[53] and are thus at risk of losing their rights as citizens and also their human rights:

> … Shamima Begum, 15, Amira Abase, also 15, and Kadiza Sultana, 16, are now feared to have reached the conflict zone….An estimated 550 Western women have travelled to join extremists in Iraq and Syria — while an estimated 600 people are believed to have left Britain to fight with IS….

---

[51]Held et al. (2005) proposed that the core challenges of the day are to address the vast differences in the standard of living between the rich and the poor. The problem is not only one of externalities that are not factored into calculations of the degradation to the environment, it is a way of thinking and 'being in the world' that shifts the extraction of profit to where labour is cheaper and where governments and citizens are less likely to complain about degradation of environment. Short-term profits are made at the expense of future generations.

[52]http://www.smh.com.au/national/emotions-run-high-for-australias-muslim-youth-risking-all-to-fight-in-the-syrian-war-20131206-2ywof.html. At that stage over 200 people have visited Syria. It is estimated at that stage over half were engaged in action beyond assisting with the wounded.

[53]http://www.abc.net.au/news/2015-03-16/syria-bound-uk-teens-held-in-custody/6321512. Updated 16 Mar 2015, 12:46 p.m.

According to the Australian Foreign Affairs Minister[54]:

> Vulnerable young Australians are being radicalized by violent extremist groups and are seeking to join Daesh (ISIL) and other terrorist groups in Syria and Iraq. While most of these foreign terrorist fighters were initially men, they are now being joined by increasing numbers of women and girls.
>
> Women account for an estimated one fifth of all foreign fighters and supporters in the conflict. Over 500 women are from western countries estimate that 30 to 40 Australian women and girls are either engaging in or supporting terrorist activity in Syria/Iraq and at home.
>
> There are a number of reasons for this deeply concerning development. One of the most pressing is Daesh's exploitation of social media to radicalize and recruit online vulnerable young women.

The development of critical thinking skills is very important in this context. Radicalization is based on a sharing of values, in other words an ideological view.[55] The issue as to whether their human rights should be set aside through denying citizenship rights has been raised by the Human Rights Commissioner Gillian Triggs[56] and defended by Amanda Vanstone[57] who 'despairs at her party's proposal to strip citizenship'.

This is of particular concern for the life chances of young people and it is indicative of a conservative turn. Despite reservations about becoming a surveillance state[58] and concerns about making people stateless, the House of Lords has conceded that citizenship rights could be revoked under certain circumstances.[59] According to an article in The Guardian:

> "the former Lord Chancellor has opposed calls for Britain to withdraw from the European Convention on Human Rights…. and has also stressed that Human Rights need to be upheld".[60]

In South Africa, I visited a range of University departments, research institutes and NGOs including University of South Africa, University of Cape Town, the Mindfulness Institute in Stellenbosch, South Africa where I attended the

[54]http://foreignminister.gov.au/articles/Pages/2015/jb_ar_150317.aspx.

[55]http://www.news.com.au/world/disillusioned-young-muslim-men-head-to-fight-in-middle-east/story-fndir2ev-1226957126037 by 2015 the numbers have escalated to at least 300.

[56]http://www.theguardian.com/australia-news/2015/jun/06/gillian-triggs-slams-scores-of-laws-threatening-fundamental-freedoms.

[57]http://www.theguardian.com/australia-news/2015/jun/08/amanda-vanstone-says-she-despaired-at-her-partys-proposal-to-strip-citizenship.

[58]http://www.theguardian.com/commentisfree/2009/feb/06/surveillance-lords-liberty-central.

[59]Deprivation of British citizenship and withdrawal of passport facilities Standard Note: SN/HA/6820, Last updated: 30 January 2015, Author: Melanie Gower Section Home Affairs Section http://researchbriefings.files.parliament.uk/documents/SN06820/SN06820.pdf, accessed 8/06/2015.

[60]http://www.theguardian.com/law/2015/jun/01/human-rights-act-repeal-tory-peer.

Mindfulness Matters Conference,[61] workshops and conversations with colleagues at Living Hope, Embrace Dignity and Africa Tikkum.

The core themes for this book start with social problems of greed and disillusionment and explore pathways for transformation. Be the change through mindfulness, listening and learning from nature and one another.

According to Eric Zeuss (2013)[62]:

> The lead research economist at the World Bank, Branko Milanovic, will be reporting soon, in the journal Global Policy, the first calculation of global income-inequality, and he has found that the top 8% of global earners are drawing 50% of all of this planet's income. He notes: "Global inequality is much greater than inequality within any individual country," because the stark inequality between countries adds to the inequality within any one of them, and because most people live in extremely poor countries, largely the nations within three thousand miles of the Equator, where it's already too hot, even without the global warming that scientists say will heat the world much more from now on…Milanovic finds that globally, "The top 1 per cent has seen its real income rise by more than 60 per cent over those two decades [1988-2008]," while "the poorest 5 per cent" have received incomes which "have remained the same" - the desperately poor are simply remaining desperately poor. Maybe there's too much heat where they live.

2.4 An Architecture for Re-generation to Maximise Changes Towards a More Sustainable Future

Participatory action research on democracy and governance to enhance sustainable living and well-being are discussed in two companion volumes (McIntyre-Mills 2014a; McIntyre-Mills et al. 2014) which explore the suggestion made by Florini (2003) in 'The Coming Democracy' that the Aarhus Convention (1998) on freedom of environmental information and participation could be usefully extended to support the nexus between sustaining human and environmental well-being and resilience. Praxis also needs to examine the impact of participation at the local level (Evans 2013) through awareness-raising about the implications of consumption choices on well-being and 'well-being stocks' (Stiglitz et al. 2010). The architecture is extended in Chap. 4 of this volume which details how to translate the rhetoric of the Paris COP into fine-grained reality to ensure that we think about how consumption choices shape our daily lives and how these choices impact on carbon emission levels and rising temperatures. Instead of merely measuring impact as a governance tool, it also helps to inspire virtue in the sense used by Aristotle, namely Eudaimoinia or being concerned about inner goodness so as to achieve public good, rather than just personal goodness.

---

[61]"From the ground beneath my feet: towards the distant horizon", Plenary paper presented by Simon Whitesman at the Mindfulness Matters Conference, Stellenbosch, South Africa.

[62]http://www.alternet.org/news-amp-politics/worlds-richest-8-earn-half-all-planetary-income?page=0%2C1.

Working across boundaries through co-creating a community of practice to address a shared concern requires:

- Developing a greater number of connections enhances consciousness (Greenfield 2000, 2003, 2008) who argues that the more we are able to think about our thinking. This can help to create closer bonds with others to foster links and to bridge differences. The approach is as relevant to education as it is to community development and in the process helps to enhance representation and accountability through exploring ideas and engaging in dialogue.
- Testing the implications of decisions with others so as to balance individual needs and collective needs.
- Enable residents to make decisions from below to inform policy at the regional level and to facilitate monitory democracy and governance.
- The capacity of members of a community to act together and to be able to modify or even transform, existing ways of life (Rose 2005; Hulme 2009). This research explores whether collaborative approaches drawing on diverse ways of knowing (Cruz et al. 2009) and user-centred governance of resources (McIntyre-Mills 2006, 2008, 2012; Podger et al. 2012) could support more systemic approaches to sustainable living, regional governance (Wear 2012) and effective environmental management.

2.4.1 Reconnecting with the Environment Through Spirituality, Oral History and Law

The metaphor of weaving together strands of experience and creating a community of practice. Indigenous Ngarrindjeri knowledge is based on understanding that river grasses enable the salinity to be drawn out of the soil to avoid toxicity during when the river levels are low. The metaphor of weaving together strands of experience through weaving artistic and functional objects such as mats, baskets and art works whist engaging in telling narratives about the dreaming and about current community rights and responsibilities enables people to think about self-others the environment including the next generation of life. The way in which excessive consumption of energy resources impacts on the size of our carbon footprint is defined in terms of the IPCC formula, namely E (Emissions) = Population × Consumption per Person × Energy Efficiency × Energy Emissions.[63]

[63]The IPCC formula suggests that the privileged lives of some could lead to existential risk for people and the planet (Bostrom 2011). This has implications for the way we live and the need to change our way of life through rethinking our relationships with others and the environment. The Intergovernmental Panel on Climate Change formula addresses the implications of polarizing 'people versus the planet' (Charlton 2011). Representation and accountability rest on showing the links across excessive consumption, unfair distribution and harm (Saul et al. 2012: 167). The issues facing the marginalised will be issues facing many in increasingly urbanised city environments.

The second part of my sabbatical was planned around preparation for this research on: How should we live and work, in order to respond to the growing challenge of meeting the needs of people living in cities? What do we value and why? What are the implications for human capacity and human capabilities? These questions are inspired in part by Stuart Hall a critical sociologist who asked: What is the social, cultural, economic and environmental context that shapes who gets what, when, why and to what effect? I explore the issue of sustainability, food, energy and water security as it relates to these questions. I visited the Schumacher Institute in Bristol, a sea port linked with slavery trade through the Royal African Company. I gave a presentation on the key points addressed in recently published Springer books at a[64] workshop organised by Dr. Jenneth Parker, Research Director with members of the UKSS and members and affiliates of the Schumacher Institute in Sept, 2014 on 'Systemic Ethics and implications for UN Development Goals'. We then engaged in conversation on future directions for research. It was suggested by the Research Director of the Schumacher Institute that I follow up on the work of Polly Higgins who suggests that a new planetary law should be passed to protect the viability of the planet and that current systems of law are inadequate to protect people and the planet. The political context is one of addressing the issue of freedom and democracy linked with social and environmental justice through balancing individual and collective needs. The participants stressed the need to foster the development of green cities whilst protecting the small farmer on the land. In the words of a farmer who participated at the Schumacher Institute in Bristol: "Remember the farmer who is outstanding in his field!" His understanding of agricultural importance was not matched with his understanding of policy implications but his points on protecting the rights of small farmers were well taken.

The participants stressed the need to educate the public, private and volunteer sectors about the importance of protecting agricultural land as urban areas are dependent on the land. Greening cities is a step in the right direction but insufficient to support growing urban populations.

The cost of living in first world nations such as the UK poses challenges for food security. For example, the BBC 1 news on the 29th August reported that families are also finding it difficult to 'make ends meet'. Middle class families are shopping more carefully whilst those with low incomes are skipping meals so they can afford to pay the rent.

(Footnote 63 continued)

The IPCC formula suggests that the privileged lives of some could lead to 'existential risk' for all forms of life on the planet (Bostrom 2011). The 'price of inequality'—national and global has escalated. IPCC formula, namely E (Emissions) = Population × Consumption per Person × Energy Efficiency × Energy Emissions. The IPCC formula suggests that the privileged lives of some could lead to existential risk for people and the planet (Bostrom 2011). This has implications for the way we live and the need to change our way of life through rethinking our relationships with others and the environment.

[64]'Transformation from Wall Street to Wellbeing', 'Systemic Ethics' and the Springer Encyclopaedia of Food and Agricultural Ethics.

2.4.2 Reflection on 'The Case Against the West and the Self-fulfilling Prophecy of the 'Clash of Cultures'

How democratic is democracy when so many fall outside the mantle of its protection? I returned from South Africa to South Australia to find that a bill had been proposed to stress that women wearing burkas be requested to sit in an enclosed part of parliament. This bill proposed by the speaker has not been supported (ABC news 3 October, 2014) which is a step forward.

In the UK a key concern was the return of UK citizens who had been fighting in Syria, because of their potential to radicalise others. This 'clash of cultures' (Huntington 1996) praxis needs to be replaced by a recognition that 'the other' is very often a projection of fears and a non-recognition of the so-called 'enemy within', namely an 'unexamined life' and unexamined values flowing from: religion, morality, politics and aesthetics (see West Churchman 1983). The self-fulfilling prophecy of the narrative myth of the 'clash of cultures' needs attention (Bottici and Challand 2006) in the wake of the Paris massacres. The inevitable clash of cultures thesis is problematic and wrong. This book critiques the thesis and proposes an alternative rooted in critical heuristics. Accordingly, it examines what is the case and explores alternative ways of being in the world linked with capacity building.

The focus of this book is to place the voiceless centre stage. The responsibility we hold as educators is that we are caretakers for current and future generations of life. In this sense, I start with the work of Nussbaum (2011) who argues that all sentient beings should be capable of fulfilling their potential for a life worth living. I also build on the argument of Donaldson and Kymlicka (2011: 33):

> the key purposes of justice, which is the protection of vulnerable human individuals. Being an I – a being who experiences – represents a particular kind of vulnerability, calling for a particular form of protection from the action of others, in the form of inviolable rights… what happens to sentient beings matters because it matters to them.

It also extend on my own work on intersubjectivity (McIntyre-Mills 2000), identity, democracy and sustainability and non-anthropocentric approaches to ethics (McIntyre-Mills 2011, 2014) which develops the argument that the zero sum approach expressed as competition across species, classes or sovereign states needs to be set aside in recognition that we are part of one web of life—that we are interdependent and that all sentient beings have rights.

As stewards—human beings have the additional responsibility to care for the land on which we are all depend.

> My awareness of the individuality of all beings, and of the capacity of at least some to respond to the individuality in me, transforms the world into a universe replete with opportunities to develop personal relationships of all kinds…. (Smuts 2001: 301 cited by Kymlicka and Donaldson 2011: 39)

As public educators, we are responsible for advocating for the rights of first nations whose wisdom and insight is vital, for young people to whom we pass on the baton of stewardship, for the disabled, for prisoners, for the destitute, for asylum seekers and those who grieve the loss of land and culture. And most importantly for all sentient beings who are commodified and treated as a means to an end, rather than as 'ends in themselves'.

We are shaped by our ability to show compassion for others. The quality of life of the vulnerable is a measure of our humanity. The extent to which our societies remain ethical and democratic is a measure of our capability to achieve a balance between individual and collective needs. The extent to which the environment on which we depend remains liveable is a measure of our stewardship.

My starting point is to think about the 'taken for granted'. As a social anthropologist and sociologist I have drawn inspiration from people with whom I have learned whilst undertaking fieldwork in a range of cultural contexts. As time passed and I was confronted by more and more social, economic and environmental issues I was increasingly inspired by West Churchman's Design of Inquiring Systems Approach 'to unfolding values' and 'sweeping in' social, economic and environmental considerations.[65]

Critical systemic praxis (adapted from Churchman 1971, 1979) helps to explore our thinking and our relationships spanning self, other and the environment. As a Meta form of inquiry, it is based on questioning boundaries of inclusion and exclusion, it examines the so-called 'enemies within' (religion, mortality, politics and aesthetics) and it considers the consequences of our choices for living systems (Wadsworth 2010).

My praxis strives to reveal 'in the small new ways of seeing the whole' (Adelman 2013: 9).[66] This has implications for public policy education and ethics, as such it strives to:

---

[65]The body of work inspired by this approach and aspects of social cybernetics (Bausch, Christakis, Flood, Haraway, Jackson, Romm, Stafford Beer, Van Gigch, Ulrich, Midgely) are also helpful in formulating more systemic research on living systems and our place in the bundle of life Most importantly the organic praxis of Shiva, Deborah Bird Rose, Max Neef and Yoland Wadsworth on Living systems is increasingly relevant to my current work on ecological footprints and social justice.

[66]Hirschman (1970) could be characterised as striving to reveal 'in the small new ways of seeing the whole' (Alderman 2013: 9). He contributed to reframing the way in which areas of concern were perceived and the possibilities for doing things differently. In times of crisis, Hirschman (1970) suggested three options—'loyalty, voice or exit'. Although it was possible for him to apply all three options to great acclaim in his lifetime—to exit totalitarian states and to demonstrate alternative ways of doing things—times have changed, because the challenges we face today cannot be addressed by working within the boundaries of a nation state or deciding to 'exit' a nation state that has become totalitarian. Currently we need to think critically about taken for granted structures. We need to take the liberative potential of small-scale projects (as suggested by Hirschman) and try to think of their potential as way to do things differently. In contrast, Hannah Arendt (1972) emphasises both potential and risks through drawing attention to evidence of the banality of evil. She considered the broad context and the structures that lead people to make unethical choices. Arendt's (1972) work Crises of the Republic on how lies and spin contribute to

- **Build** the capacity of people to think about the consequences of their choices for self, other and the environment.
- **Co-create** policy based on testing the principle of subsidiarity and Ashby's Rule of Requisite Variety (1956) and explores the policy implications for complex decision making.
- **Extend** social theory through reframing and reconsidering boundaries (conceptual and spatial) in relation to social, economic and environmental justice.
- **Contribute** to systemic ethics by: (a) expanding pragmatism through addressing 'what if' heuristics and 'if then scenarios' to enable individual self-reflection, group considerations and pilots of participatory democracy and governance. (b) Considering (i) identity and relationships, (ii) boundaries and flows and (iii) policy decisions based on drawing the ethical line through questioning taken for granted ideas about the state, market and society together with those who are to be affected by the decisions.

The conundrum is that nation states need to find ways to enhance a sense of shared ecological citizenship and responsibility for biodiversity as an ethical priority as it has social and environmental justice implications in terms of the hunger, thirst and displacement of human beings, sentient and non-sentient beings if we do not develop policies for habitat protection and rights to food, water and energy security.

First, the perceived implications of the stewardship of biodiversity will be explored. Stewardship is based on awareness that the land and biologically diverse ecosystems are a cultural heritage (Flannery 2012) on which the well-being of current and future generations depend. Resilience is defined as the adaptive capacity of the physical environment, of an individual or of a group. It concerns factors such as the capacity of members of a community to act together and to be able to modify or even transform, existing ways of life (Rose 2005; Hulme 2009).

Secondly, it explores whether collaborative approaches drawing on diverse ways of knowing (Cruz et al. 2009) and citizen centred self-governance of ecological footprints (Podger et al. 2012) could support regional governance (Wear 2012).

The premise is that the UN Sustainable Development Goals do not go far enough and that cultural and economic transformation is vital. We will try to learn more about diverse ways of knowing (Cruz et al. 2009) that could support effective environmental management by taking into consideration: (1) Non-anthropocentric

(Footnote 66 continued)

undermining democracy remains relevant as does her report on Eichmann's trial. Arendt stressed the implications of being part of an unquestioned monstrous system that becomes an unquestioned culture and a taken for granted system of bureaucracy. To avoid taking things for granted it is important to develop critical thinking based on the capability to think at a meta level about the implications of choices (Van Gigch 2003), but also to have in place constitutional structures that protect social and environmental justice for this generation and the next (Jessop 2009).

knowing drawing on the environment as well as knowing through all our senses—including empathy and intuition, (2) Respect for 'know how' developed through empirical trial and error, (3) Creativity in maintaining sustainable relationships with the land, (4) Respect for spiritual awareness, the wisdom of experience as well as the social and natural sciences, (5) Artistic expression of feelings, perceptions and emotions, (6) Caring for future generations of life, (7) 'Phronesis' or Aristotelian wisdom to match the right knowledge in context.

The research is based on:

- *Being* supported by an MOU between West Java and South Australian government.
- *Being a UNESCO project participant and reflections on narratives from young leaders.*
- *Doing* collaborative research as PhD supervisor and collaborating with staff at the University of Indonesia, Universitas National, Universitas Padjadjaran, National Islamic University, West Java and the SA Local Government Association (in particular in Alice Springs and Adelaide).
- *Undertaking* participatory action research in South Australia and Northern Territory (as evidenced by the incorporation of her research findings into Tangentyere Council Protocols and being cited at the Liquor Commission Hearing in NT).
- Facilitating workshops, e.g. Ministry of Religion, Finance and Social Affairs in Indonesia, two invited workshops for the Chief Ministers Policy Unit in Canberra in 2009 and numerous local government workshops in South Australia and research on behalf of the Attorney General's Dept. on resilience, research on behalf of the ARC and local government on well-being. *Facilitating* a leadership workshop at Flinders University together with delegates from West Java Provincial Government on the public education training and leadership. This was in line with the Paris Development Agenda (1997) that underlines the notion that development should be based on collaborative principles. The participants in this workshop stressed the importance of the need to develop leadership opportunities for women and young people. The challenges they face in West Java are associated with the impact of urbanisation as people face congested cities, lowering living standards and high risks for women and children who are unable to express their strategic rights to a voice within households, communities local government and national level organisations. To what extent can we extend solidarity to others and other species? other sources of inspiration are drawn from a sabbatical in South Africa, Indonesia, England and Japan during 2013.

Democracy and governance are in need of improvement (Hulme 2009; Giddens 2009) and lessons learned about sustainable living could help to revitalise our approaches to ethics, representation and accountability.

It seems to me that diverse ways of knowing hold the key to human and environmental security.

The argument developed in 'Transformation from Wall Street to Well-being' (McIntyre-Mills et al. 2014) and Systemic Ethics (McIntyre-Mills 2014) is that the containerist approach to science, democracy, ethics and governance is a core part of the problem. My research question is: how should we live? How can we be more sustainable in our living choices? My research tests the hypothesis that decisions that are made at the lowest level possible are more likely to take into account the complexity of the decision makers.[67]

In Australia, the gap between rich and poor has widened. The price of inequality is born by people and the environment. Most importantly, the current social contract does not go far enough to protect the needs of non-citizens, namely young people, asylum seekers as well as the voiceless who are unable to express their needs.

Critical systems thinking deals with complex wicked problems which comprise many diverse and interrelated variables which are perceived differently by different stakeholders with different values. Thus, two examples of wicked problems are climate change, poverty that pertains to our rights and responsibilities to both

[67]The interrelated aims will be addressed by asking the following questions addressed in part by some of my current and proposed projects that each address significant problems: 'The land is our mother' and stewardship for future generations underpin the philosophy of Aboriginal custodians of the land and their dreaming sites. However, there is evidence that many non-Aboriginal urban citizens wish to spend more time living slower lives, walking wherever possible, riding bikes, growing local food, recycling and reusing and consuming less, instead of living stressful, competitive lives that save time, but waste resources as they are reliant on fast food in 'throw away' containers, fast travel and a 'time is money attitude' (McIntyre-Mills et al. 2012, 2013, 2014 and forthcoming). These findings clearly link environmental health and human wellbeing and raise the question of what can we learn from mobile and egalitarian, place-attached people (Vaske and Kobrin 2001) whose history is recorded in the landscape? (Guddemi 2006; Rose 2004). The process of participation is the subject of recent research published in recent books, entitled 'User-centric policy design', 'Identity, Democracy and Sustainability' (McIntyre-Mills 2008; McIntyre-Mills and De Vries 2011) and 'From Wall Street to Wellbeing' (McIntyre et al. 2014, forthcoming) which has shown that participation enhances attachment to policy ideas. This proposed DP seeks to deepen an understanding of how people perceive, and participate in, local challenges and experiences to develop a new understanding of how to increase environmental responsibility and the stewardship of the land for current and future generations. The challenge for governance and public administration is to match the processes and governance structures to diverse interest groups—with different life chances and different socio demographic characteristics. This research explores the discourses of Aboriginal and non-Aboriginal Australians on consumption and our relationships with one another and the land. This aims to inform responsive policy decisions and service delivery to address rising temperatures and risks pertaining to water, energy and associated rise in food costs.

citizens and non-citizens. The role of theory and its relationship to practice needs to be carefully considered in terms of policy and praxis.[68]

What is needed is to roll up one's sleeves and to work on finding ways to be more tolerant more helpful and more conscious of our hubris through examination of our values and why we agree and disagree about climate change. We need overarching laws that we have a say in making and shaping locally through using and adapting currently available architectures and policies. Current challenges today such as social, economic and environmental issues associated with growing numbers of displaced people, poverty, conflict, food insecurity as a result of shrinking agricultural land, water shortages, growing urbanisation, energy shortages associated with using non-renewables, nuclear disasters on the scale of Fukushima are examples raised by the United Nations and Vandana Shiva and Maude Barlow. The issues need a new approach to the architectures of governance on a planetary scale.

In Australia, the gap between rich and poor has widened. The price of inequality is born by people and the environment. Most importantly, the current social contract does not go far enough to protect the needs of non-citizens, namely young people, asylum seekers as well as the voiceless who are unable to express their needs.

Critical systems thinking deals with complex wicked problems which comprise many diverse and interrelated variables which are perceived differently by different stakeholders with different values. Thus two examples of wicked problems are climate change, poverty that pertains to our rights and responsibilities to both citizens and non-citizens. The role of theory and its relationship to practice needs to be carefully considered in terms of policy and praxis.

Young people do not vote, because they are disillusioned. According to Deborah Mattinson, a former pollster to Gordon Brown and expert at 'Britain Thinks', believes politicians have not begun to grasp the scale of the problem. 'Voter disengagement is getting worse and worse', she says. Nobody is really taking it seriously enough.[69] I will discuss some of my findings and explain how they relate to this paper. In South Africa and Indonesia, for example, at least 65 and 75% of the population, respectively, will be living in urban areas. Urbanisation poses a systemic threat to the survival and sustainability of culture as we know it today.

What is the point of raising these concerns? The challenges pose opportunities to provide education and leadership. What is the way forward? What follows is a suggestion based in part on the work and suggestions I have gained through

---

[68]I stressed the need to reframe boundaries across a range of disciplines in Sociopedia called: Reconsidering boundaries (McIntyre-Mills 2014). Polly Higgins suggests that a new planetary law should be passed to protect the viability of the planet and that current systems of law are inadequate to protect people and the planet. Stiglitz has made the same point in relation to economics. The proposal was made by Higgins and others to the UN and Bolivia and Ecuador have succeeded in leading the way by passing a law to recognise the need to protect the earth.

[69]Mason, R. 2014 Apathetic and disaffected: millions who may never vote: the young and poor shun the ballot box but experts disagree on solutions The Guardian Weekly, 3/1/ 2014 p. 15.

working with PhD students and academics in Indonesia and South Africa doing research on a range of policy concerns. I would like to discuss the implications of increased urbanisation on quality of life and the implications for policy. Food deserts are the likely scenario if more emphasis is not placed on balance, greening cities and supporting small farmers.

This book advocates for developing education for transformational leadership and cultural change. I will compare case studies through considering the leadership of key thinkers, practitioners and community visionaries from whom we can learn. In the Participatory Education Research Journal (McIntyre-Mills 2015). I make the case for:

- Co-determination in regions—new architecture for governance democracy and ethics
- Co-learning and Mindfulness that support building communities of practice
- Supporting learning communities that help us to think across disciplines and cultures, in order to support co-determination of our future within the region
- Developing new curricula that are supported by architecture for democracy and governance
- Extending solidarity and protection to all forms of life within a region, rather than limiting protection and thus limiting human security which is dependent on biospheres not national boundaries.

It has been (wrongly) assumed that we can continue to develop and modernise the planet. It has been assumed that growth in the economy will sustain a growing population; but this is mistaken. We need to understand that the current way of life is unsustainable and that we need to rethink many of the dimensions of modern culture. Culture after all is simply a way of life and a response to the challenges that we face as human beings. The way we think shapes who we are and the sense we make of our daily experiences. Neuroscientists such as Clifford Sarron[70] and Al

[70]http://www.mindfulness.org.za/mindfulness2014/conference-speakers "Clifford D. Saron is an Associate Research Scientist at the Center for Mind and Brain and MIND Institute at the University of California at Davis. He received his Ph.D. in neuroscience from the Albert Einstein College of Medicine in 1999. Dr. Saron has had a long-standing interest in the effects of contemplative practice on physiology and behaviour. In the early 1990s, in collaboration with Francisco Varela, Alan Wallace, Richard Davidson, José Cabezón and others, he coordinated field research investigating Tibetan Buddhist mind training under the auspices of the Private Office of H. H. the Dalai Lama and the Mind and Life Institute. He has served on the Mind and Life Program and Research Council and been a frequent faculty member at the Mind and Life Summer Research Institute. Dr. Saron is Principal Investigator of the Shamatha Project, a multidisciplinary investigation of the effects of long-term intensive meditation on physiological and psychological processes central to well-being, attention, emotion regulation and health".

Kasniak[71] stress that mindfulness research has shown how thinking affects the material body that we inhabit and the way we think shapes the body and the environment on which we depend. The environment affects the body and mind. For example, the research by Sarron and Kasniak has found that people who exercise in a natural environment have higher levels of concentration than people who exercise in a built up environment. Thus, the more we decimate the environment the worse it will be for our ability to recreate ourselves when we take a break from learning, working and teaching. Another key finding from neuroscience is that the telomeres or parts of the cell that protect us from ageing are protected when we have a sense of purpose, when we use all our capabilities each day in an environment that is rich in nature and not degraded.

The will to make a difference is the motivation for change. We cannot rely on the 1% to make the difference and we cannot expect that the powerful will want to make vast changes to the status quo. The temptation to continue to focus on critique of the 1% along with pleas for transformation is great, but we also need to recognise the banality of evil. The complicity evident in entire societies that do not focus on the elephants in the room, namely, that our way of life is unsustainable.

The IPCC formula suggests that the privileged lives of some could lead to existential risk for people and the planet (Bostrom 2011). This has implications for the way we live and the need to change our way of life through living sustainably. Representation, accountability and sustainability challenges need to be met through addressing consumption choices that are currently very unequal. The gaps between rich and poor, the powerful and the powerless have become wider and wider. The greatest challenges are the consequences of inaction that will potentially pose an existential risk to humanity. These challenges include representation of the increasingly diverse populations within nation states along with accountability to ensure that resources (e.g. water, food, and energy) are used fairly, equitably and sustainably in local and regional biospheres. According to Vaughn (2014), Tax on

[71]http://www.mindfulness.org.za/mindfulness2014/conference-speakers/#alfred "Aging, Mindfulness and the Brain Biograph "Al Kaszniak received his Ph.D. in clinical and developmental psychology from the University of Illinois in 1976, and completed an internship and postdoctoral training in clinical neuropsychology at Rush Medical Center in Chicago. He is currently Director of the Arizona Alzheimer's Consortium Education Core, Director of the Neuropsychology, Emotion, and Meditation Laboratory, Faculty and Advisory Board member of the Evelyn F. McKnight Brain Institute, and a professor in the departments of Psychology, Neurology, and Psychiatry at The University of Arizona (UA). He formerly served as Head of the Psychology Department, and as Director of the UA Center for Consciousness Studies, and Chaired the Steering Committee for the inaugural International Symposia for Contemplative Studies (April, 2012, Denver, CO). He also previously served as Chief Academic Officer for the Mind and Life Institute, an organization dedicated to dialogue and collaboration between science and contemplative traditions. He is the co-author or editor of seven books, including the three-volume Toward a Science of Consciousness (MIT Press), and Emotions, Qualia, and Consciousness (World Scientific)".

meat will cut methane build up.[72] People want to eat meat but it is unsustainable because of the growing number of people. Those in denial are fighting action on climate change.

According to Goldenburg:[73]

> Brulle's study, published in the journal Climatic Change, offers the most definitive exposure to date of the political and financial forces blocking American action on climate change…the vast majority of the 91 groups on Brulle's list- 79%- were registered as charitable organizations and enjoyed considerable tax breaks. The groups collectively received more than 7bn over the eight years of Brulle's study – or about $900 m a year from 2003 to 2010.

The argument developed in 'Transformation from Wall Street to Well-being' (McIntyre-Mills et al. 2014) and 'Systemic Ethics' (McIntyre-Mills 2014) is that the containerist approach to science, democracy, ethics and governance is a core part of the problem. My research question is: how should we live? How can we be more sustainable in our living choices? My research tests the hypothesis that decisions that are made at the lowest level possible are more likely to take into account the complexity of the decision makers. The recently published volumes 'Systemic Ethics and Transformation from Wall Street to Wellbeing' provide a plea and an example of a downloadable architecture for doing things differently. Local level engagement and wide ranging goals appear to be suggesting new directions, but how do we join up the dots and become more mindful?[74]

---

[72]Vaughn, A. (2014) Tax on meat will cut methane build up The Guardian Weekly, 3/1/ 2014 p. 11. "Scientists say livestock emissions are big factor in greenhouse gas crisis. Meat should be taxed to encourage people to eat less of it, thus reducing the production of global warming gases from sheep, cattle and goats…".

[73]Goldenburg, S. 2013 Up to 1bn a year spent fighting action on climate change The Guardian Weekly, 3/1/ 2014 p. 11.

Vaughn, A. (2014) Tax on meat will cut methane build up The Guardian Weekly, 3/1/ 2014 p. 11. "Scientists say livestock emissions are big factor in greenhouse gas crisis. "Meat should be taxed to encourage people to eat less of it, thus reducing the production of global warming gases from sheep, cattle and goats…".

[74]The challenges of food, energy and water security were discussed in conversation with colleagues about the micro, meso and macro level challenges of addressing the social, economic and environmental challenges facing people living in cities that need to be supported by viable regions that make farming a less challenging experience. I discussed the liberative potential of enabling people to monitor from below, irrespective of whether they are citizens or not. Non-citizens should also be given a voice because human rights are not being adequately addressed through the nation state. My contribution to the conversation covered the need to understand that complex challenges (such as poverty and climate change) span many different variables that are inter related and that are valued differently by different stakeholders who do not agree about climate change or the causes of poverty. The systems approach begins when first we try to see the world through the eyes of another, according to West Churchman. Perceptions matter! West Churchman's so-called 'enemies within' includes religion, morality, politics and aesthetics. Paradoxically, these are the values that make us passionate and compassionate. As we face some of the greatest challenges ever facing the planet we will need to do what we have always been able to do - unite together to face 'enemies within' and the greatest external threats we all face to food, energy and water supplies. What we do upstream affects those downstream in our neighbourhood. This will affect all aspects

Nussbaum (2006) argues for social justice and for capabilities for human beings and sentients without addressing the dualism of people versus the planet and the problem of property and ownership as a core dimension of the sociocultural, political and economic problems we face today. The social contract is only extended to citizens of a nation state. It is anthropocentric and inadequate to protect the environment. Non-citizens are reliant on pleas for human rights and other sentient beings are reliant on pleas for animal rights.

De Waal makes a case for cooperation based on empathy and reciprocity as one of the bases of animal nature along with the capacity to compete. Both aspects are important for evolution. Stuart Hall on critical heuristics stresses that human identity today is not tied to the nation state. As a public academic, he posed questions about cultural identity and stressed that these days people need to give a narrative response to the question: who are you? Similarly, he has stressed in 'our mongrel selves' (1992) that we need to think critically about who, gets what when why and to what effect.

The other aspect of my argument is based on the work of Stiglitz on 'Wellbeing stocks' who argues for reframing economics to consider not merely profit and productivity, but the fabric of social and environmental life that support well-being and any economic activity. The work of Hannah Arendt 'on banality of evil' reminds us in her review of the Eichmann Trial that a single leader cannot be held responsible for making the decisions that led to many people being incarcerated and deprived of citizenship rights, it was the extensive system and the bureaucracy—voted in by the people—that supported it.

2.5 Values and Relationships: Expanding Solidarity

The need to speak out is vital if we are to maintain and extend democracy to the 50 million displaced people. This is the greatest number of people who have been displaced the UN Commissioner of Human Rights since World War 2.

(Footnote 74 continued)

of life and render all other debates pointless. In order to address these challenges it requires the ability to work across many very diverse groups locally, nationally and in post national regional contexts. It also requires the ability to evolve in ways that are quantum rather than digital. Being the change requires cooperation with others and with nature if we are to address the challenge of so-called 'existential risk' (Bostrom 2010). Monitory democracy 'from below' that supports respectful dialogue across service users and provides can be assisted through using a range of simple engagement processes that match the diverse needs of people with resources, in order to address the challenge to distribute resources more fairly. The free software that is available (McIntyre Mills et al. 2014) is downloadable. It pilots a way to enable participants to join up the dots when thinking about social, economic and environmental challenges. As it is used it updates, grows and provides pathways to address perceived challenges. The forthcoming volumes 'Systemic Ethics' and 'Transformation from Wall Street to Wellbeing' provide a plea and an example of a free downloadable architecture for doing things differently.

Judith Butler holds the Hannah Arendt chair and stresses the need to consider our humanity not only in terms of our ability to compete and to be resilient, but to also consider our vulnerability and need to cooperate. She asks what kind of world we would live in of we did not care?

Butler[75] links the need for food and employment; these along with energy and water are the basics of life that need to be protected through transnational movements. This argument is echoed by the work of Shiva on 'Flows'. She argues that instead of nationalist ideas supported by the Tragedy of the Commons Approach, we need to develop more understanding that food, energy and water flows need to be protected by a sense of our interconnectedness.

According to Ulrike Guérot:

> "The times in which the old cannot yet die and the new cannot yet emerge are the times of monsters".... Or, as the recently deceased Ulrich Beck put it: "As long as we leave authority over European integration in the hands of the states themselves, Europe cannot come into existence." [7]...Expressed more precisely: the central role of the European Council in the EU's governance system is a systematic obstacle to finding European solutions.[8] The costs of harmonised solutions are nationalised. The European solution doesn't work because the respective "national interests" are invoked.[76]

When you listen to the story of a displaced or vulnerable person, then a connection is made. The dehumanising of people and the focus on power over resources is at the heart of the issue. As is the notion that it is acceptable to treat the vulnerable and voiceless sentient beings with less respect than those who are protected by the social contract.

Solidarity with others needs to be extended from narrow association with family and friends to neighbourhoods and then to solidarity with others who are beyond our immediate human circle.

Anthropocentrism and placing the needs of the elite few ahead of the needs of the majority of human beings and the ecosystem on which we rely is at the heart of the issue.

This is addressed in the following discussion on consumption of the minority at the expense of the majority in this generation and the next. The starting point is that corporatisation and commodification pose the greatest threats to representation,

---

[75]http://www.salon.com/2011/10/24/judith_butler_at_occupy_wall_street/By viewing the video, the performative approach to social change through being the change through reframing the current socioeconomic system. "If hope is an impossible demand, then we demand the impossible. In brief remarks to the occupiers at Liberty Plaza, Butler offered her take on the continuing "demands" debate: People have asked, so what are the demands? What are the demands all of these people are making? Either they say there are no demands and that leaves your critics confused, or they say that the demands for social equality and economic justice are impossible demands. And the impossible demands, they say, are just not practical. If hope is an impossible demand, then we demand the impossible — that the right to shelter, food and employment are impossible demands, then we demand the impossible. If it is impossible to demand that those who profit from the recession redistribute their wealth and cease their greed, then yes, we demand the impossible".

[76]https://opendemocracy.net/can-europe-make-it/ulrike-guerot/europe-as-republic-story-of-europe-in-twenty-first-century.

accountability and regeneration. The most important part of the argument is that commodification and corporatisation are a result of a colonial mindset or 'apartheid thinking'. The sense that 'self–other and the environment are separate entities and that profit can be made at the expense of the well-being of others in this generation and the next cannot be limited to the social contract approach. The argument is developed by Shiva in a range of writings and in a book called: 'The rights of nature: the case of the universal declaration of the rights of mother earth' (Barlow and Shiva 2011).[77]

The rising rate of carbon emissions remains a source of concern and unless the economy changes the temperature will rise way beyond 2 °C, the safe limit for climate change rises.[78] The Paris talks have posed that the safe level should be 1.5° rise if the island homes in the Pacific are to exist[79] and that Australia and other heavy emitters will have to implement changes rather than continue with business as usual. This will require the demonstration of right living by the highest emitters, not by those who are the most affected. Singer's (2011) edition of 'the expanding circle' originally written in the 1980s remains relevant as the competition for the right to use the last of the non-renewable resources plays out in Paris in Dec 2015. But it is dated in so far as he argues that belief in religion is on the wane.

The rise in the numbers of people identifying with Islam has risen in recent years. Singer's rationale for the book was to explore the potential of sociobiology for ethics. Since writing it a number of studies by zoologists and primatologists such as Frans de Waal—underline that a sense of fairness and reciprocity are shared by many non-human species who live in groups.

The fact that Rousseau and Hobbes believed that primitive humans were solitary and savage has little support in archaeology, primatology or zoology. Human beings are capable of competing and cooperating as are many other sentient beings who live in groups. Our evolution has been supported by a sense of fairness or equity and a sense of reciprocating the sharing of food, grooming and support in times of danger. This notion of altruism has usually been extended to close kin and to one's tribe.

The notion resources should or could be shared with people outside one's own family or group can be debated by some as irrational, because it would deplete the conditions/opportunities of placing one's own family in an advantage over others. This is based on the underlying notion that we win only through competing with others. But once we realise that the zero sum approach to ethics is based on the assumption that we are in separate competing groups, tribes or nations or worse—fighting for a place in a life boat. When this approach to competition is applied to all those who are not part of our defined in group—protected by family law or the

[77]http://therightsofnature.org/related-books/http://www.ipsnews.net/2011/05/global-campaign-to-bestow-legal-rights-on-mother-earth/Polly Higgins: https://www.youtube.com/watch?v=QPUmN88htCo Earth democracy http://m.youtube.com/watch?v=Bq0lrbznsjc.

[78]http://www.theguardian.com/environment/2015/oct/10/climate-2c-global-warming-target-fail.

[79]https://newmatilda.com/2015/12/12/pacific-island-leader-slaps-down-australias-climate-claims-in-paris/.

social contract at nation state level—it renders all those human beings and non-human animals outside the boundaries of protection. The problem is that the so-called enlightenment approach—based on narrow positivism and the associated brand of rationalism—assumes that Western Philosophy and Science are right and that other ways of knowing are incorrect. As Shiva (2012) stresses reductionist approaches to testing in science and reductionist approaches to economics compartmentalise problems.

The notion that moral behaviour is the preserve of human beings (living according to the dictates of the social contract) is now no longer an acceptable premise for governance and democracy. As resources dwindle the irrationality of the assembly line approach to production, the notion that all biodiversity that does not result in profit is disposable—or can be weeded out as an externality—is also no longer an acceptable premise for governance and democracy.

The affinity we feel for another is based on emotion which can be the basis for being fair or reciprocating in a way that will help another. This is one basis for ethical choices. The other is the intuition that we feel that one option is ethical and that another is abhorrent. Group living has resulted in sentients learning that being unfair or receiving/accepting resources and not returning any form of service, favour or material resource will lead to negative consequences. But it does not explain why some animals will help others (including those from other species) just because they are able to do so. In human beings, this altruism is explained as having a good feeling, or a rush of endorphins akin to being in love, believing that their *unseen* good deed will be rewarded by some Higher Being or in some after life or that their *seen* act of goodness would result in a rise in status which would be to their own advantage or that of their immediate group.

Learned behaviour is passed on as culture, a sense of the sacred or in religious and other forms of knowledge and the evolutionary advantage of having members of the population who are capable of nurturing others and capable of defending them is self-evident.

The reductionist fallacy has slipped into many arguments that are focused on stressing that only one of the following is important for ethics: biology, intuition, emotion, culture/sense of sacred/religion or only rationality.

Intuition and the associated emotional response may sometimes prevail when rational thought is used to uphold the initial almost automatic response to news that shocks. Once we hear that someone has been charged of murdering his child, we may immediately think they deserve a long prison sentence. But when we hear that the 'murder' could perhaps be called euthanasia because the child was suffering from an incurable, agonising disease for which there was no reliable pain management. The parent decides to help the child knowing that in doing so they will be punished according to existing law. The parent does this act openly and without deceit. The act could in some contexts be seen as selfless and based on a sense of compassion. The emotional connection or love makes self-sacrifice possible in their terms. The law may see it as wrong and subject to penalties. This hypothetical instance is perhaps easier to think about than the hypothetical problem in philosophy known as the so-called 'trolley problem' (Foote 1967) posed by Foote in which she asks people to think about the difficult choice of saving five people at the cost of one person's life. In the hypothetical scenario we know that the sacrifice of a perfectly innocent person could lead to saving five or more others on a railway line from a run-away trolley.

The problem is that we cannot be sure that the sacrifice of the single person would indeed save the other five. We could in fact be actively causing another death by pushing someone to their death in the hope that his death will save five others. A normative approach would stress that one life is sacred and that violence is wrong. No hypotheticals can predict all the variables. But they can help us to think about our thinking in terms of whether we can universalize and predict on the basis of ethical principles.

In the first example, we are talking about a decision based on the emotions of a parent. In the second we are thinking about what constitutes morality when we are faced with difficult options that require rational decisions based on considering the consequences[80] but also guidance based on our emotional intuition.

Certainly, allowing many asylum seekers to live wretched lives in camps as Syria is bombed is a case in point. But the reports by NGOs and journalist give me hope that we do feel this interpretation of the social contract is unjust and ethically is wrong. Just as the post-traumatic trauma felt by soldiers, pilots and some technicians who have bombed distant villagers indicates that we have evolved with a conscience and that inflicting pain on others triggers mirror neurons that make us feel or empathise with others pain.

On the 14th December, I watched an Australian Broadcasting Corporation news report which showed a small child running as his neighbourhood in Syria was bombed. He was lifted up by a man and taken to safety, but how many other children are not rescued?

This round about discussion is to summarise some of the decades of thinking on ethics and to say that biology, intuition, culture, experience and psychology come into play in specific contexts. As resources become more and more scarce will we become amoral as we compete for resources?

Singer (2011: 25) cites the work of Turnbull and describes the way in which people competed for resources during decades of drought. The Ik apparently were portrayed (on the basis of very little and very questionable evidence) as having little empathy for the dying, but would not actively kill another. They still held to their moral beliefs that cannibalism was wrong and they continued to have a belief in the sacred mountain as a source of wonder. Thus in this very difficult situation it could be said that *they did indeed* retain dignity and that they made ethical choices that

[80]The decision to allow 1 person to die (perhaps through torture) to enable others to survive is an argument policy makers following a utilitarian approach could justify on the basis of so-called rational calculation. But it is a slippery slope as it undermines intrinsic rights. My intuitive emotional response is that it is an abhorrent choice. Some would say they could make a decision to rescue five by ending the life of one person if they could dispatch the person through flicking a switch that would lead to their demise. The more distance (emotional and physical) from the other person the easier it becomes to inflict pain on others, or so the theory goes.

demonstrated some altruism and respect for the sanctity of life. They also continued to believe in something that symbolised stability and continuity and they worshiped it.

So the ability of human beings to rise above circumstances and to make choices that may not seem to be only to one's own advantage has been recorded in times of natural disaster and war.

The axiom that this book explores is that we can be free and diverse to the extent that our freedom and diversity does not undermine the rights of future generations of life. It is based on the idea that this axiom is indeed tautological and self-evident, because if we live in profligate ways we will destroy the fabric of life on which we depend. This is the basic socio-cybernetic logic on which this book builds.

Thus, it is logical that we explore ways to ensure that the fabric of life is protected by governing the Anthropocene—based on principles of fairness and reciprocity—to underpin a new architecture for democracy and ethics.

Given that displacement is becoming more common due to both social and natural disasters more needs to be done to protect habitat to ensure that it is not destroyed. Those who are not displaced have rights and responsibilities to protect the environment by living in ways that are fair to others (including non-human animals) and the environment. All human beings have the right to expect that by living carefully and not using more than is reasonable and that upholds human and non-human capabilities to live a life worth living that others will do so too. This caretaking approach of each global citizen needs to be regarded as a norm for all global citizens.

2.5.1 How Can We Achieve Cultural Transformation on Consumption Patterns Through Balancing Individual and Collective Needs?

In an era of social, economic and environmental convergence of risk we need to accept the possibility of new forms of governance to protect those who are not protected by the nation state, because they are not citizens. The work of Ann Florini provides possibilities as does the work of Danielle Archibugi. The challenges of governance need to be addressed by working across disciplines. Gibbons et al. (1994) argue that the ability to work across boundaries is vital for 'The new production of knowledge' and vital for 'the dynamics of research' to address current complex challenges. The limited paradigmatic approach to knowledge and geographies of the mind is a reflection of human geographies.

The so-called 'Tragedy of the Commons' argument developed by Hardin (1968) is a construction informed by simplistic thinking. Ironically, it is often used as the starting point for environmental thinking, but in fact it originated as an argument developed by Locke in support of the enclosure movement and private property in Britain. The argument being that contained areas of land are cared for better than the areas of land that are held in common and shared. The example of common grazing land is shared. But ironically, it is the privatisation and commodification of land and natural resources that has led to environmental pollution and degradation. The lack of trust between nations has evoked the rhetoric of nationalism and state protectionism. This has led to 'seeing like a state' (Scott 1998), which in turn leads to the argument for competition and the zero sum approach which argues that one nation can profit at the expense of others. The organisational philosophy and governance arguments that flow from this philosophy are based on limited disciplinary paradigms that can profit at the expense of others.

Ideally, a universal respect for social and environmental justice could enable subsidiarity as a means to support freedom (Follesdale 2006; Poe 2010) to the extent that it does not undermine the quality of life of this generation or the next. Dualist thinking pervades our consciousness and reflected in socially unjust and environmentally unsustainable designs for society. Designs need to be supported by constitutions, based on *a priori norms*, and consequentialist or *a posteriori approaches*, based on testing out ideas within context and with future generations in mind. Global axioms to protect future generations and the poor of this generation need to guide legal constructs and legal decisions at the local level. The principle of subsidiarity could rest comfortably with global axioms—provided that the caveat is made that *people at the local level can be free and diverse to the extent that their freedom and diversity does not undermine the rights of others or future generations of life*. This challenge of balancing individual and collective social, cultural, political and economic concerns needs to be buttressed by regional parliaments and courts that focus on social and environmental justice at a post-national level.

2.5.2 Power of Ethical Narratives

The rise of Donald Trump who used the metaphor: 'Drain the swamp' effectively underlines how the so-called 'unprotected' and marginalised in cities feel about their lives and how identity politics has played out by those who feel abandoned by a globalised elite who benefit from 'business as usual' narratives which exclude them.

Obama has stressed that the notion of 'clash of civilisations' just helps the terrorists' and Huntington's narrative can be understood as part of the same self-fulfilling narrative. We need a new narrative of Earth Democracy.[81]

[81]We are interconnected. The narrative of 'Earth Democracy' helps to support Space Ship Earth. There is hope, because according to Evans Pritchard the Nuer (a very war like group) understood that they needed to unite against a common enemy. So they set aside their differences when facing a large outside threat. Today the threat is us. We need to understand that to survive we need to cooperate not only with one another but also with living systems of which we are a strand. Respecting and enabling diverse capabilities of sentient beings and enabling them to live a life worth living is part of Earth Democracy. Gender diversity needs to be addressed so that women are able to play an equal role in Earth Democracy where they fulfil all their capabilities—not just reproductive roles but also productive, creative and strategic roles—so that the demographic transition from over population to balanced reproduction can occur. This occurs when literacy and numeracy—or educational opportunities for all are achieved. We are interconnected—we can no

This book responds to the Noah Ark narrative. Instead of the idea of selecting a few chosen ones to protect them against the coming flood it argues for recognising that humanity faces a common threat. Just as the Nuer chose to ignore conflicts between tribes as they faced a larger enemy, humanity needs to appreciate that we face a wider threat, namely a natural, social and economic disaster as a result of current consumption choices. The challenge is to face up to our interconnectedness and to be able to hold in mind many variables. 'Us/them' need not expressed in terms of tribes, organisations and nations. Space ship earth is a metaphor used by Kenneth Bounding to help reconnect humanity's sense of geography with the planet and the universe of which we are a part. It is a plea that we should strive to achieve transcendence. Human beings face the challenge of wanting to be individuals and also to be part of group. They have evolved through ability to cooperate and compete.

If we study the work of Bentham, Mill, Hume or West Churchman who are all utilitarians who are concerned about the consequences and who base their findings on empirical research, we see that all pragmatists have in common a desire to test out ideas in context. American social psychologist, Jonathon Haidt[82] discusses moral foundation theory with an emphasis on biological determinism of culture. Haidt emphasises that rational thinking is vital for ethics but argues that often emotion and intuition prevail in shaping ethical decisions. His major problem is that he believes that human beings are guided by evolutionary drives and he confuses these with the notion of cultural choices. The theory is based on his interpretation of his empirical research. But he assumes that culture is closed system—evolution is responsive and it is an open system. I develop a critique of this approach based on the work of other pragmatists, existentialists such as Frankl (1995) and progressive idealists such as Martha Nussbaum and Judith Butler (who holds the Hannah Arendt professorship) and asks what the world would be like if we did not care for one another on a daily basis and if we did not recognise our vulnerability, hybridity and interconnectedness.

Haidt stresses that the foundations of morality are based in six basic distinctions that evolved from human beings sense of taste. The evolution of taste distinctions was useful as it promoted survival. Rotten or toxic food was bad, whereas sweet, savoury tastes are good. Social intuition is based on tastes, namely: sweet, sour,

(Footnote 81 continued)

longer address differences merely through 'exit', because the challenges are planetary and so we will need to apply ethics to enable loyalty to current and future generations through being the change not merely voicing stewardship concerns.

[82]See Haidt on moral judgement https://www.youtube.com/watch?v=vs41JrnGaxc, https://student.cc.uoc.gr/uploadFiles/179-%CE%9A%CE%A8%CE%92364/moral%20judgment.pdf, http://www.ted.com/talks/jonathan_haidt_on_the_moral_mind?language=en, http://politicsofthemind.com/2012/12/01/the-six-moral-foundations-the-real-issues-lying-beneath-the-surface-of-political-debate/.

bitter, and salt and glutamate.[83] According to Haidt, the distinctions evolved to enable people to think in terms of a range of basic distinctions: Care/harm, Liberty/oppression (supported by Libertarians), Fairness/cheating, Loyalty/betrayal (supported by conservatives/Republicans), Authority/subversion, Sanctity/degradation (supported by progressives such as Democrats).

The problem with Haidt's theory is that it supports the inevitability of cultural views and ironically despite his apparent progressive stance his belief in explaining culture through metaphors of taste reifies the notion of culture and this provides a glib theory of politics. Another more profound concern is the ability to use this theory to support Huntington's 'Cultural Clash thesis', because of its determinist approach. It is for this reason that I start by outlining the moral foundations theory before developing a critique and an alternative based on building the capacity to think critically. The role of participatory democracy is vital in this regard in order to enable people to develop capabilities for human beings and to protect the capability rights of all sentient beings.

The Social Intuitionist approach which Haidt supports explains that we respond with 'intuition' and emotion. But we need rational thinking to make moral decisions that are guided by norms to foster careful consideration of the consequences of our daily choices and to enable us to make decisions that support the virtuous life of 'Eudaimonia' or well-being. According to Kenan Malik (2015):[84]

> The right to 'subject each other's fundamental beliefs to criticism' is the bedrock of an open, diverse society. Once we give up such a right in the name of 'tolerance' or 'respect', we constrain our ability to challenge those in power, and therefore to challenge injustice. At the heart of this argument is the insistence that any form of progressive politics requires us to overcome, rather than embrace, the barriers of identity. Kenan Malik/Free Speech in an Age of Identity Politics/TB Davie Memorial Lecture that it requires us to work towards a more universalist vision of society. And that only free speech makes this possible. Free speech – proper, full-blooded free speech – is the lifeblood of any progressive politics and of any progressive transformation of society. If we treasure the one, we must treasure the other.

Furthermore, from an existentialist approach Viktor Frankl stresses that human beings have the capacity to rise above circumstances and to choose to respond differently. Jonathan Haidt and Frans De Waal make the case that intuition and emotion are important dimensions shaping ethical decisions and moral choices. But de Waal makes the case (unlike Haidt 2008) that human beings have evolved from primates who are capable of understanding fairness. Thus, he is much more optimistic about human nature and less conservative perhaps than Haidt. The conclusions reached by Haidt are problematic. From the views of the capability approach, human beings need to be supported to achieve their potential. Existentialists such as Viktor Frankl make the point that human beings need to have high standards set for

[83]http://poseidon01.ssrn.com/delivery.php?ID=652087122pdf.

[84]"South Africans must defend democracy at grass roots". Kenan Malik 30 September 2015 https://www.uct.ac.za/usr/news/lectures/tbdavie/TBDavieLecture2015.pdf.

them so that they can rise to achieve them. Viktor Frankl[85] stresses that interpretation of a situation and having a sense of purpose in life is a core aspect of being human. Neo Marxists such as Michael Hardt make the case for the adaptability of human nature and culture. Hardt (2010)[86] in conversation with Astra Taylor as part of her interview on 'the Examined Life' stresses the need for ongoing training for participatory democracy and the need to achieve transformation. His approach stresses revolution, but you cannot demonstrate empathy and democratic values through violence, so it needs to be achieved as peacefully as possible. He stresses that a great deal of the arguments about morality, human rights and transformation potential rests on a debate about whether humanity is capable or incapable of self-rule. He stresses that Bolivia and Ecuador are places to watch for examples of how to live differently—because of their constitutional protection of the planet—although they have some failings and although the rhetoric perhaps outweighs the reality of what has actually been done to protect the forests and to limit mining. But as Hardt stresses they are indeed 'places to watch', for examples, as to how to do better. But what is even more important is the world view based on a belief in 'Pachamana' and the legal transformations that could eventuate from the spiritual connection to the land. I would argue that the hypocrisy of appearance needs to be avoided through implementing ways to ensure that the ideals are translated into practice in an open and transparent manner.

In order to address areas of concern in a manner that is appropriate, it is necessary to develop ethical literacy by working across disciplines in the social and natural sciences. The argument that I develop is that by drawing on primatology, we can learn that animals (primates and many mammals) have the capacity for empathy, reciprocity and fairness and that human beings have evolved because of their capacity to cooperate and not only their capacity to compete. But the next step for transcendence is to recognise our interdependency. So the thesis I develop in this book is that the next step in our evolution is an appreciation of our interdependence.

I discuss three options for behaviour, namely cooperation, competition and an understanding that we are interdependent. Ethics is based on both rational thinking and emotions. We can start by taking into account that virtues and values can provide us a sense of two or more competing values that are equally important for ethical behaviour, such as freedom, protecting human well-being and prevention of the spread of disease. The pluralist—Isiah Berlin stresses that we need to be aware of context and to think carefully about the implications of our choices. The CSF disease or mad cows disease was caused by agriculture that did not respect the quality of life of animals. It resulted in the need to destroy farm animals and the need to confine farmers to their farms, thus limiting their movements and their

[85]http://www.ted.com/talks/viktor_frankl_youth_in_search_of_meaning, https://m.youtube.com/watch?v=Gc_COGWKKg8.

[86]https://m.youtube.com/watch?v=f0IopdH1e3s.

freedom. The concern from the point of view of the pragmatist is that CFJ disease occurred in UK and that the consequences of human animal transplants on health of humans and the consequences of inflicting pain on animals is their main concern. From the point of view of an idealist it is the sanctity of human life and the implications of transplanting across the human animal divide. When thinking about the ethical acceptability of any specific use of science and technology, utilitarians will ask: 'What are the risks and benefits?'Some criticise this as reducing ethics to nothing more than risk assessment, according to Somerville (2009: 24).

In some instances no matter how useful or risk free for some it can be said that something is wrong. Somerville stresses that some actions or non-actions are wrong, no matter how useful they may be or how much people can be enriched by exploiting others. Thus cost benefit analysis breaks down when we face moral issues, according to her. This is a vital point for morality.

2.5.3 How Can Participatory Governance Support the Self-management of Our Ecological Footprints?

The Intergovernmental Panel on Climate Change (IPCC) formula shows that the excessive consumption of energy resources impacts on the size of our carbon footprint; this footprint is defined in terms of—*E* (Emissions) = Population × Consumption per person × Energy Efficiency × Energy Emissions. This suggests that the privileged lives of some could lead to 'existential risk' for all forms of life on the planet (Bostrom 2011). The price of inequality—national and global—has escalated. According to the United Nations High Commissioner for Refugees, Guterres (UNHCR 2014), for the first time since the Second World War, the global figure for displaced persons has now passed 50 million[87] and, by 2050, this figure could be as high as 150 million (Rusbridger 2015: 13).And yet the needs of the displaced are not addressed through the current architectures of democracy, governance and education. Surely it is time to reframe the social contract and to support public education to enable people to join up the dots? Liberal democracies are increasingly criticised for not:

- **Re-presenting** the interests of diverse citizens
- **Engaging** stakeholders in dialogue
- **Building a shared sense of identity** whilst enabling individual diversity, capability and freedom to the extent that the diversity, capability and freedom *of others* are not undermined

[87]Wilkinson and Pickett (2009) argue that after a certain point, more consumption does not lead to more happiness. Thus, increased consumption based on increased economic growth does not make people happier—in fact it poses 'existential risks'. 'More equal societies almost always do better' socially, economically and environmentally, to cite Von Foerster (1979). We need to think critically about the taken for granted ways in which we live and the size of our eecological footprint.

The current difficulties in addressing representation, accountability and sustainability are theoretical, methodological and value based. The issue is that the nation state is no longer able to address the challenges that span national boundaries because poverty and climate change do not respect political boundaries. Critiques need to avoid reifying systems, rather than addressing the potential to enhance representation, accountability and sustainability through reframing the architectures of democracy and governance as detailed in 'Transformation from Wall Street to Wellbeing' and its companion volume 'Systemic Ethics' which stress that political and economic systems are constructions that can be reframed by moving beyond the recognition of well-being stocks systems to take into account cultural flows. Water and seeds can be seen as a synecdoche of a new economics and perhaps a new constitution that recognises the role of human beings as stewards.

'Towards a Planetary Passport' is an extension of this argument with an emphasis on food security as a way out of the current social, economic and environmental crisis. How can we transform democracy and governance through public education to protect the global commons and the public good? Food, energy and water provision have implications for human security that requires balancing individual and collective interests within the wider regional biosphere. Ironically, Wilkinson and Pickett (2009) argue that after a certain point, more consumption does not lead to more happiness. Thus, increased consumption based on increased economic growth does not make people happier—in fact it poses 'existential risks'. 'More equal societies almost always do better' socially, economically and environmentally. Stiglitz et al. (2011: 15) use a multidimensional measure of well-being spanning:

> 1. Material living standards (income, consumption and wealth), 2. Health, 3. Education, 4. Personal activities including work, 5. Political voice and governance, 6. Social connections and relationships, 7. Environment (present and future conditions), 8. Insecurity, of an economy as well as a physical nature.

2.6 Policy Paradoxes

> The will to power is perhaps the key concept in Nietzsche's philosophy. The will to power appears to be the basis for the geography of nation states and the way in which international relations is conducted. It is strongly connected to his concept of 'life'. So in Beyond Good and Evil Nietzsche says 'A living being wants above all else to release its strength; life itself is the will to power'[88]

[88]http://documents.routledge-interactive.s3.amazonaws.com/9781138793934/A2/Nietzsche/NietzscheWillPower.pdf.

The current way of life is unsustainable and in a bid to maintain the status quo—profit is extracted from people and the environment. Have I considered: content, context, structure and process? Have I considered a priori and a posteriori dimensions? Many ways of knowing: logic, empiricism, dialectic and pragmatism.

If there is a problem have I engaged with others and 'tried out'/piloted an alternative and engaged in ongoing learning?

The greatest challenges are the consequences of inaction that will potentially pose an 'existential risk to humanity' (Bostrom 2011). These challenges include representation of the increasingly diverse populations within nation states along with accountability to ensure that resources (e.g. water, food, and energy) are used fairly, equitably and sustainably in local and regional biospheres. But addressing the EF has implications for policy and politics, because the gaps between rich and poor, the powerful and the powerless have become wider and wider (Wilkinson and Pickett 2009).

Representation, accountability and sustainability challenges need to be met through addressing consumption choices that are currently very unequal. The gaps between rich and poor, the powerful and the powerless have become wider and wider.

Critical systemic thinking helps to explore our thinking and our relationships spanning self, other and the environment. As a Meta form of inquiry it is based on questioning boundaries of inclusion and exclusion, it examines the so-called 'enemies within' (religion, mortality, politics and aesthetics) and it considers the consequences of our choices. It examines the value placed on ecosystems (Fisher et al. 2009) and how this impacts on beliefs and intentions. This proposed research will explore the perceived implications of the stewardship of biodiversity for well-being, resilience. Well-being according to Stiglitz et al. (2010: 15) who use a multidimensional measure of well-being spans (1) material living standards (income, consumption and wealth), (2) health, (3) education, (4) personal activities including work, (5) political voice and governance, (6) social connections and relationships, (7) environment (present and future conditions) and (8) insecurity, of an economy as well as a physical nature. Leisure should also be valued. According to Stiglitz et al. the essence of the commissioned work's findings is that wealth needs to include stocks for the future—social, economic and environmental. The way to achieve this is in part through public education. Resilience is defined as the adaptive capacity of the physical environment, of an individual or of a group. It concerns factors such as the capacity of members of a community to act together and to be able to modify or even transform existing ways of life (Rose 2005; Hulme 2009). Biodiversity refers to:

> Current rates of species loss exceed[ing] those of the historical past by several orders of magnitude and showing no indication of slowing. Major drivers of biodiversity loss on a global scale are land-use changes and agricultural intensification. These processes are threatening ecosystem functioning and services on which humans depend…. In consequence, there is a deep concern that a loss of biodiversity and deteriorating ecosystem services contribute to worsening human health, higher food insecurity, increasing vulnerability of ecosystems to natural disasters, lower material wealth; worsening social relations

> by damage to ecosystems highly valued for their aesthetic, recreational or spiritual values; and less freedom ....Growing awareness of the importance of biodiversity for human wellbeing has led governments and civil society to set targets to reduce ...loss (see Convention on Biological Diversity www.cbdint) (Lindemann-Matthies 2014: 195–6)

There is little doubt that accelerated climate change will adversely affect food security and sustainability. This study looks at rapid levels of urbanisation and the implications for living in cities (Rees and Wakernagel 2008) by comparing and contrasting attitudes in Australia and Indonesia. The threats to human security will increase (Flannery 2005; Pretty 2013; Stern 2006; Stiglitz et al. 2010)—particularly if we continue to consume in rapidly urbanising cities at current rates (Davies and World Institute 2008)—resulting in significant impacts on the size of our ecological footprints.

The impact of climate change has been underestimated (Lovelock 2009; Rockström et al. 2009) and local solutions have been overlooked. The aim of the public policy and administration research is to explore the following hypothesis: The greater the level of Public Participation the greater the understanding of UN development goals. It addresses the UN Sustainable Development Goals and targets through building the strategic capacity of people to understand ecological citizenship rights and responsibilities and the implications for self-managing their ecological footprints with reference to personal decisions about food, energy and water choices as they relate to human security. The emphasis (in line with the UN) is to address healthy lives and promote well-being for all at all ages through inclusive and equitable public education through promoting life-long learning opportunities for all to address gender equality and to empower all women and girls. It strives to address ways to ensure access to affordable, reliable, sustainable, and modern energy for all in environments that are inclusive, safe, resilient and sustainable.

The emphasis is on the most marginalised and focusing on ways to protect the poor and the affected communities, focusing on women, youth, local and marginalised communities to understand mitigation and adaptation processes. The research aims to enhance strategic decision making and to develop inclusive development strategies in line with the Paris Development Agenda (OECD 2005/2008) which stresses that those on the receiving end of development aid need to be shapers of policy and the UN Sustainability Development Goals (2014). This proposed research explores perceptions on ecological citizenship and rights and responsibilities to manage the size of our ecological footprint.

The research draws on and extends Hulme's research on why we disagree about climate change (2009) and Cornelius' research on systemic approaches to understanding emotions (1996). Values, perceptions and emotions could be better understood from a systemic viewpoint that draws on interdisciplinary theories, in this case such as to 'why more equal societies almost always do better' (Wilkinson and Pickett 2009). The 'best-worst' option, namely democracy is in need of improvement (Hulme 2009; Giddens 2009). The specific challenge that this research addresses is to match forms of appropriate participation that are accessible to diverse groups and not viewed with cynicism (Cooke 2004). Significantly, these questions assess whether participatory democracy and governance enhance sustainable living and well-being. It involves public education through local governments, schools, businesses and NGO organisations. It will identify: the decision making context; constraints to achieving outcomes; elements of three scenarios (denial of the need to change, too little action too late, sustainable long-term adjustments); and key factors (variables) in facing up to the risks. It could contribute to the SD21 policy initiative 'Sustainable development in the twenty first century'. The ability of governments to secure the environmental regions—on which the cities depend—to deliver services and resources will impact on the liveability of our communities and human security within our region.

The conundrum this book addresses is that nation states need to find ways to address the biggest challenge of the decade, namely to find a way to engage large groups of people effectively and to enhance a sense of shared responsibility for environmental management and consumption. Significantly it seeks to deepen an understanding of how people perceive local challenges and experiences and develop a new understanding of how to match the processes and governance structures to diverse interest groups—with different life chances and different sociodemographic characteristics. The research will assess the extent to which participation per se could help to promote greater understanding of rights and responsibilities to self-manage the size of our ecological footprints through combining cycles of open and structured dialogue that encourage people to think in terms of the consequences of their values about how we live our lives and how we ought to live our lives and the implications of these choices on the size of our ecological footprint.

The ecological footprint will be discussed in terms of its roots and meanings and then in terms of its policy and governance implications for the common good and the global commons. The 'common good' is defined in terms of the well-being of humanity and the planet on which humanity depends. The global commons spans all national biospheres and includes the fabric of life on which living systems depend and of which we are part—such as the water we drink, the earth that feeds us and the air we breathe.

> …Henceforth, there are no merely local occurrences. All genuine threats have become global threats… (Beck 2009: 19).
>
> The current way of life is unsustainable[89] and in a bid to keep things going profit is extracted from people and the environment.

---

[89]'The coming 'tsunami of debt' and financial crisis in America Forces that caused the world economy to collapse, including income inequality and debt, are again in action, and could drag corporations down in their wake', Dimitri Papadimitriou, theguardian.com, Sunday 15 June 2014 17.58 BST. "… Forces that prompted Occupy movements protesting income inequality and financial misconduct are again in action, according to research. …The evidence demonstrates that the de-leveraging of the very rich and the indebtedness of almost everyone else move in tandem; they follow the same trend line. In short, there's a strong and continuous correlation between the rich getting richer, and the poor – make that the 90% – going deeper into debt. That the share of income and wealth to the richest has skyrocketed is certainly not a new revelation. The heralded data tabulations of Thomas Piketty and Emmanuel Saez have demonstrated just how spectacular

2.7 Debunking the 'Clash of Civilisation' Approach

The contrast between different emphases in the life of Mohammed needs to be researched carefully.[90] In 'Heretic: Why Islam Needs a Reformation Now', Ayaan Hirsi Ali (2016) argues that more understanding of Islam is needed. The early life of Muhammad is characterised as seeking a life of virtue and harmony, known as the personal journey or 'Mecca approach' to a virtuous life. The later life of Muhammad is characterised as the life of a struggle for religious conversion of others, known as the so-called 'Medina[91] approach'.

It is important to engage in a conversation as suggested in 'Philosophy in a Time of Terror' (Borradori 2003) in which Habermas and Derrida talk with a journalist Borradori about the need to develop respectful dialogue and a sense of hospitality to others. This is particularly relevant today as democracy becomes threatened with the fear of Islamic State and the violent face of Islam. Democracy, ethics and the enlightenment is threatened by closure of minds, a lack of compassion and a lack of dialogue. The potential for violence exists in both Christianity and Islam and dialogue that enables critical reflection is vital to examine the potential for the worst aspects to surface when fear of 'the other' predominates (Almond 2015).

The notion that the nation state is inevitably a bounded piece of territory that needs to be defended is an outdated idea. The nation state concept is rooted in the time when armies needed taxes to fund wars against the 'other' (Florini 2003). Sovereign nation states required loyalty to sovereign leaders and in return they protected their subjects from invading armies. The so-called containerist idea that the state is the best form of protection for people and property against the invading hordes is now in question The basis for democracy and governance needs to be revised to consider fostering relationships across conceptual and spatial boundaries to enable 'post national' constellations of states and increased roles for twinning

(Footnote 89 continued)

the plutocrats' portion became in the run-up to the Great Recession. They codified the belief that no one else can ever catch up with the very wealthiest. ...To paraphrase Voltaire's words on God, even if bubbles and debt did not exist, it would be necessary to invent them. And that is exactly what we are doing".

[90]I hypothesized in the Journal of Consciousness Studies that participation through awareness and consciousness-raising (McIntyre-Mills 2010) will influence the way in which people value relationships with others and the land. This hypothesis is based on the notion of neural plasticity in that the brain shapes the environment and, in turn, is shaped by the environment (Bateson 1972; Beer 1994; Greenfield 2000). At a practical level the greater the level of participation the better the match between service users and providers. The complexity of a governance response needs to match the complexity of the service users. In designing a response we need to ensure that the lived experiences of users and service providers is taken into account. The IPCC formula suggests that the privileged lives of some could lead to existential risk for people and the planet (Bostrom 2011).

[91]Medina is in Saudi Arabia. It is where "Muhammad fled when he was initially driven out of Mecca, and the place where he attracted his first followers. Medina currently has a population of about 600,000 people and is the home of 'The Prophet's Mosque'." http://www.religionfacts.com/medina, accessed 10/05/2016.

relationships across major cities. This is threatened by 'us them politics', as stressed by Sadiq Khan who supports remaining within the European Union.[92] The greatest threat facing humanity is the widening gap between rich and the plunder of the environment.

The challenge is to face up to our interconnectedness and to be able to hold in mind many variables. 'Us/them' expressed in terms of tribes, organisations and nations versus is a reflection of the potential to compete rather than the potential to cooperate. This has been encouraged by the western individualism and capitalism and by the evolutionary emphasis given to Darwin's thesis. But this is a misinterpretation of his work according to primatologists such as De Waal. We evolved as human beings through our ability to compete and to cooperate. Space ship earth can be seen as a metaphor to unite us. The notion of separate life boats or worse—only one life boat is problematic. But the Ark metaphor is not problematic if it is seen in the way Buckminster Fuller suggested as Space Ship Earth. Kenneth Boulding's plea is that we achieve transcendence through our ability to think about our thinking and to move from basic categorical thinking towards understanding our interconnectedness. Obama (2015)[93] has stressed that the notion of 'clash of civilisations' just helps the terrorists.

The 'Clash of Culture' thesis developed by Huntington (1996) is understood as part of the same self-fulfilling narrative. The case I make is that we need a new form of democracy, governance and ethics to reflect this new thinking.

2.7.1 Vignette: Cartoonists at Charlie Hebdo

Fraser (2015)[94] makes the point that the deaths of the cartoonists at the Charlie Hebdo magazine was a result of a sense of being affronted by the representation of Muhammad in a secular and disrespectful manner.

The ability to conceptualise a sense of the divine and sacred is cherished by many religions. Islam does not try to represent the divine through icons. This is also true of Judaism. I was at a Jewish funeral recently and the oration stressed the need to live a good life and to remember that we have to make our name through our acts. We come into the world with nothing and we leave with nothing. We also come into the world without a name. We may acquire a good or bad name depending on our acts.

The old testament of the Christian bible also gives admonitions against representation and reliance on icons:

---

[92]http://time.com/4322562/london-mayor-sadiq-khan-donald-trump/?xid=tcoshare.

[93]Roberts, D. 2015, No war with Islam, Obama says, but conflicts continue The Guardian 27.02.15.

[94]Tradition of iconoclasm: the Charlie Hebdo cartoonists were part of a history of attacking divine representations The Guardian Weekly 16/01/15 18–19.

> the danger they believe, is that we might end up overinvesting in a bad copy, just a human projection, so much better then to smash all representations of the divine (Fraser 2015, p. 18)

Interestingly, the research by Baroness Professor Susan Greenfield stresses the need to be able to think conceptually and that reading as a means to acquire knowledge is important, not only the reliance on digital media and images.

Iconoclasm means 'someone who refuses the established view of things, who kicks out against cherished beliefs and institutions. Which sounds pretty much like Charlie Hebdo' (Fraser 2015).

But it also means someone who abhors representations of the divine. Fraser suggests that both the attackers and the cartoonists have in common an undoing of representations, taken to the extreme.

> Far from being sacred, as some have claimed, freedom of speech is always contingent. All societies draw lines (Young 2015, p. 18)[95]

> The notion of polarizing ideas into black and white as if they were binary opposites denies that even if you do indeed frame events in " Manichean terms- black versus light; good versus evil – an imposed binary morality seeks to corral us into crude camps. There are no dilemmas, only declarations. What some lack in complexity they make up for in polemical clarity and the provision of a clear enemy. (Young 2015: 18)

But black and white are not separate categories. Light casts shadows; we project shadows onto others. Those who are like us and share our own norms and values are seen as good. Those who do not can be perceived to be the opposite. The connections with enemies need to be appreciated. We define ourselves in relation to others and we shape our lives in relation to experiences.

The connections with enemies need to be appreciated. We define ourselves in relation to others and we shape our lives in relation to experiences.

We need to face up to some of the reasons why our enemies oppose us.

But in so doing, we need to take care not to abandon our values and what we hold sacred.

Human beings, sentient beings and nature have value simply because they exist. This is an essentialist philosophy. It is seemingly in direct opposition to critical systemic thinking. But if we take the critical questioning process as a means to think through the consequences of our decisions for others and future generations we will realise that what we do to others we do to ourselves.

It is based on the notion that we are interconnected. Thus, narrow selfish pragmatism is expanded through thinking through the consequences for ourselves, others and the environment of which we are part. What we do to others and the environment, we do to ourselves.

At a conversation on 'Systemic Ethics' at Schumacher Institute on 26th and 28th August, 2014 we discussed the potential for doing things differently, ranging from

---

[95]The danger of polarized debate is all too real: the truth about the Paris attacks is more complex and more difficult than many would like to accept 18–19.

making use of the UN Sustainability Agenda to more profound shifts such as the declaration of the Earth Charter:

> Bolivia enshrines natural world's rights with equal status for Mother Earth: Law of Mother Earth expected to prompt radical new conservation and social measures in South American nation[96]

But testing out ideas in conversation and dialogue with others has a prerequisite, namely a willingness to engage in sincere dialogue. Power always needs to be considered in this equation and is always beneath the surface if not openly acknowledged it plays out in many unacknowledged ways that can be surfaced through discourse analysis.

It is necessary to create rapport based on many ways of knowing and by learning from all aspects of nature needs to also know where to draw the line. But boundaries need to be drawn and decisions need to be made in order to 'make cuts'. The necessity of making cuts or decisions about what is acceptable and what is not remains a challenge for current generations.[97]

2.7.2 *Vignette: ISIS*

The aim of Isis is to draw on the biblical prophesies and to make them a reality, in order 'to hasten the end of time'.[98]

Just as we cannot bargain the sacred natural environment, we need to draw the line clearly to exclude 'bureaucratic brutality' and to avoid making the same errors as those we decide are beyond the pale of what constitutes ethical policy and practice.

The BBC podcast by a double agent who once worked for ISIS now claims the role of giving information about ISIS to the British government. According to this agent, ISIS sees their role as set for them 14,000 years ago. They believe this is the final fight and they have split with other groups. The roots can be traced to Iraq and the former Al Qaida leader. The funds are from the family of Saddam Hussein and stolen from the central bank and in turn stolen by them. They became entrepreneurs and bought shops near places of power in Iraq. Helped to one-forth of Iraq and a

---

[96]The Guardian, Sunday 10 April 2011 18.17 BST.

[97]http://www.bbc.co.uk/programmes/p02gd2wjIslamic State: Bureaucracy and Brutality: "Former jihadi Airmen Dean gives a unique insight into the workings of Islamic State. Dean left school in Saudi Arabia to fight jihad in Bosnia in the 1990s. But with the rise of al-Qaeda he became disillusioned with his comrades' drift towards terrorism. He joined al-Qaeda – but working undercover for the British government. Dean has recently spoken publicly against the jihadist movement but he retains a deep network of contacts within it. Despite Dean's defection, IS supporters still debate with him? Through those discussions, Dean has gained a deep understanding of the ideology and organisational networks behind IS".

[98]"ISIS aims to establish a so-called 'caliphate', which means a state ruled by a single political and religious leader according to Islamic law, or Sharia.

one-third of Syria and also have the oil fields—oil for sale on the black market. He explains that ISIS draws on prophesies and claims that the leaders (ex-Baathist leaders in Saddam Hussein's government), professors from university, oil engineers and scientists claim that they are following god's example. According to this reference they see ISIS as 'the Ark' they argue that they are like 'Noah' the chosen ones and they see themselves as vindicated in fulfilling purges of those who do not conform to their values.[99]

The role of the double agent in striving to act as an initiator of dialogue on behalf of ISIS in BBC mainstream media to reach those of us who choose to avoid the Internet postings of ISIS also needs to be analysed using Ulrich's 12 questions.

Clearly, the dangers of scaling up terrorism beyond the capabilities of Al Qaida are a very real possibility and one that needs to be considered by all those who support the idea that the freedom to test out ideas and to voice our thinking is necessary for science, arts, democracy and ethics. Freedom and diversity ought to be fostered to the extent that it does not destroy the conditions for those freedoms to exist. This is where the line needs to be drawn and sometimes in these circumstances conflict is necessary to defend those freedoms.

2.8 Recognising Our Interconnectedness

'Surviving terrorism' is the title of a talk by Jill Hicks (2016, Flinders University) who lost both her legs in the London bombings. Her experience of being classified as 'an enemy' resulted in her life changing completely in one minute. She was completely unknown to the suicide bomber and by those who risked their lives to save her. She was named 'priority one' and rescued an hour later by people who feared a second bomb explosion.

Her every day challenge is to rebuild her life without hatred through working for peace and being thankful for the heroism that saved her:

> there is no room for ideology …it is a practical daily struggle to stand, to walk and to live with the pain…

She went on to say that all terrorists should be 're-named criminal gangs or thugs'.

Hicks also reflected that a young Muslim doctor had said to her that she wondered if she would be given 'priority one status' if she had been found that day alongside Hicks.

In conversation afterwards an Islamic student said he agreed with this view but that perpetrators of violence and war are also criminals. He commented that the perpetrators of violence do not follow Islam. For him, true Islam teaches one to greet and smile at people when they meet. He reflected however that many people not only fail to greet one another but are abusive.

---

[99]http://www.bbc.co.uk/programmes/p02gd2wj.

The same student reflected on his concerns about attitudes to Muslims in a conversation with me after the talk. This was shared on the same day that it was announced on the news that the threat of a terrorist attack was of concern in Australia.

He said that the anger felt by people who saw fellow Muslims being bombed needs to be understood. I answered that all violence is unacceptable and agreed that the ripple effects on people and their families and extended families needed to be understood.

In South Africa, I was leading a community development project on water and sanitation in Kangwane. We were targeted by Inkatha Freedom Fighters because we were labelled ANC. So the ANC who saw themselves as Freedom Fighters driven by a white SA were categorised as the enemy by Inkatha who were politically competing for power (or as suggested by some) being used by the right wing Security Police to undermine so-called anti-apartheid groups.

Fortunately, the kombi I was driving survived the sabotage and although it was filled with people (some from Kanwane and a few who wanted a lift to the station). We all survived as the wheel bolts that had been loosened did not fall off until we had reached the first stop street in the small town at the foothills of the Valley Trust. The visit to learn skills at the Valley Trust was for my PhD thesis on water and sanitation as a vehicle for job creation and transformation. I was enrolled as a student at the University of South Africa whilst employed first at the Development Bank of SA and then at the Human Sciences Research Council. At a time of Apartheid, it was important for our team to develop a so-called community of practice which addressed the practical water and sanitation needs (Levin and Greenwood 2001) of the people.

The Mixed Methods Conference in Durham provided the opportunity to share and learn from others at the workshop on doing transformational research facilitated by Donna Mertens. I spent time in Durham in the privileged surroundings of Durham University, the Cathedral and Castle where I attended a conference. The post-Brexit environment in Britain is one of contrasts. I travelled from Manchester by train to Durham. The life chances of people were starkly contrasted across the stately country homes, neat cottages in villages near green fields and the flat lands edging industrial towns where vacant blocks provided informal accommodation.[100] From the privileged precincts of Durham, I travelled to visit Hadrian's Wall to understand the history of the Durham and how it developed as a bastion against the Northern invaders. The Vikings were staved off by the Romans and later the Scots were warded off by armies funded by the vast taxes and tithes generated by this cathedral town with its large market and associated guilds.

[100]A young person hung out of a broken window puffing a pipe. At a station a young man dressed as a Goth joined the train and spent time harassing a man in a wheel chair to allow him to draw a tattoo on his hand. His offers of friendship were rebuffed.

The university which began in the monastery within Durham Cathedral is the oldest in Britain. The riches of the cathedral and the power of the bishops (known as princes) was challenged when the state felt threatened by the power of the church.

Through the quick thinking of the bishop both the cathedral and its monastery was left relatively intact. After all, the cathedral had been needed to continue to be a source of funding for the state to protect itself from the Northern invaders. The notion that universities are somehow separate from power and the market was not the case although at some stages in the development of universities they may have been self-organising with more control given to students, the tendency for hierarchical control to overtake the structures is certainly true of most.

2.8.1 Corporatism and Education: The Challenges of Dualism

Universities are in decline. We should have reformed sooner, before we were 'reformed' is the summing up of the fate of universities (Watson 2012).

The options for universities can be explored through first considering the extremes which are:

- Universities controlled by students
- Universities controlled by academics
- Universities controlled by administrators

Historically, there have been variants of all three. Apparently, the first model is relevant to the first European university Bologna (according to the call for papers I received today in my email box).[101]

Professors did however have a say in examination matters. Monasteries were the basis for universities in which the monks held sway over initiates but were controlled by a church hierarchy. Durham cathedral was set up by monks fleeing the invasions of the north and for a while they built a powerful institution which was

[101]Angliss Conference Centre, Melbourne, 17th Oct, 2016: "It is 25 years since Dawkins became the Labor Minister for Employment, Education and Training and set about remodelling the country's institutions of tertiary education into instruments of the dynamic newfangled economic order. The new economy created a new frontier and, as on the old one, the heroic doctrines were pragmatic, utilitarian, impatient, tending to anti-intellectual. By these lights, the university run by academics and according to academic values was about as far from ideal as one could get. Much closer to the model was the modern corporation and the management principles it employed. Universities became massive revenue-chasing enterprises, academics became administrators, students became customers, managers became royalty – and management's share of revenue multiplied and multiplied". Don Watson, '*A New Dusk*', The Monthly, August 2012. Don Watson's summation of the malaise in Australian universities, written in 2012 in a review of Donald Meyers' ebook *Australian Universities: A Portrait in Decline,* is perhaps the best known recent critique of the widespread malaise in tertiary education. … An air of resignation pervades our universities. Disillusionment and bitterness are palpable, but a fear of upsetting the Emperors still prevails….".

funded by tithes and taxes as well as a strong market town. These proceeds made it possible to fund an army to protect the South from the North.

The monastery, monks and church were seen as a source of wealth and plundered by the state and only by making careful deals the monastery and Durham cathedral survived.

History provides many lessons. Universities need to survive through connecting with the surrounding community and serving it. Profit without service can render universities obsolete.

The conditions under which knowledge is generated to serve society needs to be addressed. We need a sense of urgency and a commitment to serve the common good. This has not necessarily been strength in universities and it is one that needs to be developed if universities are going to make a difference and enabling us to survive the challenges of the growing disparity between the rich and the poor within growing cities.

We need to develop a collaborative learning community approach which is outlined in this chapter.

A senior administrator: 'A university does not need more democracy it can lead to Brexit.'

A junior academic: 'But a university is about testing out ideas, is it not?'

A third colleague reflected: 'Yes I agree with the junior academic but unfortunately universities were never separate from the state or the market for that matter'.[102]

'But a university is about testing out ideas, is it not?' I responded. But then I recalled how at times that too much top-down testing of ideas (based on questioning and delays) can be counterproductive. In the Development Bank of Southern Africa we called the passage of documents up and down the system the so-called 'heart attack' profile. I recall reading somewhere that it is a way to sabotage morale. This seems to be a usual process in many bureaucracies today. In the past, the universities were linked with the church and of course the state and were reliant on wealthy patrons and taxes for their survival. So what is different today? Nothing—it is just the same except that the need for independent knowledge has never been greater.

[102]But a university is about testing out ideas, is it not? I responded (sotto voce). But then I recalled how at times that too much top-down testing of ideas (based on questioning and delays) can be counterproductive. In the Development Bank of Southern Africa we called the passage of documents up and down the system the so-called "heart attack" profile. I recall reading somewhere that it is a way to sabotage morale. This seems to be a usual process in many bureaucracies today.

According to Watson (2012):

> …Donald Meyers, a former science academic, has recently published an e-book, *Australian Universities: A Portrait of Decline*, in which he argues ferociously against the view – axiomatic since Dawkins – that the old-style university administrations were inefficient. On the contrary, he says, they delivered a much better education for much less money.
>
> Then academics were teachers and researchers with minimal administrative responsibilities; now they must combine their professional duties with mind-numbing administrative tasks straight from late-model management's most depressing handbooks. …
>
> Whether management is the cause or the symptom is less the point than what Meyers believes is the sickness itself: the dumbing down of university education….

2.9 Beyond Clash of World Views Towards Leadership for Re-generation

The selfish agenda of survival of the fittest drives realist policy. But in fact the planet is not a life boat. It is a space craft and responsibility for consuming it to excess needs to be addressed as current levels of consumption pose an existential risk. West Churchman talks of the so-called 'enemies within, they are religion, morality, politics and aesthetics'. Paradoxically, these are the values that make us passionate and compassionate. It is possible that the more we think about our thinking in discursive face to face and on line dialogue in different contexts with people who are experiencing very different life chances, the more mindful we will become that we cannot win at the expense of others or future generations of life. We evolved through the ability to show empathy to others and to reciprocate and not only through our ability to compete. As we face some of the greatest challenges every facing the planet we will need to do what we have always been able to do—unite together to face (namely religion, morality, politics and aesthetics, to cite West Churchman, 1983) enemies within and the greatest external threats to food, energy and water. What we do upstream affects those downstream and with climate change the river could dry up and the ocean could be rendered toxic. This will affect all aspects of life and render all other debates pointless.

In 2015, I attended a symposium at the World Futures Institute in Cyprus where we discussed alternative architectures of democracy and governance.

Cyprus is a complicated place with a complex political place in Europe. It has a 10,000 year history that has fostered narratives of love, political and religious conflict and war. Paradoxically it is the mythological site where Aphrodite, the goddess of love rose from the waves. Archaeological sites of building devastated by war are testimony to war. But the ability to cooperate, love and to compete are part of the history of the place. In Larnaca, the castle and battlements are an everyday reminder of one side of the coin, whereas the white cliffs overlooking the azure sea are the place for mythical sightings of the symbol of love. It is thus an apt place to discuss the human potential for cooperation and conflict and ways to enhance democracy and governance. It became a member of the European Union in 2004 but it is actually closer to Syria geographically than it is to Europe. It is a country that is not recognised by Turkey and instead it is seen as a region of Turkey. Historically it was part of Greece. But recently, Cyprus stresses that it is only connected because both are European member states. Thus although Turkey is not a member of the EU the northern part of Cyprus is of course a member. In the glib but accurate summary by Tom Melzer (2013)[103] in response to common Google hits:

> The Republic of Cyprus is the internationally recognised governor of everything but two British Overseas Territories on the island, but there is also the small matter of the long-standing occupation of Northern Cyprus by Turkey. And, you know, the colossal, crushing debt.
>
> It's been ruled – over the years – by Phoenicians, Assyrians, Egyptians, Persians, Romans, Byzantines, Arabs, Richard the Lionheart, Templars, Crusaders, Venetians, Ottomans and British colonialists.[104]

The recent memories of war in 1964 was fresh in the minds of many of the participants at the symposium and the war in Syria provided the context for much of their concern about the need to develop better understanding across both sections of the island. Cyprus is a case study of social, economic and environmental challenges. As a member of the European Union, they have applied for funding to support the development of better relationships. Democratic engagement based on subsidiarity is supported by the EU and so this has fostered opportunities for SDD.

An example of democratic engagement with a wine growing district gives a sense of the social, economic and environmental challenges associated with globalisation and financialisation. The issue is highlighted by the fact that the wine growers were given conflicting information: such as 'pull out your vines as the EU is not subsidising/accommodating Cyprus any more'. This is because Italy is seen as more competitive in this industry, whereas other departments suggested ways of becoming more competitive by engaging with other wine growing areas in Cyprus, in order to form a cooperative. This cooperative agreement came about through the work of the World Futures Institute with the team led by Christakis and the local mayor who tried to find a way to help people work with one another so as to 'grow the industry', despite the difficult political and economic context.

The engagement process was important as the requisite variety of user centric policy makers with lived experience sat in the same meeting to engage in a structured dialogue with the bureaucrats. In the words of Professor Christakis: 'Suits sat alongside peasants and discovered they had something [worthwhile] to learn from each other!'

[103]http://www.theguardian.com/news/2013/mar/28/is-cyprus-a-country.

[104]http://www.theguardian.com/news/2013/mar/28/is-cyprus-a-country.

Similarly, people who lived in the same or in neighbouring villages learned that their neighbours could make helpful suggestions. Normally, considered that top-down policy making is suitable for decision making.

Cypriot shop keepers along with others in the economy are cynical about being members of the EU because debt has to be repaid.

> Greek-Cypriot political relations are excellent, reflect the closeness of the two societies and are developing constantly in every sector. Contacts are very close and are characterized by frequent meetings between the political leaderships of the two countries on all levels, with the aim of optimum coordination of actions concerning the Cyprus issue and other regional and EU issues. The bilateral agreements that have been signed by the two countries reflect this state of affairs and cover a wide range of cooperation sectors.[105]

[105]http://www.mfa.gr/en/blog/greece-bilateral-relations/cyprus/.

The problem with the clash of civilizations thesis is that it describes what is the case without making suggestions as to how the issue can be redressed through political and policy interventions. It adopts a realist approach to politics and a formalist approach to economics.[106] It assumes that the identities of opposing groups are inevitably locked in conflict. Green and Aly (2011) explain that by seeing a range of people as Muslim other, it constructs an us/them approach that makes divisions worse.

The recent shooting of Chen, a police accountant by a 14-year old, Jabar highlights the need to address pragmatic prevention through working with members of the community who show leadership and not only the police who are clearly already seen as 'the enemy' but those who have been targeted and brainwashed by Isis. Ideally, all members of the community should have equal opportunities and be seen to be treated equally and not profiled or singled out. The recent decision to heighten surveillance and the detention of young children ought not to be the focus. Working with trusted community leaders would be preferable and making opportunities to be more socially inclusive is also a priority.

The need to address why people are alienated is a priority not merely the de-radicalisation programmes as stressed by Akbarzadeh (2013). Muslims in Australia are more than three times likely to be unemployed and far less likely to be home owners than other Australians. Furthermore socio-demographically the proportion of young people who identify as Muslim is growing and the time period between feeling accepted and integrated to feeling rejected and alienated by the

[106]21–23 November 2014 plenary Session, the conference of AICIS.

mainstream is shortening as a result of on line radicalization, according to the Police Commissioner Nick Kaldas (Maley and Lyons 2015). Kaldas has stressed that:

> The key to this is families and communities, it is not just the police....its about the values that prevail at home, its about the discussions that people have with their kids – not just the actual parents of the nuclear family, but the cousins, uncles, father figures who may be friends of the family, all of them I should see as very useful influences if they're saying and doing the right things and speaking frankly about what's right and what's wrong.
>
> The missing ingredient here is listening to why young people and their families are alienated. Respectful listening is indeed the missing ingredient in democracy. This does not mean we have to start from a position of agreement. But it does start from a position of being open to hear why people are angry. Being unemployed and finding it difficult to pay the bills could be one reason for anger. Another is feeling excluded, profiled and identifying with Muslims who are experiencing war and displacements elsewhere.
>
> Social inclusion needs to start with engaging young people and their families around what they have and what they need to fulfill their rights and their responsibilities.... Radicalisation is undertaken on the internet...

The same article cites Keysar Trad, founder of the Islamic Friendship Association, who stressed that:

> "Denying the root causes is like applying a Band-Aid to an open wound before cleaning and disinfecting it," he told an anti-radicalisation forum at the University of Western Sydney on Wednesday night.[107]

Complex challenges such as poverty, climate change and terrorism comprise many, interrelated variables that are perceived differently by those who hold different values. Religion, morality, politics and aesthetics are the values that make us human. They enable us to cooperate and to compete. They are what Churchman calls, the so-called the 'enemies within'. The notion of 'wicked problems' is his concept (albeit popularised by Rittell and Webber). Power distance can be redressed by developing closer relationships with others based on respect (Romm and Hsu 2002) and this can be extended through thinking in terms of 'if then' scenarios that locate people within living systems.

Critical heuristics when applied to thinking through 'if then' policy scenarios can be shaped by everyday decision making that takes a few simple steps into account. This does require 'drawing a line' about what is ethical or unethical in specific contexts. But who draws it, what is included and excluded, where it is drawn and

[107]Olding, R 2015, Sydney Morning Herald, April http://www.smh.com.au/national/persecution-pushing-young-australian-muslims-to-the-margins-of-society-says-leader-20150409-1mhhj5.html Persecution pushing young Australian Muslims to the margins of society, says leader Read more: http://www.smh.com.au/national/persecution-pushing-young-australian-muslims-to-the-margins-of-society-says-leader-20150409-1mhhj5.html#ixzz3oPhfExia.

how it is drawn matters. Ethically, we need to consider many ways of knowing before making a decision. This is different from the language of 'taming wicked problems' or governing from above.[108] It is about stewardship based on appreciating many dimensions and many ways of knowing.

Last night, I watched an Australian Broadcasting Corporation news program (Dec 14th, 2015) which showed a small child running whilst the sound of war and falling debris provided an image of living hell. He was shown being rescued but how many children are not rescued and what will be the long-term cost to the child and society. We need to think of:

Logical coherence of an argument for war, the extent to which the perceptions and lived experience and the quantitative data are available on for example the number of civilians hurt in so-called targeted bombing raids.

Idealism in terms of what ought to be done to ensure human rights, the capabilities of all sentient beings and planetary rights. We are part of an ecosystem on which we depend.

Dialectical engagement to address alternative ways to address challenges based on considering the consequences of decisions for human and environmental well-being. Discursive engagement enables people to test out ideas and to enhance their capacity to think critically and to join up the policy dots. It also enhances their capabilities to engage in the policy and politics as to who gets what, when why, how and to what effect.

The formalist aspects of economics (the notion that classical economists know best) and that the market dictates the way in which the world system functions—is a lens through which modernist or realist agendas of western experts are set.

The political economy idea of Karl Marx explores the extent to which some people lose (time, identity and skills) as they work towards fulfilling the profit motive of their employers in the capitalist system. Global capitalism merely extends the challenge to pursue markets and profit. But the issue today is that the workers no longer have wages in the developed economies that are high enough to enable them to grow the size of the market (using disposable income). Instead they rely on debt to purchase basic needs or to enter the housing market. New market capitalism extends into developing economies where wages are lower and working conditions are of lower standard.

The substantivist economist is concerned about how people perceive their experiences of daily life and how they understand what it takes to achieve a living. Polanyi (2001) is broadly speaking a substantivist. This means that the interaction between human beings and their environment to support their livelihood needs to be understood in terms of their experiences.

[108]As an explanatory paradigm it is in line with Mary Douglas' notion of sacred and profane. Those who share the same paradigm or our own cultural narrative and those who do not share our cultural narrative are regarded as profane. Critical Systems thinking (or critical heuristics) based on the work of West Churchman and his student Werner Ulrich and others such as Gerald Midgely, Romm, Flood and McIntyre-Mills argue that systems are not closed they are open and can be redefined or reframed.

The three approaches to economics are (according to Polanyi); reciprocity, redistribution and exchange. At face value, reciprocity (give and take) may appear egalitarian as does the notion of redistribution, whilst exchange may appear to be most formal. How these concepts are interpreted matters when considering the current social, economic and environmental crisis. Narratives can be seen as formal (modernist), informed by political economy critiques or substantive (based entirely on the perspectives of people's own lived experiences).

All these lenses provide a dimension of the situation. The problem with the realist clash paradigm is that although it describes the situation it can hasten the movement towards conflict. This is the self-fulfilling prophecy argument.

The current news that Australia will take refugees from Syria is a small step in the right direction in terms of addressing the human rights of displaced people, but it will be important to ensure that the intake does not privilege some groups more than others.[109] Also it raises questions about the rights of others from Syria who remain incarcerated.

An alternative future needs to be modelled and codetermined by with and for people of good will. This is the cosmopolitan agenda. It is not naïve. It can be argued to be logical and scientific if we consider the work of Buckminster Fuller on Space Ship earth and Kenneth Boulding's work on domains of knowledge and evolution from inorganic to organic life to living systems comprising human systems that can consciously evolve in directions that they can shape with a sense of purpose. The evolutionary direction is shaped by values. Hence, the justification for codetermining and co-creating the direction of policy and politics (McIntyre-Mills 2014a, b).[110]

---

[109]http://www.abc.net.au/news/2015-09-10/neumann-lets-keep-this-refugee-announcement-in-perspective/6764150.

[110]The challenges of food, energy and water security were discussed in conversation with colleagues about the micro, meso and macro level challenges of addressing the social, economic and environmental challenges facing people living in cities that need to be supported by viable regions that make farming a less challenging experience. I discussed the liberative potential of enabling people to monitor from below, irrespective of whether they are citizens or not. Non-citizens should also be given a voice because human rights are not being adequately addressed through the nation state. My contribution to the conversation covered the need to understand that complex challenges (such as poverty and climate change) span many different variables that are inter related and that are valued differently by different stakeholders who do not agree about climate change or the causes of poverty. The systems approach begins when first we try to see the world through the eyes of another, according to West Churchman. Perceptions matter! West Churchman's so-called 'enemies within' includes religion, morality, politics and aesthetics. Paradoxically, these are the values that make us passionate and compassionate. As we face some of the greatest challenges ever facing the planet we will need to do what we have always been able to do—unite together to face 'enemies within' and the greatest external threats we all face to food, energy and water supplies. What we do upstream affects those downstream in our neighbourhood. This will affect all aspects of life and render all other debates pointless. In order to address these challenges it requires the ability to work across many very diverse groups locally, nationally and in post national regional contexts. It also requires the ability to evolve in ways that are quantum rather than digital. Being the change requires cooperation with others and with nature if we are to address the challenge of so-called 'existential risk' (Bostrom 2010). Monitory democracy 'from below' that supports

Social media and digital networks have made a difference to the way we engage, but they carry the risk of our not being sufficiently rooted in the everyday here and now, face to face moment to moment communication.

The future is quantum and not digital is a profound statement made by the physicist Turok. In simple terms it is being the change and making paths through walking and engaging. It is possible that human beings will evolve through their ability to hold in mind more variables as a result of external and internal digital and then quantum tools and then to use these to enable them to leap to a higher level of consciousness. But the challenge will be how do we use this technology, will we use it to include or exclude people en route? The potential for the 1% who are now able to protect their interests will be the 1% who survives the natural disasters and the social disasters in enclaves whilst those outside perish.

(Footnote 110 continued)

respectful dialogue across service users and provides can be assisted through using a range of simple engagement processes that match the diverse needs of people with resources, in order to address the challenge to distribute resources more fairly. The most powerless and voiceless are those who do not have the vote… they are young people, children, asylum seekers, the dis Abled, sentient beings who are commodified and traded and of course the planet which has no rights at all. So the focus needs to be on the inadequacies of the nation state and the social contract. The recently published volumes 'Systemic Ethics and Transformation from Wall Street to Wellbeing' provide a plea and an example of a free downloadable architecture for doing things differently at the meso level. These volumes are extended in the following chapters. Local level engagement and wide ranging goals appear to be suggesting new directions, but how do we join up the dots and become more mindful?

Chapter 3
Planetary Passport for Social and Environmental Justice to Address an Increasingly Vulnerable and Unequal World

3.1 Introduction

The participatory action research discussed in this chapter aims to contribute to discovering and piloting new forms of architectures to address the increasingly diverse world views of people living in nation states that do not represent their needs. As such it addresses (a) engagement to support user-centricpolicy, expanded pragmatism ethics and law to (b) enable multilevel and multiagency approaches for social and environmental justice based on the axiom that people can free and diverse to the extent that their decisions do not undermine the rights of others through negatively impacting food, energy and water supplies for other living systems. As such it enables people to think through the consequences of their choices and to make a posteriori decisions that are informed by norms but which take into account the many views that need to be appreciated when making policy and governance decisions:

> Representative government is in crisis today, partly because it has lost, in the course of time, all institutions that permitted the citizens' actual participation, and partly because it is now gravely affected by the disease from which the party system suffers: bureaucratization and the two parties 'tendency to represent nobody except the party machines. (Arendt 1972: 89)

Cosmopolitan democracy and governance requires a multilevel and multiagency approach within and across overlapping regions. We need to work within a common framework for co-determination that operationalizes a form of Global Covenant that protects social and environmental justice through monitoring social, economic and environmental indicators of wellbeing in a responsive and contextual manner. This chapter extends the argument developed in *Transformation from Wall*

J. McIntyre-Mills, *Planetary Passport*, Contemporary Systems Thinking,
DOI 10.1007/978-3-319-58011-1_3

Street to Wellbeing (McIntyre-Mills et al. 2014)[1] and '*Systemic ethics and non-anthropocentric stewardship: implications for transdisciplinarity and cosmopolitan politics*' (McIntyre-Mills 2014). It builds on the previous research to enable people with complex needs to consider the consequences of continuing to live the same way as before or the consequences of making changes to the way that they live. It is hoped that this research could make a contribution to resilience and wellbeing. People face increased risks associated with climate change, such as rising food, energy and housing costs. Part of the inquiry involves a study on emotional connections to particular kinds of life. We can continue as ostriches denying our responsibility or we can strive to engage in cultural transformation by making everyday decisions that support the re-generation and sustainability of our habitat.

The Anthropocene is characterised by rapid urbanisation and unsustainable development. According to the United Nations World Urbanization Prospects (2014) in 2014, 54% of the world population was urban and by 2050 66% of the world's population is projected to be urban if current trends continue. Better urban governance needs to ensure that cities remain liveable and sustainable during and after the development process. Another central concern that urban governance schemes need to consider is the balancing of resources to meet both state and individual needs and goals. Land usage, either for agricultural production or for urban development, strategies and policies needs to be well informed to ensure not only optimal production is achieved but also elements of justice and equity prevail for a balanced development. Thus the critical systemic approach addresses production, consumption, evaluation and policy based on participatory democracy and governance that is user-centric.

This chapter[2] makes an effort to re-conceptualise cross-boundary democracy and governance to protect human beings and the planet. The challenge of implementing control from above and below—to limit the size of our carbon footprint whilst providing protection for sentient beings and the living systems on which they depend—will require a new form of governance based on nation states within interdependent biospheres that protect biodiversity. As emphasised in

---

[1]This chapter develops a means to implement an alternative form of governance that weighs up the social, economic and environmental indicators of wellbeing, in order to enable equitable distribution of resources and to ensure that some are not living at the expense of others and future generations of life. It moves beyond a critique to suggesting an alternative form of governance and democracy that spans communities of interest at a planetary level.

Dialogue, conflict resolution and a new architecture for cosmopolitan governance and communitarian approaches need to support democracy on a broader scale across overlapping regions of planet earth. But reality is business as usual.

[2]The chapter is based on an expanded version of a paper delivered at ISA conference in 2013: entitled 'Challenges for social and environmental justice in an increasingly vulnerable and unequal world posed by the global panopticon, penal states and disappearing states'. The research discussed below has been published as '*Transformation from Wall Street to Wellbeing: Joining up the dots through participatory democracy and governance to mitigate the causes and adapt to the effects of climate change*'. This section discusses the process as requested by Prof. Romm who reviewed it and requested more elaboration on the process of public engagement.

McIntyre-Mills et al. (2014), the starting point is to address the five areas of priority highlighted by Archibugi as:

(i) 'Control over the use of force;
(ii) Acceptance of cultural diversity by ensuring internal sovereignty based on constitutional rules that enable freedom and diversity to the extent that the freedom and diversity of others is not undermined. External sovereignty is 'replaced by global constitutionalism';
(iii) Strengthening the self-determination of people based on participatory democracy and the 'absence of domination' over other states;
(iv) Monitoring internal affairs, based on a form of monitory democracy and governance;
(v) Participatory management of the global commons as the fabric of life' (to cite and paraphrase Archibugi, in Wallace and Brown 2010: 321);

The implications of the new global accord on Climate emissions on 12 December 2015 is worth celebrating. But it will need to be worked through to ensure that trade agreements are fairly implemented. For example how will trade agreements support the accord given the lack of transparency in some agreements? For example the ABC 9 am radio report (8/7/2015) stressed that the TPP had not engaged the broader public and that the exclusion of the public from public comment at the time was considered problematic.[3] Concerns about trade arrangements were raised during the recent American election.

3.1.1 Aims and Logic of the Data Collection Process and Passport

The aim of this research is to enhance the awareness of local residents and to help local government decision-makers to understand the needs and perceptions of local voters. The software is populated by data on services that are available to enable more sustainable living choices. Thus it aims to help with raising consciousness which can provide a step towards the creation of a new more sustainable culture that supports 'wellbeing stocks'. This research explores people's perceptions of what works why and how, but also assesses whether the process of engagement enables them to move from denial towards taking steps towards a sustainable future. Soft Systems Mapping and System Dynamics Modelling techniques were applied as compatible methods. Soft Systems Mapping Methodology was used to enable

[3]This echoes the comments made at http://www.aflcio.org/Issues/Trade/Trans-Pacific-Partnership-Free-Trade-Agreement-TPP/Consumer-Rights 'The United States and 11 Pacific Rim nations—Australia, Brunei Darussalam, Canada, Chile, Malaysia, Mexico, New Zealand, Peru, Singapore, Vietnam and Japan—are negotiating the Trans-Pacific Partnership (TPP) Free Trade Agreement' (see http://www.aflcio.org/Issues/Trade/Trans-Pacific-Partnership-Free-Trade-Agreement-TPP/Consumer-Rights). These arrangements will be reconsidered under a Trump administration.

conceptual model building of the diverse perspectives of the stakeholders. By enabling service users and providers to use the software it will enable them to think about their thinking, in order to inform decisions. These first steps were used to populate the software and to enable it to display 'if then' scenarios that enable the involvement of end users in modelling to respond to complex social, economic and environmental challenges that are interrelated.[4]

Stiglitz et al. (2010) in '*Mis-measuring our lives*' introduce the concept of 'wellbeing stocks' and they argue for reframing economics to consider not merely profit and productivity, but the fabric of social and environmental life that support wellbeing and any economic activity. But the problem is that their work can also be used by neoliberalists to commodify the environment.

Dobson (2003) and Shiva (2002) emphasise the essentialist nature of the environment and they advocate virtuous behaviour through engaging people in thinking about the consequences of their choices and demonstrating ways to live lives based on the sufficiency principle. The work of Hannah Arendt on 'the banality of evil' reminds us of the extensive system that supports everyday choices. Judith Butler holds the Hannah Arendt chair and stresses the need to consider our humanity not only in terms of our ability to compete and to be resilient—but to also consider our vulnerability and need to cooperate. She asks what kind of world we would live in of we did not care?

---

[4]This chapter develops a new architecture for governance and democracy that enables a more organic approach to intervention based on 'being the change' in our daily lives. Reforming economics and accounting and accountability needs to be driven systemically (McIntyre-Mills 2014b). Wicked problems comprise many diverse variables that are interred related and that are perceived differently by different stakeholders (Rittel and Webber 1984; Flood and Carson 1993). As the nation state increases in size it is increasingly difficult to respond to the diverse needs of citizens and non-citizens. The nation state only protects those who are citizens through the social contract. But non-citizens are not protected and the environment is not protected by nation states that complete for resources. Nation states are indeed no longer the only dimension of democracy and governance. Democratic and governance interventions also need to consider cultural identity which is many layered—as stressed by critical theorists such as Stuart Hall in his work on representation and identity in which he reflects on his own experience as a Jamaican living the life of a public academic in the United Kingdom (see Hall 1992, 1997; Akomfrah 2013). In this chapter I map out three basic discourses on the notion of truth along a continuum from one truth based on modernist enlightenment to multiple or no truths (based on extreme versions of post-modernist thinking) and I explore a middle ground based on co-creating meanings within context. Each discourse is unpacked in terms of policy discourses and the implications for justice (drawn from McIntyre-Mills 2000: 22). The notion of one truth supported by divisive us/them categories is no longer possible as the movement of people and ideas intensifies in an increasingly interconnected world. Extreme forms of post modernism go beyond respecting diversity to suggest that all perceptions are of equal merit and thus extreme forms need to be avoided, because they undermine the notion of truth. Striving for truth is vital if we hope to achieve social and environmental justice. The middle ground of testing out ideas and policy options by considering the views of those who are to be affected provides a more fertile ground for cooperation, informed by Ashby's Rule of Requisite Variety and by the principle of subsidiarity, namely that people who are to be affected by a decision should be part of the process and that the decisions should be made at the local level possible and that the rights of the voiceless and the environment should be central to the decision-making process.

Butler[5] links the need for food and employment; these along with energy and water are the basics of life that need to be protected through transnational movements. This argument is echoed by the work of Shiva on 'Flows'. Shiva (2002) argues that instead of nationalist ideas supported by the tragedy of the commons approach, we need to develop more understanding that food, energy and water flows need to be protected locally by a sense of stewardship.

Six principles of sustainable governance underpin the Lisbon principles (1998) these principles are: 'responsibility' (individual and collective), 'scale matching' (at multiple levels), 'precaution', rather than narrow minded profit, 'adaptive management', based on integrating as much social, economic and environmental information as possible, 'full cost allocation', based on considering the inter connected dimensions of social and environmental considerations and 'markets should be adjusted' and 'participation' based on engagement using discursive dialogue.

Previously the work of Beck (1992, 2005, 2009) discussed the so-called 'boomerang affect' of poverty and pollution, Beck (2009: 96) stresses the need for sub politics 'from above' and 'below'. The first mention of climate change in sociology is by Lever Tracey (Pidgeon and Fischhoff 2011).

The implications for climate change and the ecological footprint caused by human society has not progressed—partly because of the lack of interdisciplinarity applied to a study of power and the *lack of will* to change in the interests of the majority in this generation and the next.[6] The Brundtland Report (1987) on 'Our Common Future' stressed our interconnectedness and the rights of future generations. Clarke (2000) stressed that the planet is a system, but all the policy documents and conferences to date have been limited by the containerist philosophy echoed by the rights of sovereign states (UN 1992; Copenhagen 2010; Cancun 2011; Johannesburg 2012) to make decisions on the use of resources—most particularly in connection with the carbon-based economy. If we are to balance the interests of the planet as well as ensure the rights of less developed nations to redress poverty—then we must find a way not only to innovate—but also to limit the misuse of resources and to enable the fair distribution (Rawls 1999) of rights and responsibilities to ensure food, energy, water supplies that underpin quality of

---

[5]http://www.salon.com/2011/10/24/judith_butler_at_occupy_wall_street/ '…we demand the impossible—the right to shelter, food and employment are impossible demands, then we demand the impossible. If it is impossible to demand that those who profit from the recession redistribute their wealth and cease their greed, then yes, we demand the impossible'.

[6]For example: rising sea levels destroying some nation states. The issue of climate change has been discussed by Sydney Peace Prize Winner, Vandana Shiva, a scientist, feminist and environmentalist who has raised concerns about commodification, capitalism, climate change and the implications for rising sea levels on human lives (see Sydney Foundation Peace Prize http://sydneypeacefoundation.org.au/peace-prize-recipients/2010-dr-vandana-shiva/ for Lecture: Making Peace With the Earth).

life and wellbeing.[7] The United Nations declaration of human rights will remain rhetoric unless new forms of governance that respect diversity and freedom to the extent that the diversity and freedom of the next generation of life is not undermined.

Democratic governance requires listening to perceived needs and responding within the architecture of rights and responsibilities that are in line with cosmopolitan law to protect human rights and the human environment, as outlined in 'Our Common Future' (Brundtland 1987) and the 'Earth Charter' (1997) that outlines world environmental citizenship, namely to: 'Build rapport and trust' in order to *counter* the realist approach to international relations through:

- Deciding how best to use scarce resources
- Drawing on local wisdom
- Drawing on creative ideas of local people
- Enabling participation to match responses to perceived need but also through
- Enabling people to blow the whistle on misuse.

The research was based on previous research on social inclusion with First Australians (McIntyre-Mills 2010), in order to protect the global commons[8] and the collective good (Eder 1996). The philosophy underpinning the approach grew out of the research in Alice Springs and in the Southern Region with Aboriginal Australians who stressed that 'we are the land'. Olive Veverbrants, an Arrerente Australian, stressed 'the earth is our mother' and we depend on it for our survival.

---

[7]The planet is not a credit card to be used to benefit some—whilst others pick up the tab! The responsibility weighs differently on developed and less developed nations and also differently on the rich and poor, this generation and the next. The architecture of governance (Archibugi 2008) needs to respond to environmental needs that protect the wellbeing of global citizens (Haydon 2005). Consequently sovereign nation states need to consider regional rights and responsibilities in a series of overlapping organisations that apply the rule of law to ensure equitable and fair distribution of rights and responsibilities for food, energy and water. The research is about developing post-national organisations to ensure human rights. Disaster management, movement of people fleeing disaster or injustice, securing food, energy and water are concerns that span boundaries (spatial, conceptual and temporal). We need to solve problems across generations across disciplines and across nation states to create good neighbourhoods. This requires trust and better communication so that we understand one another better.

[8]Initially the concept of global commons was narrowly defined as 'assets outside the national frontiers such as oceans, space and the Antarctic' OECD definition http://stats.oecd.org/glossary/search.asp. Our area of concern is the extent to which resilience could be achieved through attitudinal and behavioural change fostered through participatory democracy and systemic governance that incorporates structured 'if then considerations' (Christakis and Bausch 2006; McIntyre-Mills and De Vries 2011) to develop clear priorities as to how to respond to contextual challenges (Ng 2011). The theme of global stewardship, democracy and governance frameworks to support wellbeing is addressed in this programme of research by asking questions guided by a Design of Inquiring Systems (DIS) (adapted from the work of West Churchman (1979, 1982; McIntyre-Mills 2006a, b, c, 2010; McIntyre-Mills and De Vries 2011). The DIS refers to addressing an area of concern by 'unfolding' human values and 'sweeping in' contextual considerations and many domains of knowledge spanning logic, empiricism, idealism, the dialectic and pragmatism before making any decision.

The weaving together of stories of what works and why shapes the approach. Participants are invited to reflect on their lives in terms of what they *already have* and what *they still need*, in order to enhance their wellbeing. They are asked to consider what they would be prepared to add to their 'In baskets' and what they would be prepared to discard to their 'out baskets'. These are metaphors—not just in the sense of managing our lives—but more poetically, they are Murray River grasses woven into baskets that symbolise our connections to one another and to the land. The more we weave together strands of experience and respect the other—including the land—the more chance we have to achieve wellbeing. This is the assumption of the project that builds on a previously funded ARC linkage project with the SA Department of Health and Neporendi to address complex needs by making better decisions to match services and to enhance wellbeing. It is also informed by the work of thinker in residence Susan Greenfield, a neuroscientist. The more we are able to 'join up the dots' by making more neural connections the more conscious or aware we become. The more we engage in 'thinking about our thinking' and engage in testing out ideas with others the closer we get to truth. The engagement process is to help achieve 'a change of heart' and 'mindfulness' about our interconnectedness. It is guided by the ethical axiom for democracy and governance that 'we can be free and diverse to the extent that we do not undermine the rights of others or the next generation of life'.

The aim of the model is to help key stakeholders to make strategic decisions to support sustainable and regenerative living choices at the household, organisational and regional level. The passport is detailed under the case study section at 3.4. The passport is based on the following logic:

1. Individuals addresses personal wellbeing in terms of social, economic and environmental factors that impact on food, energy and water security. Those with heavy footprints who consider only their own wellbeing at the expense of others will be required to self-contract to make social, economic and environmental adjustments.
2. This user-centric evidence-based approach addresses the social, economic and environmental overlaps in Venn Diagrams to create pathways to move from (a) business, (b) making small steps towards (c) living harmoniously in ways that enhance wellbeing.
3. The individual, neighbourhood, community, regional post-national biosphere comprises a series of overlapping spirals that are interconnected. The carbon footprint of each person is considered as an individual weighting made up of (a) what they have in their lives, (b) what they need, (c) what they are prepared to add or discard, (d) the turning points for the better and worse, (e) the barriers and services they are prepared to contribute to as well as (f) the services they need, (g) a drop-down list of post-coded socio-economic and environmental wellbeing services are provided.
4. Exel Spread sheets of the domains for each individual, organisation and region span boundaries to protect the individual and the common good through

creating shared pathways that provide sustainable futures through regenerating relationships that span living systems.
5. Thus it re-frames democracy, governance and economics based on the synecdoche of water ripples and water spirals.

3.2 The Challenges for Social and Environmental Justice

The argument that a 'clash is inevitable' and that 'the global commons is best protected through containerist states' (see Hardin 1968) becomes a self-fulfilling prophesy. The mythic narrative has been the focus for realist international relations needs to be changed through alternative narratives that are co-determined and co-created through dialogue. This is the argument I develop in this chapter. Currently, my practical concern is that there are more refugees and asylum seekers now than ever before and the current form of democracy and governance does not work. Policy researchers, practitioners and educators need to work with many forms of knowledge and across conceptual and spatial boundaries to address the big issues of the day, namely peace, gender mainstreaming and social, economic and environmental security within increasingly densely populated cities that need to respond to increasing poverty and climate change.

This has implications for the quality of life from a human, non-anthropocentric and environmental perspective. We need education and a new form of participatory governance that enables:

- Mobilising people to redesign the way we live through the appreciation of living systems of which we are part
- Co-determination of our future and that protects both human and environmental wellbeing
- Extension of the social capital thesis to include natural capital as the basis for all living systems.

The research in this chapter:

- Draws on and extends Hulme's research (2009) on perceptions and Cornelius' research on systemic approaches to understanding emotions (1996). Values, perceptions and emotions are better understood from a systemic viewpoint that draws on interdisciplinary theories.
- Explores voluntarist approaches to ecologically responsible living and explores structuralist options for encouraging people to make better ecological choices through participatory public education—using scenarios as a means to assist understanding and engagement for democracy and better governance linked with local government. The intention is to provide the background material for creating and supporting a curriculum for public transformation.

- Explores the wider potential of Max-Neef's 'Being, Doing, Having and Interacting' Index as it relates to extending the UN Sustainable Development Goals (2014) to include re-generation, human capital and capabilities (Nussbaum 2011) locally and regionally. Dobson (2003) distinguishes between environmental citizenship versus ecological citizenship.[9] The former relates to the role of the state through law and control, the latter refers to the personal choices made by citizens for the common good as they relate to the ecological footprint. Human rights underpin all relationships within the nation state and within the broader region.[10] This has implications for social and environmental justice. Brundtland (1989) and Kaldor (in Held 2005: 177) have stressed that human security rests on creating a peaceful world, enshrined in rights expressed in the United Nations Charter (Article 55, 56) and in the constitution of the European Union.
- Re-frames ways of knowing to preserve a sense of our connection with the earth and waters—on which we depend—to protect future generations through remembering our heritage, reconnecting to others (including sentient beings) and regenerating ecological citizenship.[11] The move towards remembering and reconnecting with the land echoes the voices of Indigenous First national movements and the implications for the way we live, our relationships, our sense of identity and for new forms of governance and democracy.
- Constructs the notion of a planetary passport as a means to rethink our relationships.[12]

---

[9]The aim is not to avoid detailing the problems, but it also aims to explore alternatives and to discuss some ways of achieving wellbeing. The lens through which I am working starts with discrimination against the majority of this generation and the next.

[10]The argument I have developed in 'Reconsidering Boundaries (2014) and in Systemic Ethics (2014)' is that Shiva critiques Hardin's 'Tragedy of the Commons' approach. It applied John Locke's defence of property and the enclosure movements of the seventeenth Century in Britain. Hardin's core argument is that in the absence of private property there would be lawlessness. Hardin's classic argument supports nationalism and private property.

[11]Human rights, discrimination and outlining the notion of a planetary passport is the starting point for transforming our lives through recognising our shared hybrid needs and appreciating space for diversity. This is quite different from on the one hand, the problems associated with universalising culture or on the other hand, associated with the 'clash of cultures' thesis. Some of the best examples of structuralist agendas are constitutions that give rights to nature such as the Ecuadorian constitution. Some of the best examples of voluntarism include the Future Worlds Organisation and the World Futures Institute that encourage people to become more politically active in transforming what is valued through 'being the change'.

[12]We are part of one region and we need to understand cultural flows and connections that will be tested through our cooperation in managing ocean space, airspace and the fair share of food, energy and water. Cosmopolitan democracy and governance requires a multilevel and multiagency approach within and across overlapping regions. We need to work within a common framework for co-determination that operationalizes a form of Global Covenant that protects social and environmental justice through monitoring social, economic and environmental indicators of wellbeing in a responsive and contextual manner. This chapter is drawn in part from '*Transformation from Wall Street to Wellbeing*' (McIntyre-Mills et al. 2014) and 'Systemic ethics and non-anthropocentric stewardship: Implications for transdisciplinarity and cosmopolitan

- Makes a case for the regeneration of aspects of Indigenous stewardship (without lapsing into a naïve belief in cultural relativism or that one culture has all the answers) and not just sustaining the current ways of life, because the so-called enlightenment has delivered modernist capitalist, global society that is part of the problem and not part of a solution. Sustainability also does not really go far enough in protecting gendered ecological citizenship or in co-creating new ways to regenerate and reverse the extinction process.[13] The Sydney Peace prize winner Shiva (1998, 2002, 2011) sums up the challenge for this century as one of preventing the commodification of life and the importance of thinking about the interconnections across a number of issues. Ensuring a fair use of resources will require not only a Global Covenant but a means to implement local management that respects regional benchmarks and that honours the spirit of the following to source a new architecture:
- The Aarhus Convention, based on the notion of participation, freedom of information and the right to be heard.
- The Tuvalu test, based on the notion that all decisions need to be tested in terms of their impact on our neighbours, as detailed in the above publications and in this chapter. Measuring the size of a carbon footprint begins with the choices made by the public and private sector, but major emitters such as USA, China and India will be pushed in the right direction by requiring that they measure their footprints. The evidence of 'right living' is already demonstrated by the very poor (in terms of having low carbon footprints). The heavy emitters will need to work out ways to progress through a green economy with subsidies going from the richest nations to the newly developing and poorest nations.

3.2.1 Climate Change Refugees and Climate Change Displacement

Off shore processing of asylum seekers in states whose economy is reliant on aid from more powerful economies such as Australia leads to the development of

(Footnote 12 continued)

politics' (McIntyre-Mills 2014). It strives to build on the previous research to enable people with complex needs to consider the consequences of continuing to live the same way as before or the consequences of making changes to the way that they live. It is hoped that this research could make a contribution to resilience and wellbeing. People face increased risks associated with climate change, such as rising food, energy and housing costs. Part of the inquiry involves a study on emotional connections to particular kinds of life.

[13]In 'Transformation from Wall Street to Wellbeing' (McIntyre-Mills 2014), 'Systemic Ethics' (2014) and 'Reconsidering Boundaries' (McIntyre-Mills 2014) the case is made for co-creation and co-determination within regional biospheres that respect diversity and freedom based on stewardship for current and future generations.

oppressive camps along the lines of convict colonies, such as Manus Island. The propensity for riots has already been demonstrated along with the reaction of increased control and increased distancing of displaced people as 'other'. The IPCC formula suggests that the privileged lives of some could lead to 'existential risk' for all forms of life on the planet (Bostrom 2011). The 'price of inequality'—national and global has escalated.[14]

Each of the concepts will be defined in turn. I refer to the recent Sydney Peace Prize Winners as indicative of some of the key social and environmental justice concerns:

1. Global panopticon—the surveillance approach that has been revealed by the Guardian through the work of Greg Snowden and Julian Assange (2011).[15]
2. Penal states—the incarceration of asylum seekers because of the containerist approach to social justice in nation states (Julian Burnside 2014).[16]

---

[14]Don't expect Indonesian support for tow backs by Damien Kingsbury http://www.abc.net.au/news/2014-10-27/kingsbury-turning-back-the-boats/5844926.

[15]http://sydneypeacefoundation.org.au/?s=Julian+assange&submit.x=-696&submit.y=-246&submit=Go "'*The Director of the Sydney Peace Foundation, Professor Stuart Rees, has stated: 'Assange's work is in the Tom Paine Rights of Man and Daniel Ellsberg Pentagon Papers tradition—challenges the old order of power in politics and in journalism. Assange has championed people's right to know and has challenged the centuries old tradition that governments are entitled to keep the public in a state of ignorance. In the Paine, Ellsberg and Assange cases, those in power moved quickly to silence their critics even by perverting the course of justice'". The Sydney Peace Foundation Gold Medal was awarded ... at a ceremony in mid-2011...Presented by veteran Australian broadcaster Mary Kostakidis, the Sydney Peace Medal recognises Assange's leadership, courage and tenacity in journalism and publishing, and pays tribute to his enduring conviction that truth matters and justice depends on it.Kostakidis, a former member of the Australian Human Rights Consultation Committee, praised WikLeaks as an 'ingenious and heroic website that has shifted the power balance between citizen and the state by exposing what governments really get up to in our name'*.

Acknowledging that 'exposing secrets can be dangerous business', Kostakidis thanked Assange for his 'heroic courage' as a whistle-blower to take 'great risks for our benefit'.

[16]Citation: 'for his brave and principled advocacy for human rights and for those wronged by government, for insisting that we respect our international obligations toward those seeking asylum, and for his unflinching defence of the rule of law as a means to achieve a more peaceful and just society'

http://sydneypeacefoundation.org.au/peace-prize-recipients/2014-julian-burnside-ao-qc/ *'I am deeply honoured to be selected to receive this year's prize, especially in a year when Australia's reputation is being tarnished by its intentionally harsh treatment of asylum seekers. I have long admired the work of the Sydney Peace Foundation. It continues to advocate for human rights and promote peace. It helps keeps our focus on the importance of the rule of law, the need to treat all human beings with compassion and the need to ensure that human rights are respected', says Julian Burnside AO QC. 'Australia is currently marooned at a moral and legal crossroads over policies towards asylum seekers and refugees', says the founder of the Foundation, former Director, Chair and current Vice Chair, Emeritus Professor Stuart Rees AM, 'Julian Burnside's decades long advocacy of the of the human rights of some of the world's most vulnerable people makes him an extremely worthy recipient'.*

3. Disappearing states—the rising sea levels destroying some nation states (see the work of Vandana Shiva,[17] a scientist, feminist and environmentalist on commodification, capitalism, climate change and the implications for rising sea levels.

3.2.2 States, Boundaries and Biospheres: Redesigning Governance and Democracy

A case is made for two approaches to addressing the inadequacies of the social fabric and the social contract for protecting food energy and water through *balancing individual and collective needs (or collective co-determinism).*[18] The process involves monitoring for social and environmental justice:

> '*From below*' based on subsidiarity and enabling people at the local level to have a say in protecting the local environment. The Max-Neef Human Sustainability Index applied to valuing, measuring and managing environmental and human assets. A new architecture and process for governance and democracy is needed in order to balance individual and collective needs.
>
> '*From above*' *based on the arguments of* strong cosmopolitans such as Archibugi (2010: 319) who stress the need for global democracy to support global governance by means of a constitution spanning nation states and thus creating 'overlapping spheres of power'. Just as decisions are made at the nation state level by legal systems buttressed by constitutional law:
>
> At the global level, similar institutions should also have the task of deciding on the decision-making clout of the stakeholders and of assigning competencies (Archibugi 2010: 320).

Unlike the federalist or confederal model Archibugi argues that global constitutions replace sovereignty at the national level and instead of 'one state one vote', the principle of 'one individual one vote' prevails for global citizens. Archibugi envisages that citizens and their governments would participate in decisions and that an international court would make decisions and that 'the member states would accept the compulsory jurisdiction of the international courts'. Archibugi stresses that each state would 'retain their own armed forces', but that humanitarian interventions could draw on the resources of many states and could be 'managed by cosmopolitan institutions'. According to Dean (2009), the Marxist argument against

---

[17]Winner of the Sydney Foundation Peace Prize http://sydneypeacefoundation.org.au/peace-prize-recipients/2010-dr-vandana-shiva/"Sydney Peace Prize Citation: '*Recognised for her courageous leadership of movements for social justice—the empowerment of women in developing countries, advocacy of the human rights of small farming communities and for her scientific analysis of environmental sustainability, Sydney Peace Prize Lecture: Making Peace With the Earth'.*

[18]Full Bright public Lecture on recasting international relations by Professor Victoria Farrer-Myers.

cosmopolitanism is that normative and legal arguments are inadequate to address the class inequalities within and between nation states. The cosmopolitan response is one of promoting social democracy and redistribution of energy resources through governance. The Marshall Plan approach is suggested by Chevalier (2009) as a possible response to the energy crisis. The problem is that that the greatest emitters such as USA need to accept their responsibility to developing nations who need to consider development as a way out of poverty.

In the light of the above, how then can we develop a way forward to create transnational webs of shared meaning that are buttressed by governance to protect the global commons and the public good? Held proposed that the core challenges of the day are to address the vast differences in the standard of living between the rich and the poor through reforming world trade and through using the market to address global warming by means of a global covenant that prevents unilateral interventions.

Stewardship governance is necessary to maintain the balance across individual and collective interests. The current architectures for democracy, economics and governance are fatally flawed and the time to reframe it is long overdue. How democratic is democracy if it does not foster food and human security? Liberal democracies are increasingly criticised for not representing the interests of diverse citizens, engaging stakeholders in dialogue, building a shared sense of identity, whilst enabling individual diversity, capability and freedom to the extent that the diversity, capability and freedom of others is not undermined.

The current difficulties in addressing representation, accountability and sustainability are theoretical, methodological and value-based and it has implications for the size of our ecological footprint. The privileged lives of some are at the expense of the majority and at the expense of future generations of life. The issue is that the nation state is no longer able to address the challenges that span national boundaries, because poverty and climate change do not respect political boundaries. Critiques need to avoid reifying systems, rather than addressing the potential to enhance representation, accountability and sustainability through re-framing the architectures of democracy and governance.

An emphasis on food security is a way out of the current social, economic and environmental crisis. The governance of the footprint 'from below' through subsidiarity—means that decisions need to be made at the lowest level possible and that freedom and diversity should be fostered—with the caveat that these decisions should not undermine the rights of others or future generations of life (Table 3.1).

David Held's work is criticised by Dobson (2003) for making too much of the global interconnections and that he pays too little attention to the reality that for many local lives are all they will ever experience. The hyper mobility and interconnectedness of elites is a world away from the experience of those who are outside the digital world of 'Networked Society' (Castells 2001). De Waal (2006, 2009) makes a case for evolutionary cooperation based on research on primates and other animals that have demonstrated that empathy and reciprocity are one of the bases of animal nature along with the capacity for compete. Cooperation is important for evolutionary transcendence to protect the ecological footprint. This

Table 3.1 Architecture to protect 'wellbeing stocks' and minimise existential risks: combining participatory democracy (based on taking decisions at the lowest level possible, subsidiarity and post-national governance of regional ecological footprints/biospheres)

	Structure	Process	Action
Micro-level individuals	UN local Agenda 21 (1992) and Aarhus convention (1998)	Questions raised and posed to local government by individuals	Local government, NGOS and individuals
Meso States and regions	Aarhus convention linked to global covenant	On line monitory democracy and governance to address state/market/civil society concern	Networking NGOs and INGOS to address representation and accountability
Macro Cosmopolitan governance	Legal structures to support the global covenant, Aarhus convention and Biospheres convention	International Criminal Court United Nations	Global action to pass laws to protect social and environmental justice in overlapping biospheres

Source Adapted from Florini (2003) and Archibugi in Wallace Brown and Held (2010: 322) cited in McIntyre-Mills et al. (2014:92) and McIntyre-Mills (2014: 7) in 'Reconsidering Boundaries', Sociopedia

requires the capability to hold in mind many variables and not to think in terms of 'us/them' expressed in terms of tribes, organisations and nation states.

Giddens (2009) in 'the Politics of Climate Change' warns that localised efforts, whilst important for prefiguring change are insufficient to hold the market to account, unless they are applied regionally and internationally.

Habermas argues that we can no longer limit democracy within the boundaries of a nation state. This is a relevant point and builds on the work of Dobson (2003) who argues that discursive engagement is important for changing mind sets. Dobson also believes that engaging at a cosmopolitan level needs to be more than just about dialogue. It also needs to be about distribution.

Flyvbjerg (1998), an urban planner who succeeded in developing Aalborg as a green city asks whether Habermas (a Universalist) and Foucault (a contextualist concerned about power) are both vital thinkers for civil society? Flyvbjerg argues convincingly that we need both approaches to balance individual and collective concerns. The social sciences argue that we need to enhance our capability to make connections. Digital communications are already widely used, but their potential is under tapped as stressed in research on wellbeing, representation, accountability and sustainability (McIntyre-Mills and de Vries 2011, 2012).

A new narrative of 'Earth Democracy and Governance' is urgently needed according to Shiva and Barlow (2011).[19] According to Evans Pritchard the Nuer

[19]http://www.democracynow.org/2011/4/22/earth_day_special_vandana_shiva_and.

understood that they needed to unite against a common enemy so they set aside their differences.

As stressed at the outset, whereas in the past concerns were at the level of the organisation or nation state, they are now at a post-national level. We are interconnected—we can address differences through 'loyalty' and 'voice', but exit—as described by the development economist Albert Hirschman—is no longer viable because the problems are planetary.

The level of analysis needs to be extended by working on transdisciplinary capacity to address the convergent challenges (social, economic and environmental) that do not allow exit from the problem which is in global.

Similarly, we need to understand that space ship earth is the new 'ark of the covenant'—planet earth and not the ark of 'us versus them'. Thus the clash of cultures (Huntington's 1996) narrative needs to be reframed from a 'zero-sum' approach to an appreciation of the planetary scale of stewardship. Governance of the Anthropocene requires new architectures for democracy and governance.

Nussbaum (2006) argues for social justice and for capabilities for human beings and sentients. Robert Costanza in '*An introduction to ecological economics*' (2015) stresses that the magnitude of the Anthropocentric human footprint has implications for new forms of governance:

> Satisfaction of basic human needs requires a balance among social, built, human and natural capital (and time). Policy and culture help to allocate the four types of capital … as a means for providing these opportunities. One institution that helps build social capital is a strong democracy…. where all citizens are free (and expected) to participate in all political decisions affecting the community. Interactive discussion plays an important role…preferences and opinions are not fixed and hence can be changed/shifted through a deliberative process. The process of deliberative democracy involves 1. Sharing of information and policy options among a set group of individuals, 2) the process of deliberation among the group, and 3) the socialization and dissemination of the results of this process (Held and Hervey 2009; Herbrick and Isham 2010 cited in Costanza et al. 2015: 294–295).

Stuart Hall stresses that human identity today is not tied to the nation state. As a public academic he posed questions about cultural identity, colonisation and post-colonial mindsets and stressed that these days people need to give a narrative response to the question: who are you, because people identify with many spaces and places, even if they are rooted in one place. He applied critical thinking in 'our mongrel selves' (1992) about who, gets what when why and to what effect. By encouraging people to give a narrative of what they value and who they are, it could help to address the issues raised by Wilkinson and Pickett about consumption and status. People consume to be accepted in a globalised world where status indicators become a way of life.

3.3 Collective Co-determinism to Protect Human Security: Better Forms of Participatory Democracy and Distributive Governance

In 2014 Professor Victoria Farrar Myer (Fulbright Distinguished Chair) echoed this plea for greater emphasis on rebuilding relationships and emphasised the need to foster collective determinism through public engagement processes. The sovereign nation state has treated the planet as zero sum—when in fact the notion of externalities is quite wrong.

3.3.1 Alternative Architectures to Enhance Representation, Accountability and Sustainability

The greatest challenges are the *consequences of inaction* that will potentially pose an existential risk to humanity. These challenges include r*epresentation* of the increasingly diverse populations within nation states along with *accountability* to ensure that resources (e.g. water, food, and energy) are used fairly, equitably and sustainably in local and regional biospheres. Better forms of e*ngagement* are needed to enhance the capability of people to understand that attitudes towards consumption have profound implications for social and environmental justice. The rationale for the research is that we face risks that could undermine quality of life and our capacity to achieve wellbeing.[20] Droughts, floods storm surges are regional concerns that affect human security. Many of these risks are the result of climate change.

- How should we live? The challenge is to enhance engagement and address the question about our relationships with others and the land; in other words to address the implications for post-carbon economies. According to Rann (2012) integrated, intelligent design needs to respond to the long-term planning needs of cities within the region. Thinker in Residence, Architect Professor Laura Lee has stressed that cities do not grow organically they need to be designed to meet the need of people living a low carbon life style. Thus public transport and cycle

[20]Part of the rationale for trialling the software was to establish to what extent computing per se is a useful way to engage different interest groups and whether it could be useful to enable managing large, diverse data sets spanning many diverse interrelated variables about which residents have strong values. A diverse range of values, emotions and opinions are associated with wellbeing, quality of life and climate change. These can be regarded as a wicked problem, in the sense that it has many, diverse and interrelated variables that are viewed differently by people with different values (see the West Churchman Series 2006). How to enable democracy to balance rights and responsibilities in increasingly diverse nation states remains a core challenge. How to enable governance within post-national regions poses a similar challenge.

paths are important (Fred Hansen). The choices we make will be vital for our wellbeing in the future.
- How do we enhance or maintain well being through acknowledging existential risk and considering the rights of next generation of life (Young 2000; McIntyre-Mills 2000; McIntyre-Mills and de Vries 2011)?
- How can we develop a greater ability to think about the bigger picture[21]?
- How do we enhance engagement?
- How can we achieve cultural transformation on consumption patterns?

This section discusses a participatory action research project in a place I will call 'Middleville' in South Australia.[22] The area of concern is to take into account social, economic and environmental factors that make a difference to wellbeing and our ability to mitigate or adapt to climate change and to develop climate change policy and actions, using a participatory process with local residents to inform policy makers and service providers. The addictive culture consumption is unsustainable and the research aims to enable people to understand that decisions to support the mitigation and adaptation to climate change are rooted transforming daily living choices.

It is rational to take policy precautions on whatever side of the scientific debate we are located. This requires wellbeing and sustainability decisions and reconsidering our democratic rights and responsibilities by using participatory engagement aimed to match services to needs—within limits—so that local communities make the best decisions that minimise social, economic and environmental risks and that shape 'policy from below' by enabling people to engage in a process of thinking about what *material and non-material* resources they have to support wellbeing, what they still need, what they are prepared **to do** to enhance their social, economic and environmental wellbeing, what they are prepared to discard; and then to consider what they personally consider to be the social, economic and environmental turning points for the better and the worse. Participants were also asked to consider

[21]De Grasse in conversation with Richard Dawkins explains that we need to see horizons as always changing, but we need to strive to think about larger assemblies and to make use of conceptual tools to enhance evolutionary consciousness. The limits are beyond us and always relative to where we are located physically and conceptually. http://www.youtube.com/watch?v=9RExQFZzHXQ&feature=related.

[22]The research has been funded by the Centre for Research into Aboriginal Health (CRCAH) Australian Local Government Association and is currently in progress (McIntyre-Mills and de Vries 2011). The transdiciplinary team, comprises: Janet McIntyre (Social and Policy studies) to conceptualise and implement the project with Dr. Denise de Vries (Informatics, Flinders University) with the assistance of Sun Binchai (Hons graduate, Informatics, Flinders University), Ivantia Mokoginta (Ph.D. student, Economics and Public Policy), Adib Mohamned Shomad (Ph. D. student, Public policy and management), Barbara Dickson (MA student, Public Policy and engagement specialist for an urban renewal department, public sector, Adelaide), Dr. John Mugabushaka (Ph.D. student and public policy and youth justice officer, public sector, Adelaide) and David Hope (accountant and risk management consultant for local government). Details can be found at: The log in is user name = test, code = test.

the barriers that they experience in achieving wellbeing. All the decisions are based on participant's perceptions about material and non-material facets of their life.

The public engagement software updates as they add their new ideas and these help to inform others who see only the list of de-identified social, economic and environmental factors and the particular local resources that are available to assist them to move from (A) Business as Usual, to (B) Making Small Changes and (C) Sustainable wellbeing.

In an era of social, economic and environmental convergence of risk we need to accept the possibility of new forms of governance to protect those who are not protected by the nation state, because they are not citizens.[23] The so-called 'tragedy of the commons' argument developed by Hardin (1968) is a construction informed by simplistic thinking. Ironically, it is often used as the starting point for environmental thinking, but in fact it originated as an argument developed by Locke in support of the enclosure movement and private property in Britain. The argument being that contained areas of land are cared for better than the areas of land that are held in common and shared. The example of common grazing land is shared.

But ironically it is the privatisation and commodification of land and natural resources that has led to environmental pollution and degradation. The lack of trust between nations has evoked the rhetoric of nationalism and state protectionism. This has led to 'seeing like a state' (Scott 1998), which in turn leads to the argument for competition and the zero sum approach which argues that one nation can profit at the expense of others. The organisational philosophy and governance arguments that flow from this philosophy are based on limited disciplinary paradigms that can profit at the expense of others.

Ideally, a universal respect for social and environmental justice could enable subsidiarity as a means to support freedom (Follesdale 2006; Poe 2010) to the extent that it does not undermine the quality of life of this generation or the next.[24]

---

[23]The work of Ann Florini (2003) provides possibilities as does the work of Danielle Archibugi. The challenges of governance need to be addressed by working across disciplines. Gibbons et al. (1994) argue that the ability to work across boundaries is vital for 'The new production of knowledge' and vital for 'the dynamics of research' to address current complex challenges.

[24]Dualist thinking pervades our consciousness and reflected in socially unjust and environmentally unsustainable designs for society. Designs need to be supported by constitutions, based on a priori norms, and consequentialist or a posteriori approaches, based on testing out ideas within context and with future generations in mind. Global axioms to protect future generations and the poor of this generation need to guide legal constructs and legal decisions at the local level. The principle of subsidiarity could rest comfortably with global axioms—provided that the caveat is made that people at the local level can be free and diverse to the extent that their freedom and diversity does not undermine the rights of others or future generations of life. This challenge of balancing individual and collective social, cultural, political and economic concerns needs to be buttressed by regional parliaments and courts that focus on social and environmental justice at a post-national level. A Community of Practice needs to develop curricula that enable students to identify with others across the differences and to recognise our interconnectedness and to reframe the nation state. The Springer Encyclopedia of Food and Agricultural Ethics (McIntyre-Mills 2013) provides a brief outline of the landscape of this complex field, defined by intersections spanning social, cultural, political, economic and environmental contributions from the social and natural sciences.

We have designed and tested an alternative approach using critical heuristics—engaging in questions that explore if/then scenarios with local service users and providers. I hypothesise that participation through awareness and consciousness-raising (McIntyre-Mills 2010) will influence the way in which people value relationships with others and the land. This hypothesis is based on the notion of neural plasticity in that the brain shapes the environment and, in turn, is shaped by the environment (Bateson 1972; Beer 1994; Greenfield 2000). The liberative potential of aspects of Sunstein's 2013 book on simpler government is a starting point for enabling people to think differently through providing scenarios and then getting people to make choices that are transformative:

- The Opt out or redistributing saved resources to others (food, energy, water) and organs for those in need
- Rules against using private transport where public transport is available, rules about texting or phoning when driving, wasting food, energy and water
- Rules against waste of food, energy and water
- Rules about distribution of saved resources to those who opt in for help at the local level.

Designing public education requires enabling participants to address representation, accountability and sustainability. Why start with West Churchman's (1979) 'Design of Inquiring Systems Approach'? It helps to think about our thinking and our relationships. It spans self, other and the environment. It is based on questioning boundaries of inclusion and exclusion, it examines the so-called 'enemies within' (religion, mortality, politics and aesthetics) and it considers the consequences of our choices. It helps to address a core issue such as Food security in order to unpack it and to demonstrate that it is a systemic issue that poses both challenges and opportunities. We need to work with many forms of knowledge and across conceptual and spatial boundaries to address the big issues of the day, namely poverty and climate change within increasingly densely populated cities.

Developing learning organisations and learning communities could help to create bridges across diversity and strengthen bonds of friendship through enabling people to feel that they have a voice and a right to a voice. Participation in vertical democracy through elections, whilst important needs to be extended to include

(Footnote 24 continued)

A case is made for those who are not protected by the social contract, including young people, non-citizens, the disabled, sentient beings and the environment on which we all depend. The contribution highlights the empirical contradictions and theoretical tensions that have implications for social and environmental justice.

other forms of participation.[25] Polly Higgins suggests that a new planetary law should be passed to protect the viability of the planet and that current systems of law are inadequate to protect people and the planet.[26]

3.4 Case Study of Engagement to Address Low Carbon Ethical Living

The research began in the NT and South Australia with Aboriginal Australians on ways to enhance quality of life and wellbeing. The relevance of this definition is that the NT and SA are, respectively, one of the hottest and driest parts of Australia and thus most affected by climate change. Furthermore sustainability thinking is now more relevant than ever for planetary wellbeing. Most importantly it is relevant because it is likely that the sorts of challenges faced by the poorest on the planet—are the sorts of conditions that we face in the future, if we continue our addiction to consumption of resources.

3.4.1 Design

The design and iterative approach to research on wellbeing spans the following stages:

1. The first step began years ago when I worked in Alice Springs on quality of life on behalf of the Northern Territory Local Government. I was asked to undertake

[25]Discursive discussion and direct votes on policy ideas could help to foster connections and a shared ideology for a sustainable socially and environmentally just future. Opportunities exist to include creating a better balance between rural and urban areas, greening cities (Dryzek 1999, 2000, 2010) through the creation of urban agriculture in open spaces, on top of, attached to and within building, co-determination in regions—new architecture for governance democracy and ethics—learning, Mindfulness to support communities of practice within learning communities. These aspects will be discussed below.

[26]See https://www.youtube.com/watch?v=QPUmN88htCo and Earth democracy http://m.youtube.com/watch?v=Bq0lrbznsjc. Stiglitz has made the same point in relation to economics, whilst the need to reframe boundaries across a range of disciplines has made similar points as I stressed in a recent entry in Sociopedia called: 'Reconsidering boundaries' (McIntyre-Mills 2014). Whilst the proposal was made by Higgins and others to the UN, Bolivia has succeeded in leading the way by passing a law to recognise the need to protect the earth. This could help to ensure that elected leaders remain more connected to the people they are supposed to represent. So the future lies in participatory democracy and more direct interventions. What is needed is a greater emphasis on core principles and engagement with the people in discursive democracy. Participatory governance needs to test the implications of decisions with others so as to balance individual needs and collective needs by enabling residents to make decisions 'from below' to inform policy at the regional level and to facilitate monitory democracy and governance.

research on the quality of life of rate payers. This was reframed as the quality of life of all residents as a requirement of my undertaking the research. The findings have been discussed elsewhere in a Monograph (McIntyre-Mills 2003) in which the details of the mixed ethnography are discussed. But I mention it here only because Kathy Abbott and Olive Veverbrandts (my Aboriginal mentors along with Peter Turner of the Stolen Generation) stressed that health and wellbeing could only be achieved through working on ways to care for country. Caring for country involves protecting earth, water and living systems. Olive and Peter share Hawke Dreaming and have a view of 'the big picture'. When I began the work with Neporendi in South Australia funded by the South Australian Dept. of health and by an ARC Linkage grant it was stressed that weaving strands of experience in South Australia (one of the driest states in the driest continent) was a precursor to wellbeing. This research has also been described in a Monograph (McIntyre-Mills 2008) and when the work on complex health, housing and social inclusion was completed members of the Neporendi board launched the project the day before the Rudd Apology to the Stolen generations and Doug Morgan made the request that the learnings be scaled up and applied not just by one small NGO set up to provide services by, with and for Aboriginal Australians living in the South, but to be applied more broadly by others in South Australia.

2. This led to the next step in the project, namely research funded by The South Australian Local Government Association on complex decision making to address Climate Change through mitigation of the causes and adaptation to the effects through addressing social, economic and environmental indicators of wellbeing. The research was inspired in part also by the work of non-Indigenous theorists in a range of disciplines such as Joseph Stiglitz who had recently addressed the Australian Productivity Committee in which he stressed the need to emphasise wellbeing as the economic bottom line—rather than profit. The findings of this research have been detailed in *The Contemporary Systems Series* (McIntyre Mills et al. 2014) and the methodology was discussed in Systems Research and Behavioural Science (McIntyre-Mills and De Vries 2012). The development of an alternative form of engagement was not explained in sufficient detail, however, and the mixed methodology used to foster engagement needs to be documented in more depth. The next step was to engage the population living in Middleville in a series of focus group discussions that were advertised by the library and in a newsletter and co-facilitated by the community development officer at the council. The three focus groups were well attended and were held in the morning, afternoon and evening to enable different groups to attend. The discussion around what constitutes wellbeing in Middleville combined narrative accounts, group discussion and Soft Systems Mapping of shared concerns noted on yellow stickers and then combined in a map on large sheets of white paper placed on a central table. Other interviews were also held in the community garden, the library, the shopping centre and with significant members of the community who play an active role in shaping the local government agenda. This then led to the development of three scenarios which

became the basis for the development of an interactive engagement platform (McIntyre-Mills et al. 2014). The platform was populated by adding material to the software on the basis of the interviews and then a team of post graduate researchers helped to interview participants who created over 300 pathways to wellbeing.

Box 1: Building capability to think critically and systemically
The aim was to:

- Develop software to enhance decision-making.
- Enable service users, policy makers and providers to develop clear priorities in response to contextual challenges through careful consideration of the consequences of social, economic and environmental decisions.
- Understand the diverse perceptions on mitigation and adaptation. Mitigation means reducing our impact on climate change. Adaption involves changing our daily decisions and behaviour.
- Enhance awareness of our interconnectedness and dependency on nature.
- Build the capacity of participants to realise that they can live elegantly and well and that they do not need to polarise productivity versus the planet (Charlton 2011) by building wellbeing stocks for the future.

The core point made by the participants in South Australia to date is that people need to take time or find time to make a difference to the way they live.

The architecture for engagement could have potential if applied as a means to implement monitoring 'from below' as detailed below. The book discusses the testing of a pilot project to develop raft of systemic wellbeing indicators to address the notion of wellbeing stocks and being the change globally.

- It tests the principal of subsidiarity and Ashby's rule and strives to implement Murrays's notion of a Tuvalu test for decision-making. The software is designed and extended by Denise de Vries and Natasun Binchai. Details can be found at: https://wellbeing.csem.flinders.edu.au/ and now http://wirasoft.com/pathways-to-wellbeing/.
- The log in is user name = test, code = test.

This book has two main objectives. The first is to make the case for social change through exploring post disciplinary and post materialist frameworks to address greed, zero sum competition for resources, the commodification of the powerless and the environment. Secondly, it develops a reframed approach to measuring wellbeing—not productivity—as a sign of economic success. Thus the book considers the challenge posed by Stiglitz (2010) to the Australian Productivity Commission, namely to foster an understanding that the wellbeing of humanity is dependent on protecting the global commons and ensuring that some do not live at the expense of others and future generations of life.

- Makes a case for cosmopolitan approaches that scale up local engagement and that enable monitory democracy 'from below'. Instead of Big Brother controlling from above, the people are encouraged to monitor the use of resources from below at a local level.
- The process of monitoring needs to be supported by means of the principle of subsidiarity and buttressed by international law spanning post-national regions.
- It discusses participatory action research to prefigure a means to hold the market to account—to ensure that the use of resources that are necessary for the common good are accessible and equitable—implications for distribution and sharing.

3. We concluded that wellbeing requires re-conceptualising our identity and relationships to others and the environment. It is achieved through systemic intervention, based on 5 axial themes. Research on wellbeing requires re-conceptualising our identity and relationships to others and the environment through systemic interventions.[27] The 5 axial themes mentioned above are universal. (McIntyre-Mills with De Vries 2008: 307). Similarly, Stiglitz et al. (2010: 15) use a multidimensional measure of wellbeing spanning: '(1) Material living standards (income, consumption and wealth), (2) Health, (3) Education, (4) Personal activities including work, (5) Political voice and governance, (6) Social connections and relationships, (7) Environment (present and future conditions), (8) Insecurity, of an economy as well as a physical nature'.[28]

The idea is to use the scenarios as starting points to explore ways to enable people to live healthy, happy ethical lives that are not at the expense of others or the next generation of life. The scenarios are for Australia and we need scenarios from other parts of the world that reflect social, cultural, political, economic and environmental diversity.

---

[27]The definition of wellbeing is developed on the basis of research firstly in the Northern Territory with Aboriginal Australians and then in South Australia in the Southern Region of Adelaide on complex needs.

[28]Together with vulnerability to job loss, many people across the world also face risks associated with fires, flood and drought. Arguments for and against the national, transnational and supranational organisations have been raised by idealists, pragmatists and realists. A way to bridge the divide is through considering the consequences for current and future generations through addressing wellbeing and developing governance systems to ensure that stocks for the future are developed through limiting the way in which production processes and distribution practices benefit the minority at the expense of the majority.

It could help people to realise their dependency on their relationshipwith one another and the land and on one another for peace and security. The problem with current forms of governance is:

- Spatial—Limited to the nation state but the resources to support quality of life are beyond the boundaries of the nation state.
- Temporal—Limited to electoral cycle and this generation.
- Conceptual—Wellbeing not profit (economic) is the problem—addressed through education and transformation of ethics and law.

The outcome of the research with non-government and human service organisations was to explore ways to scale up the engagement process to the local government level to enable wellbeing through wide-ranging social, economic and environmental changes. The research makes the case that developing connections with others could provide a way forward. Decisions need to be informed by connecting many kinds of knowledge. If people can make conceptual connections through thinking through scenarios it could enable those of us who are not physicists to think about the big picture. Our wellbeing depends on the wellbeing of the planet. Our home or sense of place is thus both local and global. At the moment our software enables people to choose from three optional scenarios. Then to select one with which they identify and to consider what they are prepared to add or discard from their lives, in order to make a change. The data informs them of their choices and helps to inform local government of the patterns of policy choices. What it does not do, however, is help people to see to what extent their thinking and practice matches the characteristics for making small adjustments for the long haul or sustainable living. This is the next step. Many people deceive themselves that they are living sustainably when actually their range of behaviour comprises many different choices– some of which support change and some of which undermine change. Overall little change is made to the size of their carbon footprint. We need to obtain more insight into why this group thinks the way that they do. What we need is for people to be able to see how they can move closer towards sustainable living by enabling them to see the implications of their choices. This is a precursor for the transformation of cultural 'habitus' (Bourdieu 1977).

4. The case study at Middleville helped to develop and adapt the software for a wider audience. It involved 'teaching' the software a wider range of social, economic and environmental issues through adding the data obtained through focus groups that mapped and created typologies of social, economic and environmental concerns and some initial pilot interviews. The next step involved populating the software and testing it out before linking it to a range of websites. The research made use of many forms of engagement to help participants to consider the potential consequences of their choices. These approaches span narrative (to explore complexity and to enable a creative flow of ideas), dialogue and questioning to reformulate boundaries. Scenarios were also explored spanning 'business as usual, 'small steps towards a sustainable future' and achieving 'wellbeing stocks' (Stiglitz et al. 2010). The research

approach at this stage comprised: (a) Gathering data through focus groups using basic scenarios and extending these through in depth conversation. (b) Identifying patterns through mapping ideas and connections. (c) Designing and Piloting the software through testing the logic—it updates and grows as data is added. It is an open system geared an enabling people to have a say using their own words and expressing their perceptions and emotions. (d) Populating the software—now more than 300 pathways on an interactive data collection system located on the Flinders Website.[29]

5. The software enables thinking through 'if then' scenarios to enable people to consider the consequences of one choice versus another. The aim is to enable the management of social, economic and environmental resources, such as energy or water usage. The research considers the potential for decision-makings software to address social, economic and environmental factors that support or undermine adaptation to the challenges and mitigation of the effects of climate change. It evaluates the extent to which the approach enables the community to consider their perceived assets and risks and the implications of their consumption choices.[30] The demo with illustrative EXCEL sheets can be viewed at https://socsci.flinders.edu.au/fippm/pathways_demo/.
6. Each personal pathway to wellbeing can be printed out and kept for personal review. Pathways could help service providers provide better responses to social, economic and environmental challenges. This is a dedicated website which shows the software linked with the research (see McIntyre-Mills and de Vries 2011) on representation, accountability and engagement to enhance wellbeing. The aim is to enable the more effective matching of services to meet diverse needs and to enable the more equitable use of social, economic and environmental resources at a local level. Requirements: Recent web browser with JavaScript enabled (i.e. Internet Explorer 7, Mozilla Firefox 3.5, etc.). Plug-ins for a media player (avi files) and PDG reader, such as Adobe Reader. There is no installation required as all data is stored on our server. Application

---

[29]This will be linked with the local government website. It is also currently with a number of relevant professional and public sector organisations. The members of the research team ensured that Aboriginal stakeholders were included as a means to extend the initial research funded by the ARC (McIntyre-Mills 2010) on user-centric design to meet complex health, housing and social inclusion needs. This is because as climate change accelerates it is likely that living sustainably will need to draw on the lived experience of the first Australians. It is also likely that the living conditions that many in urban areas have taken for granted will need to change in response to more post materialist values.

[30]The design was developed through focus groups and through 90 face-to-face interviews. The next step involved populating the software and testing it. To date there are more than 300 participants in total. Of which 65 are aged 18–25 years, 31 are aged 26–35 years, 21 are aged 36–45 years, 86 are aged 46–55 years, 74 are aged 56–65 and 24 are aged over 65 years. To date female informants have generated 159 pathways as to how to enhance quality of life and 140 men have generated pathways. Those aged from 26 to 45 participated less than other age groups as a result of other responsibilities.

for local government: https://wellbeing.csem.flinders.edu.au/.The log in is user name = test, code = test.

7. The evaluating approach is ongoing and iterative as more data is collected and we connect to more websites. The de-identified data can be seen by those who have been given administrator rights.

3.4.2 Enabling Local Communities Through Participatory Design

Participatory design has a long history in the International Systems Sciences, International Sociological Association, the Action Learning and Action Research Association, for example. The rationale is that testing out ideas with people who have lived experience as well as by people with professional expertise it is possible to enhance the quality of the decisions and to achieve a greater match. It is based on the notion that the complexity of the decision needs to be matched by the complexity of the decision makers. This enhances the robustness of the decision and it is also more likely to be supported politically by those who are affected by the decisions.

The process is vital to develop appropriate indicators of wellbeing that are linked to an interactive map of the perceived haves and needs of the population in the context of a post-carbon world in which the cost of resources is affected by the cost of oil. This is where the collaborative work on participatory democracy and governance is located. The role of the good citizen is to: (a) enhance wellbeing of communities, (b) support the demographic transition and (c) enhance life chances for all through securing food, energy and water. The good global steward builds stocks for future generations and does not exclude the cost to the environment or future generations. Poverty, pollution and carbon emissions ought not to be shifted to future generations or other people. This requires post-national and post disciplinary cooperation and new forms of governance. Could we start to see patterns in the statistics and in the stories that we co-create as shared narratives that become increasingly harmonious song lines for a new society that is indeed post-materialist?

The research explores whether using a computer-aided model could enhance data collection, analysis and ongoing engagement to address complex social, economic and environmental issues associated with climate change.[31] The software

[31]At the heart of democracy, science and ethics needs to be an appreciation of stewardship not commodification. As stewards, at best we can strive to consider social, economic and environmental indicators to prevent destroying the environment and considering options 'from the point of view of future generations. The programme of research begins with the issue that governance and democracy have to deal with three options pertaining to truth (Crowder 2003; McIntyre-Mills 2000, 2011a, b, c): (a) One truth (monist) responses defended by conflict and grand, (b) No truth (postmodernist) approached defended by relativism, (c) Mediated (harmonised) responses based on stewardship. Each approach has specific implications for governance and democracy. The

was loaded onto the Flinders Website,[32] in the first instance. It is now co-located on the Wirasoft website, and has the potential for links with local governments elsewhere at a sub and post-national level within the region and elsewhere that we are currently developing. Technology needs to serve social and environmental justice, not curtail it. Our approach to design is an essential aspect of creating a livable planet. As tool makers we have the potential to use technology in diverse ways.

The research design comprises a multi-method approach. The first stage included focus groups spanning age, gender and areas of concern to address wellbeing and climate change in a middle class suburb in South Australia, Adelaide. The initial data collection on perceptions with regard to three scenarios (gained through face-to-face engagement and focus groups) provided the basis for formulating the question schedule. One of the major challenges for the research was to find a way to engage the attention of the residents of Middleville on line. Face-to-face engagement by members of the team was the most successful means of eliciting responses, followed by focus groups facilitated by D.R. Time taken to develop rapport was crucial for obtaining useful information. Nevertheless, even the question schedules that were left for collection in the Middleville library provided some useful (albeit mostly incomplete data) by those who took the time to fill in the form. The initial questions were piloted and critical feedback was incorporated.

The first pilot followed the initial focus groups and a total of 90 one to one interviews to populate the software. Three informants did not understand the connections across climate change and wellbeing and some of the informants did not complete the initial pilot of 28 questions, because they did not understand the questions. This was remedied by developing more explanations and examples. The attempt to enlist participation via a Google monkey question schedule was also unsuccessful, in part because the issues are associated with personal issues, rather

(Footnote 31 continued)

monist approach lends itself to top down interventions and autocratic interventions. The post-modernist approach lends itself to arguments about non-commensurable notions of truth and justice and thus national silos, whilst the harmonisation approach lends itself to working across conceptual and spatial boundaries. Engagement to enhance the capability to think critically and systemically is one of the major challenges if democracy is to be enhanced and if people are to be able to understand the consequences of their decisions. The signing of the United Nations Declaration of Human Rights has not lead to detailed covenants or the means to implement the noble ideas. Transforming legal frontiers is a necessity for human security. The legal frameworks to address greed and the commodification of the powerless and the environment need to be informed by narratives and public debates on ethics to enhance human reasoning capability, empathy and humility.

[32]Until broadband is made widely available the software will run less effectively at peak hours. Participation is thus effected by the lack of broadband. This point was made strongly by the Thinker in Residence, Genevieve Bell (2009), who was invited to review the potential for ICTs in South Australia in her report ‘Getting connected, Staying Connected, Exploring South Australia’s digital futures’. Another challenge was the migration of data from one part of the server to another and the right to obtain access to log into the data systems by both users and administrators. The next hurdle will be to link the Flinders website to the Local Government Association (LGA) website. It is currently linked as a demonstration to the Wirasoft Website.

than overtly public policy concerns and partly because of time constraints and the lack of clarity in explaining how to use the software.

The scenario that they developed follows the prompts that were provided to enable people to think about the future and their choices.[33] An approach to democracy, governance and ethical decision-making based on considering both the existing a *priori norms* linked with a particular time and space through contextual, 'if then' scenarios could enable participants to consider the implications of their decisions. The structure and the process need to support respectful dialogue.

The research at the local government level begins by exploring the discourses to establish whether mitigation of consumption can be encouraged through enabling people to think through the consequences of their choices in scenarios. Collins (2004, 2010) discusses the need for overarching legal structures as well as emotional commitments made when taking an oath or communicating an idea and receiving positive affirmation from others. The engagement process of participatory democracy provides an everyday ritual to enable people to think about the implications of their choices. In this study people are invited to do mental 'walk throughs' to think about the implications of climate change on their lives and what they are prepared to do or to change, in order to meet their own daily needs and those of their families. The aim is to enhance capability through conversations and thinking about our choices,[34] in order to develop an 'ecology of mind' (Bateson 1972) and to develop a greater ability to process large amounts of information—which is needed to avoid projecting problems onto others (Baruma and Margalit 2004) and to develop the capacity for cosmopolitan approaches (Beck 2009) and systemic governance (McIntyre-Mills 2006b). Beck stresses that we learn from the consequences of our choices and that we cannot anticipate all the risks nor can we prepare to plan for them in a linear, controlled manner, but we can try to think in terms of scenario of risk. For example, the research in this vein was done by asking people to consider a variety of scenarios (as outlined below).

---

[33]C.H. heads the environmental Community Advisory Group and R.I. and J.F. are active members who are committed to the issue. The youth advisory group via K.M. also contributed along with a number of community members who gave their time to participate in the focus groups that were followed by one to one conversations following a set of guided prompts, then the questions were piloted iteratively in stages at the community garden, outside the library, at opportunity shops and in a range of community groups through networking. Critical feedback leads to ongoing modifications during the pilot stage of the research.

[34]One way to do this is to assess the quality of conversations and to find out what kinds of conversations can be used to enable people to make changes in their lives through helping them to think through the implications of choices. Collins (1981: 1001) argues: 'It is insufficient to address a single conversation; instead he suggests the value of analyzing conversations over time to assess the way in which power is garnered in symbolic interactions'. 'Emotional energies-emotions affect ritual membership in several ways. There must be at least a minimal degree of common mood among interact ants if a conversational ritual is to succeed in invoking a shared reality. The stronger the emotional tone, the more real the invoked topic will seem to be and the greater the solidarity of the group (see Collins 1975: 94–95, 153–154)'.

What example do we want to set for our children and grandchildren? How can we make a difference today in the way we live? Please would you suggest ways in which Middleville residents could minimise their carbon footprint in their everyday lives? What ideas do you have for living sustainably? What ideas do you have to promote wellbeing in the community?

What are the implications for transport, cycle paths, walking paths? What are the implications for housing, heating, cooling, building codes? What are the implications for gardening, water use, recycling water, recycling household waste?

Participants were asked to identify ways to live sustainably by participating in a focus group and a meal. The information they provided was loaded onto the website. After participating in the focus group they were invited to add to the ideas by logging on and by inviting others in the community to contribute as well. It is an open site that is administered by Flinders University, linked with the participating local council. The link has also been added to other professional and social networking sites. Some training was made available to those who were interested in logging onto the site. People were asked if they were interested in participating in a training programme and to register their interest with contacts undertaking this research.

Scenarios along a continuum to address climate change and wellbeing	
(a) Business as Usual, (b) Doing too Little Too Late, (c) Sustainable Future and wellbeing	(a) Not coping[a], (b) Just keeping it together (c) Wellbeing

[a]This reflects the underlying science, but from the point of view of the some/perceptions they probably don't perceive themselves as not coping. The software is aimed at enabling participants to make the connections

Participants were asked to think about health, housing, employment, education and social inclusion within their communities. Our conversational structure to address complex needs is to consider the following:

What do I/we have?
What do I/we need?
What do I/we want to add to my life—the resources and networks?
What am I prepared to discard from my life—the resources and networks?
What are the turning points for the better and worse
What are the barriers?

The reason for engaging people in work on the programme is in order to make an emotional connection with what they are thinking about. The symbolic action of talking with a sympathetic 'other' is one aspect of the process for engaging in a therapeutic conversation.[35] The second aspect is the commitment to do something about the decision and to make a personal contract to do things differently. The idea is that people will be invited to think through the implications of doing nothing to change one's life and the implications of making small (cumulative) adjustments over time.

In order to help the thinking process, people will be prompted by means of questions and '*if then*' scenarios that enable them to think through the implications of micro decisions. What difference will it make if I sell my car and use public transport? Will using a car help me to solve my problem better than if I make use of public transport? Perhaps I can live closer to work? Perhaps I can use the car only in instances where public transport is unavailable? The next step is to take action and to think through what I choose to add to my life in the forms of actions or what I am prepared to discard from my life. This is extended by making a contract which spells out what individual users are prepared to contribute, in order to make a difference. The local resident who chooses to engage in a conversation about how s/he will live in the future will make a choice and then consider the implications for their pathways. The de-identified choices can be mapped and viewed by everyone. The identity of individual user is indicated by an anonymous user name.

Scenarios and focus groups were used along with one to one interviews. The focus groups explored three scenarios. The aim was to build a shared sense of identity, whilst enabling individual diversity, capability and freedom to the extent that the diversity, capability and freedom *of others* are not undermined. These were inspired by the Mont Fleur Scenarios that were used to enable people to think about what a future in South Africa could be like in the year 2002 (Kahane 1992). Similarly scenarios were shared with participants who were asked to think about how they would envisage life in Middleville and Australia if business as usualprevailed, or if small changes were made, or if progress was made towards living sustainably. They were asked to consider three typical scenarios and then to reshape them according to their own perceived ideas. The first step was to address the question: How do we address the many social, economic and environmental variables with different value dimensions? Scenarios can enhance our capability to join up the dots. Three scenarios were introduced to address climate change and wellbeing, namely: 'Business as usual, Small Adjustments for the Long Haul and Sustainable Futures'. Scenarios such as these can be used as a starting point for an

[35]By studying interactions from local to global we can see how positive responses build energy and emotional attachment and how negative responses deplete energy and create divisions. 'The most basic common emotional ingredient in interactions, I would suggest, is the minimal tone of positive sentiment toward the other. The solidarity sentiments range from a minimal display of non-hostility toward mutual liking and enthusiastic common activity. Where do such emotions come from? They originate in previous experiences in IR chains. An individual who is successfully accepted into an interaction acquires an increment of positive emotional energy. This energy is manifested as what we commonly call confidence, warmth, and enthusiasm. Acquiring this is one situation; an individual has more emotional resources for successfully negotiating solidarity in the next interaction. Such chains, both positive and negative, extend throughout every person's lifetime'. (Collins 1981: 1002).

engaged conversation at the local or transnational level. As human beings we need to re-learn how to be resilient. The scenario that they developed follows the prompts that were provided to get people thinking about the future and their choices. C.H. heads the environmental Community Advisory Group and R.I. and J.F. are active members who are committed to the issue. The youth advisory group via K.M. also contributed along with a number of community members who gave their time to participate in the focus groups that were followed by one to one conversations following a set of guided prompts, then the questions were piloted iteratively in stages at the community garden, outside the library, at opportunity shops and in the wider Middleville area through networking.

3.4.3 Potential Value of the Case Study

What engagement processes could be effective to address the policy and governance challenges of mitigation and accountability? This study addresses the programme of research on representation, accountability and engagement to adapt to and mitigate the extent of risks. Thus its emphasis is not merely to describe what the case is, but to discuss participatory action research that strives to design and prefigure alternative praxis as a seed for transformation. The research tests the principle of subsidiarity, namely that people ought to have the right to make decisions by drawing on their own experiences at the local level, but they have the responsibility to follow norms and laws that protect the interests of the public. In order to address convergent social, economic and environmental questions it is necessary to address complexity, by making sense of issues through narrative and dialogue in order to understand the diverse, interrelated variables each of which is perceived differently by different stakeholders and then to work out socially and environmentally just decisions (Talpin and Wojcik 2010).

Perceptions and emotional dimensions to quality of life (Nussbaum and Sen 1993) could change as a result not only of laws to manage energy, water and food resources, but through engagement with a form of interactive software that helps to develop consciousness of our interconnectedness.[36]

[36]It strives to enable a greater awareness of consciousness of our rights and responsibilities as global citizens. It is practical in that it attempts to acknowledge that localised, slow lives that are sustainable could lapse into the Orwellian nightmare of control from above—without surveillance from below to hold the elites to account. It is for this reason that the digital future needs to be shaped not only by the state military complex—but by those who wish to act as caretakers for the next generation. Members of the public need to be protected by a scaled up version of the Aarhus Convention that gives residents within post-national regions the right to access information and to the right to be heard by the state (1998, see McIntyre-Mills 2011), thus this report (a) discusses the theory of participatory democracy—based on the principles of subsidiarity and Ashby's rule of Requisite Variety, meaning that decisions need to reflect the complexity of the area of concern and (b) makes a case for new forms of accountability to support regional governance that is capable of extending the social contract beyond the nation state and thus to protect biospheres and the

It is likely that a greater balance will need to be achieved between individual and collective needs through emphasising the determinants of wellbeing for this generation and the next. This requires a new approach to public education, democracy, governance and ethics that places food security at the centre of decisions. The role of local participation and governance thus becomes increasingly important (Schweber 2012). Our research strives to enable people to have a say about how scarce resources should be used, in order to inform elected representatives of their perceived social, economic and environmental needs. It is hoped that by thinking about what they have and what they need and what they are prepared to change (by being prompted to consider the consequences of our choices for others and for the environment). The notion of engagement in 'if then scenarios', is helpful to enable people to develop sustainable social, economic and environmental decisions (McIntyre-Mills and De Vries 2011).

A sustainable local community is determined by a sustainable region in which food, energy and water supplies are considered as major determinants for wellbeing. No community can be expected to transform from a high carbon life style (or aspiring to this life style) without being part of the design process and owning the decisions as to how resources should be used.[37] The exercise of 'thinking about their thinking' is an attempt to build new neural pathways in the mind—to rehearse a different way of 'being in the world', based on thinking about the consequences for themselves, others and the environment.

The research aims to establish if it is possible for cultural transformation to be scaled up to enable a shift towards sustainability. According to Barbara Dickens from the Renewal SA State Government urban renewal authority (see McIntyre-Mills et al. 2014):

> Pathways to Wellbeing software offers a unique avenue for ongoing, long-term engagement between local residents and local government. It facilitates expanded pragmatism based on considering the consequences of our choices for ourselves, others and the environment. These considerations are used to shape evidenced-based policy to inform decision-making at the local level. By asking residents to make choices from Pathways to Wellbeing scenarios, participants are invited to make changes in their lives. It aims to help them think through the implications and consequences of their choices for themselves and future generations.

The objective is to make a case for social change to address wellbeing (not merely productivity for profit). By working in middle to upper income area, the aim

(Footnote 36 continued)

diversity within them. The governance decisions will need to be upheld by rule of law based on the axiom that we can be free and diverse to the extent that freedom and democracy is not undermined (McIntyre-Mills et al. 2006).

[37]This research emphasises the creation of multiple paths to form webs to wellbeing through critical, systemic engagement to ensure that local people get to know about local resources (that are listed in directories created by both service users and providers) and that their identity is protected. They are able to see the typical scenarios of what people have, need, are prepared to add or discard from their lives and what they think are turning points for the better and worse.

was to establish to what extent this group of people (the majority of whom are employed and own their own homes) are able to think in terms of enhancing their wellbeing whilst lowering their carbon footprints. The central challenges for democracy, governance and ethics are as follows: (1) Representation, (2) Accountability, (3) Sustainability and (4) Engagement (McIntyre-Mills and de Vries 2011).

3.4.4 *Options and Potential for Engagement*

It is worth bearing in mind that 93% of Australians have access to the internet in 2015.[38] This contrasts with the official statistics by the Australian Bureau of Statistics in June 2016 which states there are 13.3 million internet subscribers in Australia at the end of June 2016 and that this marks a 4.2% increase since June 2015. But only 66% of households had access to computers, according to the census in 2006. The rapid increase has been extended by mobile phone connections which appear to be replacing traditional dial up connections. The way forward will be tested through linking with a range of sites and moving gradually towards social networking sites, once we are satisfied with the testing of the software through smaller pilots in other contexts. For example, in Indonesia[39] there are 100 million internet users which makes 40% of the population. In South Africa 52% of the population use the internet and the numbers of users are growing (see Footnote 40).

3.4.4.1 Engagement for Those Who Prefer Face-to-Face Engagement

Participants were asked to consider the following scenarios[40] as a starting point for answering the questions listed in the proforma below. Each participant will have new ideas and they are just a starting point for asking you to share your own ideas about how we should live:

A. **Business as usualscenario**

We continue to believe in economic arguments that others believe ignore the social and environmental dimension. We continue to think that our way of life is sustainable and are not prepared to manage the perceived risks of climate change by changing our way of life. We attribute drought, bush fires and floods to one-off unrelated events or natural cycles, and deny that climate change can trigger rising

[38] Internetworldstats.com, accessed, 23rd Dec 2016.

[39] www.internetlivestats.com 2016.

[40] The three scenarios need to be suitable within cultural contexts where the research is to be tested. Also the updated design needs to have language translation built in. If the software is to run on an iPhone it needs to be colourful but simple and not take up too much bandwidth.

temperatures in some areas and plummeting temperatures in others as melting ice effects the ocean currents.[41]

B. **Small changes scenario**

People make slow annual progress towards goals which they meet for the benefit of their children and grandchildren. People of all ages and from all walks of life who are able to join up the dots between the economic, social and environmental dimensions help to motivate movement towards a better future. We do not perceive these small changes as being too slow to sustain beyond our grandchildren, or we envisage that something else will happen by then to reverse the current trend. Governments and non-government organisations take the initiative. They hold workshops to demonstrate how people can make a difference. They listen to the people and help local groups to respond to local challenges. Together they undertake model projects that demonstrate how it will be possible to live differently. They model different ways of thinking and through living the changes show that it

[41]The detailed scenario is as follows: We continue to believe in economic arguments that ignore the social and environmental dimension and we continue to think that our way of life is sustainable and are not prepared to manage the risks of climate change by changing our way of life. We blame the increasing risk of drought, bush fires and floods are on 'one-off' unrelated events or deny that climate change can mean rising temperatures in some areas and plummeting temperatures in others as melting ice effects the ocean currents. The sea is used as a dumping ground and it no longer helps to regulate our climate. More and more of us suffer from viruses and food poisoning. Animals are diseased. Most of our rivers are polluted and many have dried up. We fight over the last of energy and natural resources. We export our waste material to poorer nations who 'offer' to store it. 'The government' and 'the economy' are blamed for the problems, but we do not make any changes to our personal lives, because it is too hard, or not our problem. The corporate business sector continues to tell us that the market self-regulates. We compete with one another and are proud to wear designer labels or to carry designer packages. We engage in fund raising activities and give money to charity. These small gestures are to enable us to pretend that we are making a difference. We refused to an agreement in Copenhagen, because it is bad for the state of the economy. We continued to wrangle for after the inconclusive Climate Change talks in Copenhagen (18 Dec 2009). We cannot achieve agreement internationally as to how we will go about reducing our emissions and changing our way of life. We feel anxious, stressed or depressed and we use drugs, alcohol and shopping to provide temporary relief. We flop in front of television and watch mind numbing programs or endless DVDs so that we ignore the problems in our neighbourhood. We withdraw and do not know the people in our street. We feel we cannot be bothered, because we are too busy making a living and worrying about our own problems. We base our decision on narrow pragmatism, because we think about the consequences only for ourselves and not others. We believe that our power and profit must be driven by self-interest and the bottom line, namely ensuring our powerful positions and our profits. We think that social and environmental considerations are 'externalities', rather than imbedded in the current system.

is possible to balance individual and collective interests, because we are not selfish nor are we unable to create alternative ways of governing at a regional level.[42]

C. **Sustainable future scenario**

We live in an environment that can support this generation and the next. Housing is affordable and made of sustainable materials. We have faced up to the convergent social, economic and environmental challenges and we are resilient, because we live in clusters of homes, share rain tanks and solar grids that are subsidised by local governments. Our living and working areas are powered by alternative energy. The new status symbolis the environmentally friendly lifestyle. Public transport is green. The green economy supports a vibrant job market spurred by subsidies to enable packaging goods, housing people, and transporting people, educating and entertaining the public. The carbon economy is replaced through innovative inventions. All members of the public are encouraged to share their experiences and ideas for living sustainably. The futures market has been reconstructed to take into account the air, water and earth we need to grow organic, safe food. We have thought carefully about the implications of treating people, animals and the land as commodities and we strive to care for ourselves, others (including the voiceless) and the land.[43]

---

[42]The detailed scenario is as follows: People make annual progress towards goals which they meet for the benefit of their children and grandchildren. But they do not move quite fast enough. People of all ages and from all walks of life who are able to 'join up the dots' could help to motivate cultural transformation towards a better future. They are motivated by concerns for others and the environment and are becoming increasingly less selfish and more concerned about the common good. They empathise with others. Local governments and non-government organisations take the initiative. We hold workshops to demonstrate how people can make a difference. We listen to the people and help local groups to respond to local challenges. Together we undertake model projects that demonstrate how it will be possible to live differently. We model different ways of thinking and through 'living the changes'. We show that it is possible to balance individual and collective interests, because we are unselfish nor are we unable to create alternative ways of governing at a regional level.

[43]The detailed scenario is as follows:

A. **Best case scenario: an inclusive, creative society that balances individual and collective interests**
We live in an environment that can support this generation and the next. Women, children and men have a say in local public agoras. Those who wish to attend the face-to-face meetings in the local town hall. Others log on to the website to follow the debates and to add their own ideas (using freely available software such as this software). The summaries of ideas update automatically and are shared horizontally with other local governments and with the state government who in turn shares the findings at joint meetings spanning all the states in Australia. These findings are shared by local government representatives at a regional forum every three months which is held to address quality of life concerns are being met in a timely, transparent manner. Quality of life is defined in terms of social, economic and environmental indicators by the people in terms of what works, why and how. Housing is affordable and made of sustainable materials. We have faced up to the convergent social, economic and environmental challenges and we are resilient, because we live in clusters of homes. We share rain tanks and solar grids that are subsidised by local governments. Poor local governments and rich local governments form twinning arrangements to support each other. No one in our

3.4.4.2 Exploratory Focus Groups (See McIntyre-Mills and de Vries 2012: 355)

The participants summed up their vision as follows:

(Footnote 43 continued)

community is homeless, because those who are ill or unable to work are assisted in community housing where they make a contribution in many different ways to public projects spanning home construction, food production, repairs, art projects and recycling materials. Our living and working areas are powered by alternative energy. The new status symbol is the environmentally friendly lifestyle. Public transport is green. Off road vehicles are used only when appropriate. The green economy supports a vibrant job market spurred by subsidies to enable packaging goods, housing people, transporting people, educating and entertaining the public. The carbon economy is replaced through innovative inventions. All members of the public are encouraged to share their experiences and ideas for living sustainably. The futures market has been reconstructed to take into account stocks for the future, such as the air, water and earth we need to grow organic,safe food. People develop new economies and new trading systems that enable them to have time to enjoy many activities. The clothes and shoes we wear are made of renewable resources. The windmill and fabric shoes are the new chic! People understand that their carbon basket can be stretched further by growing their own veggies. Most waste is recycled locally and used for building or composting. Packaging is designed to ensure that waste is minimal. Animals live in a carefully monitored environment to ensure their quality of life and ours. We are better off because we respect ourselves, one another (as one of many other sentient beings) and the environment on which we are all dependent. Bird flu, swine flu and bovine disease are unheard of in this scenario, because governance standards for animal husbandry and farming are monitored and applied (Horvath and Odor 2010). The Union after and International Commission on the Future of Food and Agriculture 2007). We no longer take too many antibiotics, because we take time to recover from illnesses. We live in harmony with the people of our region and our economy prospers through being able to work in one another's countries. We learn many languages. We are enriched by the diversity of language and culture. We are free and diverse in our neighbourhood, sub national region and supranational region, to the extent that our freedom does not undermine the freedoms of others. Each local area enables each resident to be heard. The concerns that they raise about living in the hills or on the plains or near the coast are given careful consideration when making complex policy and planning decisions. Community networks are formed to enable people to discuss their fears about bush fires, drought and the inundation of coastal properties. People who are worried or stressed are able to access specific services to address their mental health needs and their practical concerns about building regulations and the safety of their neighbourhood. The ideas of local people are scaled up through interactive democracy and governance software. People have a say in ensuring social and environmental justice. We are happy and creative, because we have time to sleep, make slow food, talk to our neighbours, work in communal gardens, irrigated by water harvested and saved in many ways. We have green parks where the trees look healthy because they thrive on grey water. We play sport and express ourselves in a range of art forms. We have hope for the future. We do not commute long distances to work. We teleport to virtual communal areas and congregate in streets. We have technology that is inexpensive. Our desire for recognition and status is supported through being rewarded for innovation that supports the next generation by living in ways that not only sustain, but regenerate resources. We live not only for ourselves but for others and the environment. We understand and remember what the first nations have taught us that we are the land and we are the earth to which we return. We base our decisions on an expanded form of pragmatism based on thinking about the consequences for our own family and neighbours and also for the next generation of life. We understand that what we do to others and to the environment, we do to ourselves.

A. **Sustainable future: stewardship and wellbeing**

Social

- Everyone on public transport
- Shared spaces safe for children
- Grow food
- Share skills and Car share
- Bike sheds Riding bikes or Rickshaw taxis on Bike streets al la Copenhagen
- Decentralised community owned renewable energy production
- Volunteering, community groups—more input by indigenous people in Middleville—as the traditional land owners they have the right to know what is going to happen to their land

Economic

- Buy less—live more
- Simplicity
- Work lives—meaningful employment
- Community education and Volunteers
- Energy—rickshaw taxis and meaningful jobs
- Community owned energy

Environmental

- Shared spaces, shared assets big play grounds
- Dry zones
- Simplicity, Volunteering sharing skills cheaper renewable transport
- Environment—recycling water and Shared access Big playgrounds
- Buy less Education—awareness of environmental issues—primary, secondary, community
- Paperless business and schools
- Solar panels
- Less expensive public transport
- Wider bike lanes
- Green spaces and more trees—use less plastic when constructing play grounds
- Lots of people about in open spaces so we feel safe and become more active

B. **Small changes for the 'long haul'**

Participants viewed incremental change as too little too late and summed up the characteristics of this scenario with the image for this stage was summed up as 'scared rabbits frozen in denial'. They raised to following concerns:

- Carbon pricing gives 'mediocre false security'
- Taking away fossil fuel subsidies—investing in renewables
- Plastic bag ban
- Sustainability officers at coal mines
- Business products need integrity and not more than green washing

- Education need awareness about Nuclear
- Uranium is unhealthy—Transparent government—Closer to lobby groups
- Reduce the distance between government and community
- More input by Indigenous people in Middleville—as the traditional land owners they have the right to know what is going to happen to their land
- Problem is that dollars [are] placed first then the rest second
- Sustainable futurists are isolated and in the minority—they are regarded as hippies and this leads to burnout.

C. **Worst case: Business as usual and a large carbon footprint**

This is how the participants perceived 'Business as usual' Business—all making more money:

Environmental

- Need smarter use of land for waste management and food
- Too much reliance on fossil fuels and plastics
- Traffic jams and poor air quality
- Degradation leads to less outdoor activity
- High CO_2 smog—need more trees and green space
- More cars—lack of responsibility
- Push for sustainable infrastructure
- All buildings a concrete jungle—push for sustainable future
- Smaller subdivided properties
- Mass produced commodities through
- Outsourced labour
- Lack of control of weeds
- Immoral trade practices—support local businesses–Adelaide showground and farmers market
- Continued investment in cars, roads and infrastructure
- Average public transport
- Water restrictions and not enough recycling and preservation of water—less green space less recreation
- Heavy traffic—cars—carbon pollution
- More tax on pollution

Social

- Less resources leads to poverty
- Big gap between rich and poor
- Industry growth—lack of responsibility
- Insufficient incentives to do the right thing—smog
- More cars—lack of responsibility—cyclist danger
- Uninclusive (sic) political system—insufficient to only engage during the election
- Building codes continue to be voluntary/weak and too many exemptions
- People are overworked and selfish and no time for the community

- Individualism—Population growth ignorance
- No creativity
- Smaller houses—many stories, subdivided land or no gardens all building in a concrete jungle—[leads to] more isolation and depression
- Mass production—no handcrafts lack of creativity Minorities stifled—no motive, no collectivism, no voice
- Isolation—loneliness/lonely people—depression
- More isolation between people—individualism—selfishness = depression bad social health
- Increased depression leads to less community interaction, no new friends
- Laziness [need] encouragement for sustainable transport—bikes
- If less outdoor activity—leads to obesity—need community exercise programmes, such as Samba dancing or yoga
- Not getting outdoors and active—no love for nature

The results can be displayed in excel sheets and by means of word queries to cross-tabulate concepts such as:

Independent variables:

1. Age
2. Gender
3. Level of education (primary, secondary, tertiary)
4. Employment status (employed, unemployed, retired)
5. Number of dependent children
6. Number of dependent elderly
7. Place of birth (Australia (post code), Africa, Asia, Europe, North America, South America, Pacific Islands, Other)
8. Aboriginal/Torres Strait/non-Aboriginal
9. Religious affiliation (Aboriginality, Agnostic, Atheist, Baha'i, Buddhist, Christian, Hinduism, Islamic, Jewish, Jainism, Sikhism, Spiritualism, Taoism, Other)

The following systemically linked variables:

1. Scenario with which they identify—A (business as usual), B (small adjustments for the long haul), C (sustainable futures)
2. Social, economic and environmental factors drawn from what participants consider that they have in their lives (material and non-material resources that make them resilient)
3. Social, economic and environmental factors drawn from what participants consider they need (resources to enhance their lives)

   (a) What they wish to add
   (b) What they wish to discard

(c) Turning points for the better or worse
(d) Barriers
(e) Services and actions

The extent to which demographic variables (age, gender, level of education, place of birth, home language, Aboriginal/non-Aboriginal, religious affiliation, type of occupation or employment status) make a difference to:

- With which scenario they identify
- What they perceive they need to **add to** their lives to make a difference to mitigating or adapting to climate change
- What they perceive they need to **discard** from their lives to make a difference to mitigating or adapting to climate change
- What they perceive are the **turning points** for the better or worse, what the barriers are and what services make a difference.

Telling a story and thinking about what we have and what we need and what we are prepared to add or discard from life is part of stepping into another conceptual space.[44] The evaluation of the level of importance of multiple and a simultaneously important issue is important by reflecting on one's life in terms of different scenarios and the consequences of these choices, for example:

• *I have* the following things in my life—understanding of human rights, respect for biodiversity, fear for the future/hope for the future, a confidence, or lack of confidence, loss of home due to natural or other disaster, no family/community support, responsibility to care for others and very high levels of stress
• *I need* in my life—a home, a sense of safety, a place near public transport and hope for the future
• *I will add* to my life—more community supports from a range of services and/or more community engagement to lobby for resources, more connection to nature
• *I will discard* from my life—a sense of hopelessness, a sense of entitlement, excessive consumption
• Self-reflection on *the turning points* for the better or worse—hope that consumption can be replaced with greater sense of attachment to others and the environment
• *Consideration of the barriers* that currently exist and consideration of what could be done to transform society and our relationshipto the environment

[44]This needs to be a therapeutic process and cannot be rushed. The time taken is part of the emotional connection with a different *way of being*. It involves thinking about being a different person. It is the same kind of process as taking an oath or making a vow (Collins 2004), or entering into a contract to undertake certain actions.

3.4.5 Data Collection System

- People rank what they have in their lives to support wellbeing and what they need to make their lives better in terms of quality of life and wellbeing. This requires building 'stocks for the future' and developing links with one another and the environment.
- Accountability rests on showing the links across excessive consumption, unfair distribution and harm (Saul et al. 2012: 167).
- The planetary changes associated with increased risk of drought, famine, floods, fire and tornados is becoming more pervasive. These are not one-off events. They are a way of life that poses a human (not merely a national security).

The software collects people's responses and perceptions to the following:

- *I have* the following things in my life
- *I need* in my life
- *I will add* to my life
- *I will discard* from my life
- Plus, self-reflection on:
- *The turning points*—for better or worse
- *Consideration of the barriers*—for self and society

The following are extracted from the user guide (McIntyre et al. 2014):

Wellbeing

- **Achieved by** making time to connect with people and place; a sense of community
- **Enhance access** to sustainable public and private infrastructure for housing such as affordable transport, safe cycle paths and accessible public transport
- **Agencies/**facilitators; willingness to use and reuse resources carefully; give up greed and waste; sense of community and respect others.

Constraints

- Lack of time and money; sense of hopelessness; inappropriate infrastructure
- Related to/closeness of fit; a sense of community and belonging; connection with native; family friends; beautiful streetscapes; few cars.

This information was provided to local government to help them to plan and deliver services to residents. The administrators are able to access summarised data in excel spread sheets.

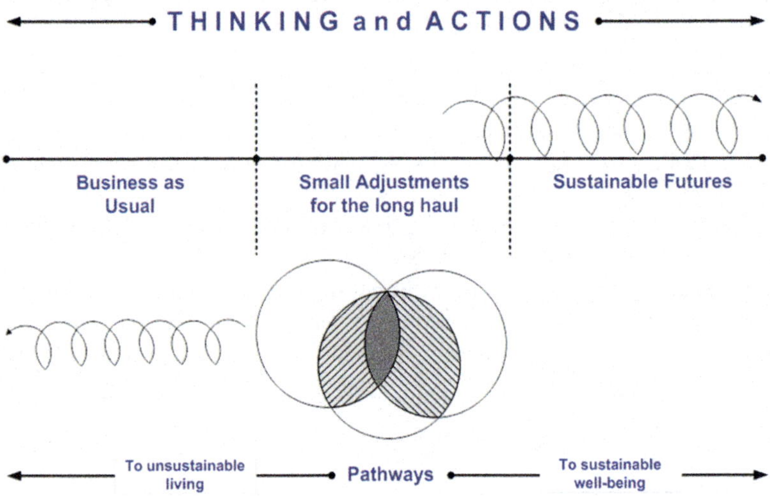

Diagram 3.1 Conceptualisation of the continuum of praxis from Business as usual to Sustainable Futures (See McIntyre-Mills et al. 2014: 57)

3.4.6 *How the Engagement Process Works*

- Personal stories based on addressing the question 'how should we live in order to achieve wellbeing' reveal patterns.
- Patterns of adaptation to effects and mitigation of causes provide residents and service providers the opportunity to explore pathways to wellbeing.
- The software lets residents build on other people's experiences.
- The software summarises personal experiences and perceptions for service providers (see http://wirasoft.com/pathways-to-wellbeing/ User name = test Password = test) (Diagram 3.1).

Pathway To Wellbeing: example of screen shot

Home Log Out

Report Name: **In Basket Factor per Domain 33by User**

Set selection criteria:

All	☑				
Start Date	31	Dec	2012	Start of Records	☑
End Date	1	Feb	2014	End of Records	☑

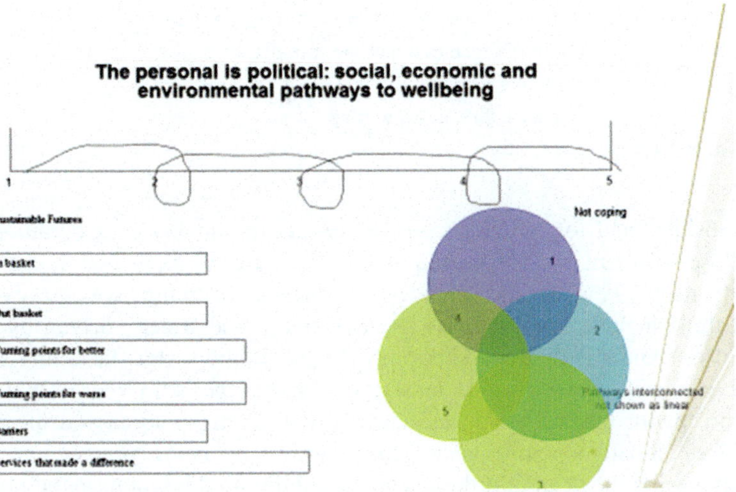

Typological characteristics

'Business as usual' is characterised by the perception that: (a) Physical space is segregated, whereas (b) Conceptual spaces are individualistic and competitive. Typically informants stress that 'time is money' and life is considered to be stressful and competitive. Adaptation is through increased expenditure and technological innovation, rather than a willingness to give up their standard of living.

Small Steps towards sustainability are characterised by: (a) Making more use of energy efficient appliances and cars and a 'less is more' approach to living elegantly and ethically. At the time this report was written the responses from participants about what they perceived to be positive about their quality of life (Nussbaum and Sen 1993) are as follows (Table 3.2):

Sustainable living extends 'the less is more' approach to supporting wellbeing for this generation and the next by 'being the change'. Physical and conceptual space is considered to be integrated. Time is perceived to be constructed as 'we are the land'. Those without the means to install new technology stressed the need to

Table 3.2 Highlighted concerns as the basis for policy intervention

Social	Lack of Public transport and poor bus service—buses not running regularly Cannot discard petrol car, less time to pursue interests if cannot take car to work Social and work commitments make slow living difficult
	Unwillingness to change way of life, Remembering to take reusable shopping bags
	Loss of inspiration, apathy, doubt about ability to make a difference, People 'bringing you down
Economic	Unwillingness and greed
	Cost of adaptation and lack of money
	High cost of local produce
	Lack of Time to walk or catch multiple buses
Environmental	Native Vegetation and fire hazards
	Extreme heat and drought
	Lack of time to grow food

See McIntyre-Mills et al. (2014: 58)

provide subsidies and for building codes to ensure the quality of accommodation for all. Those who rent and are unable to afford to purchase their own homes—are at a disadvantage. Few fit into the ideal of sustainable living. But the overlaps across all three include a willingness to save energy and water—driven by price, fines and user standards that are determined by the public sector. Installing energy saving devices and expenditure on new technology is the main way in which the groups who favour business as usual adapt to rising temperatures and the rise in energy prices. Small adjustments are characterised by being prepared to make material changes if they can afford to do so. Capability to think in terms of multiple factors is vital for enhancing our human capabilities to adapt to climate change and to mitigate the worst effects—thus making us more resilient, whether we live in areas threatened by drought, flood or other dramatic climate change events (Diagram 3.2).

3.4.7 Cultural Ambivalence

In the first instance the data are analysed according to how each informant wishes to live. The majority think that they are living sustainably or are prepared to 'make small adjustments for the long haul'. The factors include the material and non-material conditions that each informant perceives that they have or need, in order to experience a sense of wellbeing and so-called quality of life and that they are willing to add or discard from their lives, in order to enhance their quality of life in the future and the conditions that they consider to be turning points for the better

Diagram 3.2 Inter-relationships that support Wellbeing and Quality of Life (see McIntyre-Mills et al. 2014: 59)

and worse. The challenge is that most participants seem to combine a range of thinking and actions, some of which lead them towards sustainable living and some which are unsustainable. Only a minority of respondents have chosen 'business as usual' as their preferred way of life. But a running list of those who have refused to participate indicates that a willingness to engage is also indicative of openness to change. Furthermore, once the actual choices are analysed in terms of what they are prepared to do to make a difference, it is clear that the majority are clustered closer to the mid-point in the continuum. This is because so many of the small actions such as recycling, etc., are counterbalanced by the other choices made. Nevertheless with the increasing direct and associated costs associated with the use of energy and water will mean that local living will have increased benefits. The savings associated with living locally, engaging in local food production, sharing the resources and developing local recreation and volunteering opportunities will make investment of time in the community more attractive. For more wide ranging changes to occur there will need to be transformations that are driven from below and above.

To date a preliminary analysis of the factors shows that the following are the most frequently mentioned non-material responses to change: water conservation through introduction of rain tanks, installing solar panels and using air conditioning more responsibly. But the cost of making these changes are a barrier for some. Only those who are in full time employment or with access to capital are able to make these changes. The more affordable options are using public transport more frequently and walking or cycling. The barriers are lack of time to spend waiting for public transport and concerns about safety—due to inadequate pathways for cycling or walking and inadequate public lighting. Other non-material changes are community based gardening and home based gardening leading to the sharing of produce. For some this is about growing seeds and also transforming people's attitudes to one another and the earth. In the process of engaging in community gardening they learn to share and to grow more connections with the land and one another. The keeping of livestock on a small scale was raised, in Middleville, a suburban area, but bylaws will need reconsideration in Middleville.

The state needs to be persuaded of the need to support changes through subsidies to households and incentives to innovators in all sectors. More regulations that give market signals need to support those who are prepared to make changes to their homes, or who are prepared to make use of public walkways, cycle tracks and public transport. Their community gardening philosophy and emphasis on slow transport (cycling and walking), getting to know neighbours and growing vegetables together has made a tangible step towards a 'less is more' approach to community life. Some emphasise wellbeing changes—a life style of growing local food and developing greater concern for those beyond the immediate family to include the wider community and global concerns. An appreciation of the importance of kindness to others was emphasised and a few questioned their sense of entitlement to a standard of living that some can only dream about. But the predominant cultural shift or 'habitus' or way of being (Bourdieu 1977) is merely at the level of implementing changes to the housing stock, in terms of installing rain tanks and insulation or suggesting policy changes to infrastructure. To sum up, the changes required are social in order to free up time to spend on building the community and fostering links with others. The desire to build a creative, active and renewable lifestyle through growing vegetables, living locally and relying more on local produce. The processing of rubbish, reusing, recycling, growing vegetables, making compost, sharing community harvest are the basis for growing community in Middleville. A future is possible where local, sustainable living is a lifestyle choice will coexist with those who are forced to remain in one area as a result of poverty (Diagrams 3.3 and 3.4).

Adaptation usually means adapting to *the effects* of climate change at the local level, which requires awareness, whereas, mitigation usually means responding *to*

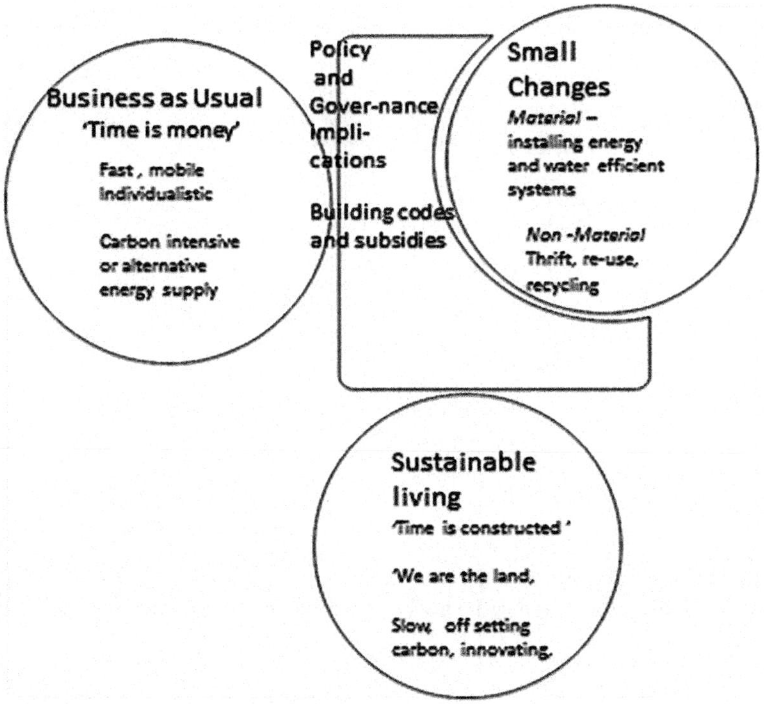

Diagram 3.3 Arenas for policy change

the causes through changing policy on emissions. But the latter requires adaptive responses to new policy and governance requirements. Adaptation is through reducing the effects of emissions through spending money or living differently, whereas responding to heat, for example involves installing air conditioning. Adaptation to fires requires changing designs for living and our expectations about consumption and the commodification of life. For example, in the words of an informant: 'Even taking out the rubbish is a spiritual matter'. He explained that sorting out the rubbish carefully is a matter of respecting others and the environment. A strong sense of community is also displayed by the organisers of the community garden and the Opportunity Shop linked to St Middleville's Church. The notion of reusing and recycling is demonstrated in the way in which all waste is recycled from the community op show. All items that are not sold are sent to State-wide Cleaning Cloths, Australia. Surplus is also sent to the 'Anglicare NGO or to Korea and Africa'. This contrasts with the exasperation expressed by some informants who said that 'Climate change is natural' but 'we do care about the environment'. Clearly the politicisation of the carbon tax that became official on 1 July 2012 could play out in the next election. This research strives to make a contribution to gathering data on perceptions that can be summarised without losing the fine grained detail and the emotions of what people are saying. It is vital to be

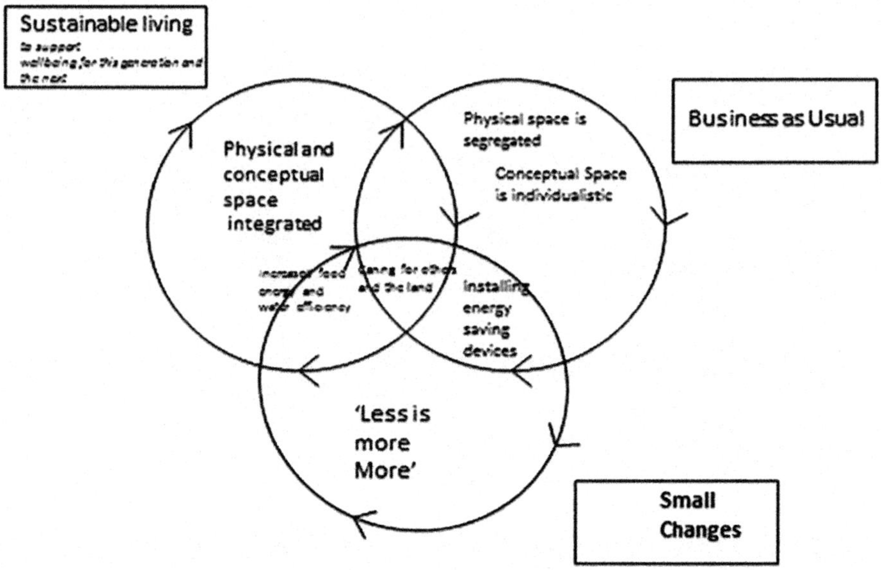

Diagram 3.4 Overlaps and turning points for the better and worse (see McIntyre-Mills et al. 2014: 61)

able to gather large data sets on perceptions if policy engagement and community education is to be achieved. Clearly, the desire to engage more in the community and to have more spare time is shared by the majority of participants. Energy awareness does not, however, translate into a willingness to change greed and wasteful habits. Poverty was considered a barrier by those without financial security or facing poverty. The weightings given to perceptions were obtained by analysing data from a matrix spanning the following snapshot of factors downloaded from the publicly accessible software in June: for example, these are the findings on 1/07/2012, but the research is ongoing and the data set continues to grow as it is extended (Table 3.3).

Sustainable living extends '*the less is more*' approach to supporting wellbeing for this generation and the next by '*being the change*'. Physical and conceptual space is considered to be integrated. Time is perceived to be constructed as 'we are the land'. Those without the means to install new technology stressed the need to provide subsidies and for building codes to ensure the quality of accommodation for all. Those who rent and are unable to afford to purchase their own homes—are at a disadvantage. Few fit into the ideal of sustainable living. But the overlaps across all three include a willingness to save energy and water—driven by price, fines and user standards that are determined by the public sector. Installing energy saving devices and expenditure on new technology is the main way in which the groups who favour business as usual adapt to rising temperatures and the rise in energy prices. Small adjustments are characterised by being prepared to make

Table 3.3 Weightings of perceived areas of concern based on the number of responses per category

	Social Inclusion needs + turning point	Less waste and more recycling needs and turning point	Energy Aware-ness as a need or turning point	Financial and job security as a basis for change	Greed and waste-habits as a barrier	Spare time	Sustain-able transport as a need	Information and education on climate change adaptation	Food security	Poverty as barrier	Housing quality
Business as Usual	9	2	7	8	1	2	0	3	0	1	1
Small changes	9	6	28	4	3	7	5	11	7	0	1
Sustainable future	63	23	52	17	16	35	20	7	9	8	11
Total	81	31	87	29	20	44	25	21	16	9	13

See McIntyre-Mills et al. (2014: 62)

material changes if they can afford to do so. Capability to think in terms of multiple factors is vital for enhancing our human capabilities to adapt to climate change and to mitigate the worst effects—thus making us more resilient, whether we live in areas threatened by drought, flood or other dramatic climate change events.

3.4.8 Critique of Version 1 of the Software and Future Directions for Representation Accountability and Sustainability

The challenge is to enable people to think not only about rights, but also about their responsibilities to enhance their own resilience and to adapt to risks within their neighbourhood and the wider national and post-national region in which they live.

Strengths

The software works as a prototype. It enables people to think about whether they identify with:

A. Business as usual
B. Small changes
C. Sustainable living that support wellbeing now and in the future.

It enables people to rank what they have in their lives to support wellbeing and what they need to make their lives better in terms of quality of life and wellbeing. Each of the items needs to be thought in sequential terms so people get a sense of the consequences of having for example a large status symbolcar that guzzles fuel or a large, badly insulate home and to give each item a weighting.

Then people are asked to consider what they need to help them improve their lives. These items also need to be thought through in sequential terms so people get a sense of the consequences of using a more energy efficient vehicle or using public transport.

People are prompted to think through what they are prepared to add or to discard. Linked with this process are services to help them achieve their perceived needs.

Then they are asked to think through turning points for the better and worse, the barriers and they are able to print out their own report.

The benefit is that people are able add concepts to an ever growing list of social, economic and environmental indicators of wellbeing.

The administrators are able to summarise the data on excel sheets.

Weaknesses

Whilst the software is robust, it is not sufficiently intuitive as yet. The identified weaknesses are iteratively tested by members of the team with volunteers with varying levels of skills.

Some participants will never engage on line, but the software needs to be as easy to use as possible on as many types of computers, iPads and iPhones as possible. In addition the engagement process needs to enable people to: (a) connect face-to-face in groups and then to share ideas with a facilitator who loads their ideas on line on their behalf, or (b) email their responses. Current forms of democracy and governance need to enable more engagement with diverse people with diverse needs and computing skills.

Opportunities

The long-term policy aim is to extend the testing of the research in post-national regional biospheres. The open democracy forum was the starting point for research. The Local and Regional score cards could be developed by:

- Scaling up the process to respond to social, economic and environmental needs.

Threats

People tend to deal with information or fear through denial. McLuhan and Powers (1989) expressed reservations about the ability of people to override the sense of overload as a result of exposure to images and information by cutting off from the world. But with increasing exposure to a sense of global issues through a range of media not only the elite global knowledge workers have a sense of being global citizens.

3.5 Second Case Study: Planetary Passport as a Means to Ensure the Fair Distribution of Resources for Locals and Displaced

The engagement process needs to develop indicators that reflect the perceptions of the service users and providers. So they should be able to frame the agenda items and then develop the indicators. Human security should thus no longer be tied to nation states. Instead regional overlapping regions would undertake to ensure the right and the responsibility to support representation, accountability and sustainability.

- For service users the architecture could provide a pathwayto wellbeing that can be further developed after reflection
- For service providers a spread sheet, listings and cross-tabulations of the inputs from residents providing information to assist in developing strategies to deal with poverty and climate change
- Respect for self, others and the environment requires time to live slowly and to re-connect with others.

3.5.1 Piloting a Way to Do Things Differently by Translating the Theory and Policy into Practice: Equitable Distribution of Resources and Protection of the Global Commons

The work of Shiva (2002) stresses that global markets and mainstream research supported by corporate profits have disempowered local producers and in particular women farmers whose knowledge of plants was challenged by patriarchy and the 'enlightenment' during the witch hunts in Europe and America and then extended through the commodification of seeds through patents that prevent the collection and protection of biodiversity (Shiva 2010: xvii).

> The destruction of the self-reproducing character and genetic diversity of seeds was the beginning of seeds becoming a source of profits and control. This was an attack on the feminine principle. The hybrid 'miracle' seeds are a commercial miracle because farmers are forced to buy new supplies of them every year. Hybrids do not produce seeds that reproduce the same result because hybrids do not pass their traits on to the next generation. With hybridization, seeds could no longer be viewed as a source of plant life; they were now a source of private profit only.

My area of concern is: what combinations of social, cultural, political, economic and environmental factors make a difference to wellbeing and our capability *to adapt to climate change at the local level and mitigate its effects* by scaling up local interventions? The premise is that the UN Sustainable Development Goals do not go far enough and that cultural and economic transformation is vital. We will try to learn more about diverse ways of knowing (Cruz et al. 2009) that could support effective environmental management by taking into consideration: (1) Non-anthropocentric knowing drawing on the environment as well as knowing through all our senses—including empathy and intuition, (2) Respect for 'know how' developed through empirical trial and error, (3) Creativity in maintaining sustainable relationships with the land, (4) Respect for spiritual awareness, the wisdom of experience as well as the social and natural sciences, (5) Artistic expression of feelings, perceptions and emotions, (6) Caring for future generations of life, (7) 'Phronesis' or Aristotelian wisdom to match the right knowledge in context.

The idea is to take the Global covenant idea developed by David Held to protect the global commons and to find a way for this cosmopolitan approach to be applied in practice through drawing on the liberative potential of the Aarhus convention that supports access to information and the right to participate and to be heard on environmental matters. Participatory democracy is needed to protect local citizen's right to a safe environment. The application of policies such as the UN Local Agenda 21 and the Aarhus convention provide potential for a scaled up approach to managing resources through considering what people have, what they need and what they are prepared to add or discard from their lives in order to use resources carefully and to feed forward resources for future sharing. This works along the lines of the veil of ignorance approach underpinning the social contact but it is expanded to the entire human population on the planet. As we do not know where the next disaster will be located within the region, human security and hospitality is extended across the region. Resources are managed carefully so that resource rich areas share with those in need. Stocks are not stored and used within contained areas or interest groups. They are shared according to a system of accountability that measures resources in terms of a planetary passport and score card. This has implications for complex decision making within and across regions.

People who are displaced need to be able to track and link with other people in their local area. The software could provide a means to do this by enabling displaced people to continue to link with people in their geographical space and with those with whom they connect beyond their immediate community. In other worlds it provides a means of tracking resource use and networks of people.

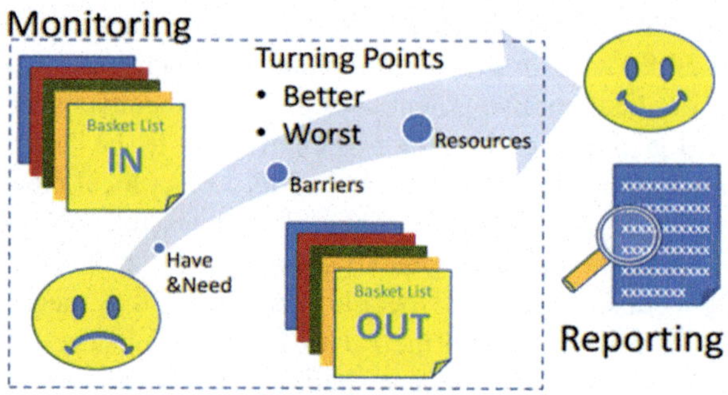

Source McIntyre-Mills and Wirawan (2017)

Iterative Process Pathways To Wellbeing

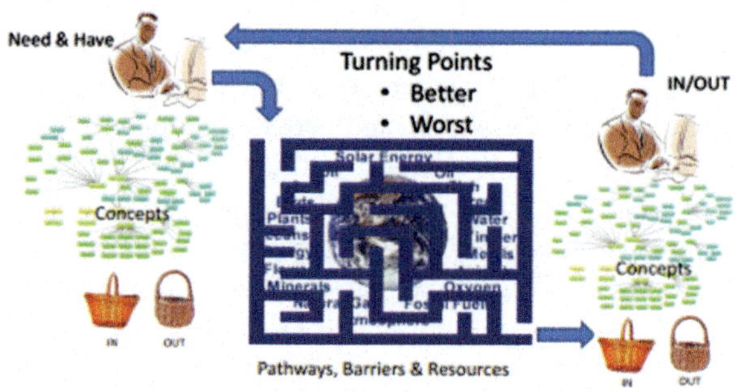

Source McIntyre-Mills and Wirawan (2017)

Core Data Collection

- I **have** the following things in my life …
- I **need** in my life
- I will **discard** from my life
- I will **acquire** for my life
- Plus, self-reflection on:
  - The turning points – for better or worse
  - Consideration of the **barriers** – for self and society

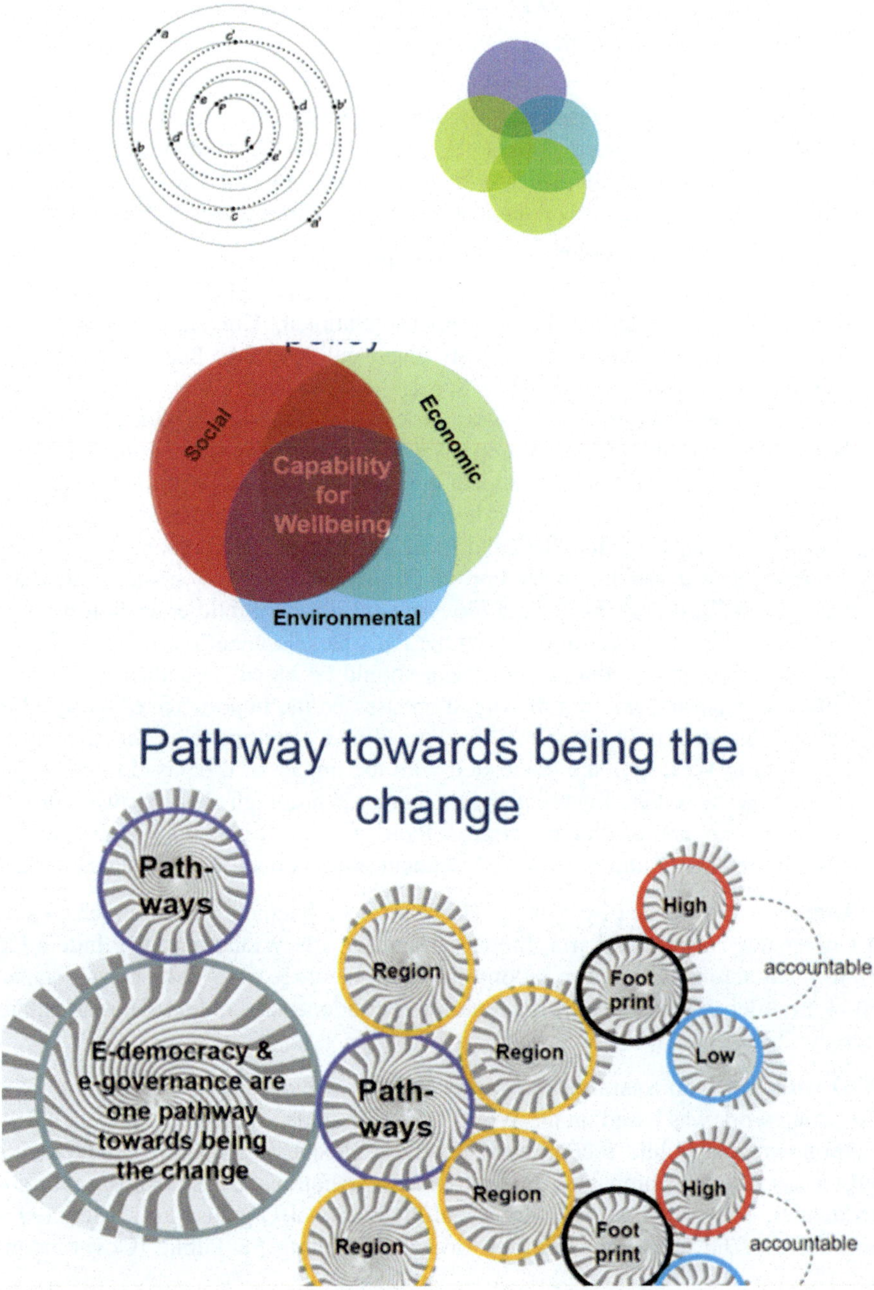

Source McIntyre-Mills and Wirawan (2017)

3.5.2 *Transformative Work in Progress Research for Re-generation*

The current team comprises Assoc. Prof. Janet McIntyre (Social and Policy Studies) and Pak Rudolf Wirawan (Wirasoft, CEO, leader of Indonesian diaspora, Australia). Dr. Ida Widianingsih (Assoc Prof at Padjadjaran University) and Ibu Novieta Hardini (lecturer at Universitas Nasional). The research is based on:

- *Being* supported by an MOU between West Java and South Australian government.
- *Doing* collaborative research as Ph.D. supervisor and collaborating with staff at the University of Indonesia, Universitas National, Universitas Padjadjaran, National Islamic University, West Java and the SA Local Government Association (in particular in Alice Springs and Adelaide).
- Facilitating workshops, e.g. Ministry of Religion, Finance and Social Affairs in Indonesia, two invited workshops for the Chief Ministers Policy Unit in Canberra in 2009 and numerous local government workshops in South Australia and research on behalf of the Attorney General's Department on resilience, research on behalf of the ARC and local government on wellbeing.
- *Facilitating* a leadership workshop at Flinders University together with delegates from West Java Provincial Government on the public education training and leadership. This was in line with the Paris Development Agenda (1997) that underlines the notion that development should be based on collaborative principles. The participants in this workshop stressed the importance of the need to develop leadership opportunities for women and young people. The challenges they face in West Java are associated with the impact of urbanisation as people face congested cities, lowering living standards and high risks for women and children who are unable to express their strategic rights to a voice within households, communities local government and national level organisations.

McIntyre was assisted by Novieta Hardeani Sari, Lecturer and Research Fellow in Universitas Nasional Jakarta, Indonesia and Dr. Ida Widianingsih (Flinders Ph. D.) to facilitate the research in Indonesia across the three case study areas detailed below in order to adapt the programme for the Indonesian context (Human Ethics Protocol 7213).

FGD with UN[45] on Sustainability with Private Sector

The group work was based on the World Cafe technique. The participants included: Government and public sector representatives, Businesses, including advertising and market research companies, Private sector associations, Consumer associations, Academics, NGOs, think tanks, Regional/International Organisations, Media representatives. The technical workshop on ECO-labelling for Public (Government)

[45]The UN workshops addressed eco lables according to ISO 14024 classification.

and Private Sector Participants (4 October 2016). The participants included the Ministry of Environment and Forestry, Republic of Indonesia, Foundation for Sustainable Development (Yayasan Pembangunan Berkelanjutan or YPB and United Nations Environment Programme (UNEP), EU-Funded SWITCH-Asia Programme.

The need to protect agricultural areas abutting Jakarta is vital. Land is being lost to development and forests are being chopped down to provide for food. So pristine habitat which also protected Jakarta from flooding is being lost to agriculture and production resulting in worse run off and flooding.

The packaging of food in biodegradable plastic packed into shopping bags has not reduced the amount of waste which is dumped in the river. This too adds to the risk of flooding. The challenge facing the current governor is to move the small urban villages along this river where the water is rising to higher levels than previously due to higher run off and more dumping associated with higher levels of migration into Jakarta.

In Jakarta people tend to buy food en route to and from work so in fact they provide work opportunities for street traders and this is one way of reducing the need to package food in plastic![46] Jakarta has 99% of its plastic shopping bags made out of biodegradable plastic. But the Jakarta Post reports that consumers are now allowed to purchase plastic bags (Ribka 2016)[47]:

> Rp 200 (15 US cents) per bag might seem like nothing but I do think it has an impact on reducing plastic waste', a university student informed the Jakarta Post.
>
> the government's call to reduce waste was followed up by a ministry circular requiring retailers to charge Rp200 per plastic bag during a trial period from Feb 21 until May. The circular's policy was enforced in 23 cities. In June, the ministry announced the trial would continue until it issued a regulation in July and expanded the coverage to 514 cities and regencies, however no regulations have been issued….
>
> Retailers will continue giving free bags until the ministry issues a regulation.
>
> Ribka describes how the Indonesian Consumers Foundation shows consumer acceptance and that the majority of customers (87.1%) use their own bags when shopping….plastic bag production dropped 25 to 35% in the first half owing to the pay for plastic policy…

[46]The Malaysian businessman stressed that a decade ago he was in Bali and his father spotted a plastic wrapper with his company logo lying on the beach. He and his son committed company funds to produce biodegradable plastic which takes two years rather than 500–10,000 years to degrade. The reliance on plastic was stressed as unavoidable by the Malaysian and Indonesian companies who stressed that food loss would be grater without plastic. I commented that if people ate local food and minimised the food miles this could help.

[47]"In 2015 alone, the country produced 64 million tons of waste, with plastic shopping bags accounting for 14% of the trash, according to data from the Environment and Forestry Ministry'. The governments call to reduce waste was followed up by a ministry circular".

The discussion on recyclable and biodegradable products caused some controversy and many thought that the two routes were confused and that more government subsidy should be given to the production of biodegradable packaging.

A Malaysian bioplastics researcher mentored an Indonesian company 'Tessa Tissues' based on his own work at Tirta Marta.[48] He developed the first company to make wrapping and packaging out of cassava and tapioca plants.

Participants were invited by administrators at the following organisations: Universitas Nasional, Padjadjaran, the Ministry of Education and Ministry of Religion and through the West Java Provincial Council. Once the pilot has been completed we will then use the data collection tool to explore: (a) awareness of scarcity and sensible use of scarce resources as keys to sustainability; (b) how living more simply promotes sustainability and can minimise inequality; and (c) views on consumption-related ethical decision-making.

The Tati Erianti is the Food Security Representative from West Java Provincial Government and Elli Wasliah (City level) discussed urbanisation in terms of the need to foster ways to prevent the continuing subdivision of land, such as the policy in Japan because it leads to the increased inability of people to make a viable living from agriculture. Other key themes are the need to create opportunities for young people in rural and remote areas as cities attract young people who seek a foothold in the cash economy and often face limited opportunities and vulnerability to exploitation. The lack of security for workers in casual employment was underlined by participants. These themes will be addressed in a forthcoming publication. The focus groups with the public, private and volunteer sectors[49] deepened this understanding and will be used as a basis for developing the software.

The mission of the rector of the University of Padjadjaran is to serve the community and has set up so-called Team 11 comprising staff members who are university leaders to help address the needs of the community. The rector is a medical doctor and has suggested working with the mayors of regions to sponsor students who will return to the village to work for 2 years. They pay no fees at UNPAD. His mission is expressed in the new university batik cloth. The flower symbolis a lotus (blossoming means rebirth and immortality), the pentagon stands for the philosophy of Pancasila and a better Java, the torch stands for light and enlightenment and the wheel stands for community support dynamic ready for change.

[48]Sugianto Tandio, CEO PT. Tirta Marta—*Social Entrepreneur* dari *World Economic Forum*—From Zero to Hero—*The journey towards biodegradable plastic (ecoplas)*, Bong Edison, Business and Operation Director, President Director PT.Graha Kerindo Utama (Tessa Tissue)—Market Trends for environmental products.

[49]A forthcoming publication: "Getting Lost in the city" will explore the themes in more depth.

Participant type/group	Population pool	Numbers to be approached	Expected/required no.
Pilot	40 post grads The participants include staff and students who will be recruited via the research office through a general letter sent to all students and not directly via class room contexts	40	30
Focus groups	The first in Jakarta with a population of 1.898.537. It has a low unemployment rate of 2.4%. It will be conducted in they urbanised area of Depok City at the University of Indonesia	Purposive judgmental sampling	60 a purposive sample spanning FG1 public sector (20), FG2 private sector (20) and FG3 (20) NGOs
Interview	As above	Purposive judgmental sampling	10
Focus groups	The second in Jatinangor, a suburban area that is becoming increasingly affected by loss of land as urban sprawl extends, with a	Purposive network sample of participants at the local government level will be invited through the Ministry of Education and Ministry	50 a purposive sample spanning public sector (20), FG2 private sector (20) and FG3 (20) NGOs

(continued)

(continued)

Participant type/group	Population pool	Numbers to be approached	Expected/required no.
	population of 4407. The poverty rate is 11.85% with an unemployment rate of 6.5%	of Religion and through the West Java Provincial Council 10 in depth interviews	
Interview	As above	Purposive judgmental sampling	10
FGD	**A third case study** with residents from Ciangur, a food production area		60 A purposive sample spanning public, private and volunteer groups
Interview	As above	Purposive judgmental sampling	10

FGD held at Padjadjaran, Bandung Campus

One of the university leaders, facilitated the fieldwork and we visited Jatinagor Campus of the UNPAD on the 10th of October and gave a workshop for West Java Provincial Government attended by staff of the university. One of the staff is working on coffee production and agricultural opportunities. Pak Rudolf Wirawan, CEO of Wirasoft and president of the Indonesian Diaspora presented his idea of a systemic response to coffee production based on a non-anthropocentric approach to working with free ranging civet cats to ensure that the coffee is ripe so that only the best and ripest coffee is picked. Then we met the students and they gave overview responses to the challenges of living in an increasingly urbanised area. They focused on the impact on both the village and the city as a result of travelling to the city. Loss of land is caused by subdivision into uneconomic units that are often sold off to small businesses. The concentration of people in the city results in an increase in crime and a decrease in social connections based on a sense of neighbourhood.

Only 6 of the 80 students at the Jatinagor campus lived in a village environment, the majority were born in Bandung or Jakarta and had little or no knowledge of the plants and animals in the region.

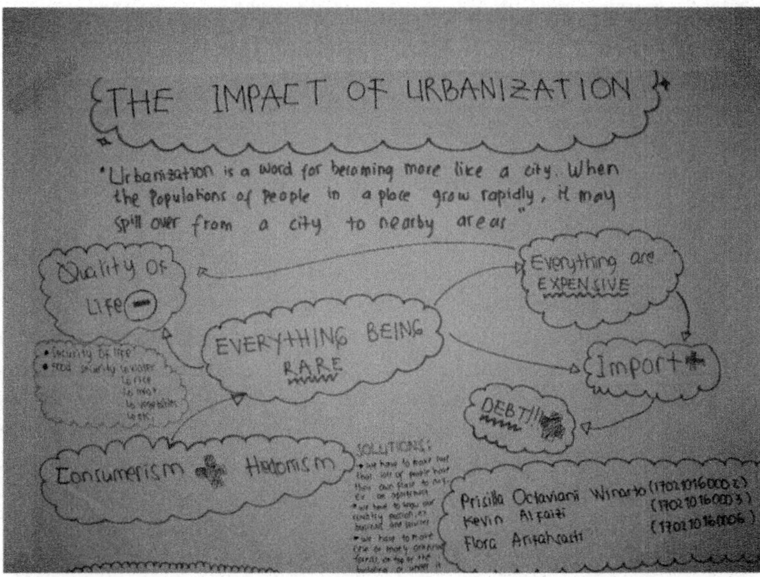

We then met with the Train the Trainer leader Dwi Purnomo who has set up several small incubator projects to enable students to work with community on a 50/50 basis to share the profits associated with value adding to mangoes through creating pulp at harvest time, so that mangos would not be wasted. The mangos are packaged and marketed by the students and further training on safe production, business and marketing is given to those who wish to join the project. Maize rice is also produced by the project and marketed. They are interested in selling to Australia and the gluten free market is of interest to them. Other products are shoe production which they have facilitated in the flood prone area of Garut.

The project has is housed in four houses which are rented at approximately Australian dollars $15,000 per year.

The training is conducted through facilitation to plan and manage each stage of the project and then to enable the small business to move from a room in a shared house to rent their own property when they have become financially viable and able to pay the rent. This incubator project is self-funding but supported by the university. However UNPAD subsidises free initial training courses and this is part of the UNPAD cares for the community approach.

Jatinagor is in the eastern part and it is close to this region. Travelling to Jatinagor highlighted that each district has a different water authority. It used to be a rubber production area, but now it is a development hub. The Bandung Institute of Technology is also located here. The National Institute of Government was also located here in 1980. They graduates become government officers. The President and the Ministers come once a year visit for graduation which is why the area developed.

The Institute of Cooperatives is a Private Institute established by the Ministry of Cooperatives and originally founded by Suharto.

In line with Pancasila model. It is managed by Padjadjaran University. From Jatinagor we travelled to Cibodas and visited a successful biodigestor project discussed in Chap. 2.

Ciangur is the rural rice growing area close to Bandung and is a more relevant site for agricultural production. This region is a neglected area that borders the ocean. Despite its agricultural potential it is one of the poorest area particularly in the southern area of West Java. Rice, coffee vegetables are produced here.[50]

The interviews in Ciangur the following week required travelling en route through villages named after the type of work for which the village is known. One is named for stone masonry, another for marble mining another for pottery, for example. Most areas continue to farm rice in between a growing number of light

[50]Drawing on a data base for Ciangur Local Government statistics 2015: "The rural area of Cianjur a food production area with 32 sub districts and 360 of which 6 are city villages and the rest are rural. The local government officials comprise 55% are male and 45% female officials and 99.15% Islamic and the rest are Catholic and Protestant. There are 11 political parties the biggest Democratic Party, Golkar, Indonesian Democratic Struggle Party. Only 9 women are elected out of 50 legislators. The population is over 2 million and 17% live in the 3 subdistricts and rest are rural. Only 55,000 are over 65 years which is indicative of a high mortality rate. Only 600,000 attend primary schools and 148,000 attend junior high and senior high 144,000 and university 41,393 go to tertiary education out of a population of 2 million. 2,235,418 pop total. Women comprise 1,084,036, Men comprise 1,151,382. People work in agriculture, forestry, hunting and fisheries 342,422 and most male 96,000 are employed in industry, whereas small business including hotels, restaurants and markets employ 182,356. The rest 119,925 are employed in a large category comprising industries ranging from: Mining, electricity, gas and water, construction, transport, storage, insurance, rental property, land and company service. 1,031,622 is the number of working age people of which 878,215 working and 153,407 are unemployed. 94,208 have lost jobs, 59,199 never worked. Number of women at home 369,748. There are only two hospitals with 400 beds to serve 2 million people and only 1 pharmacy for the whole, 140 GPs for the region, 12 clinics, 10 dentists, 11 specialists and 273 midwives and 8 remote maternity services in the area are in the city and they have 45 community health centres. In Ciangur the 8889 people have TB viral infection and 4206 deaths this comprises 50% are exposed to this disease. This is one of the areas with the greatest number working in Saudi Arabia from Indonesia. 350,148 acres for rice and non rice fields 284.239 for other farming and 46,421 of rice is well irrigated. Cianjur is an area where Australian troops were located for border protection and it is an area where a high rate of out migration occurs because agricultural production is declining," (Indonesian Stats 2015).

The following are the ten most common diseases according to the Provincial Government:

1.	TB
2	hypertension
3	Gastric illnesses
4	Diabetes
5	Schizophrenia residual
6	Otitis media
7	epilepsy
8	Bronchitis
9	Control Luka
10	Cataracts

http://pusdalisbang.jabarprov.go.id/pusdalisbang/berkas/jabardalamangka/747Provinsi-Jawa-Barat-Dalam-Angka-2016.pdf

and heavy industry combined with small warungs selling food, gas, water in bottles and larger containers for varying prices (shops), mobile hawkers on bikes and horse taxis.

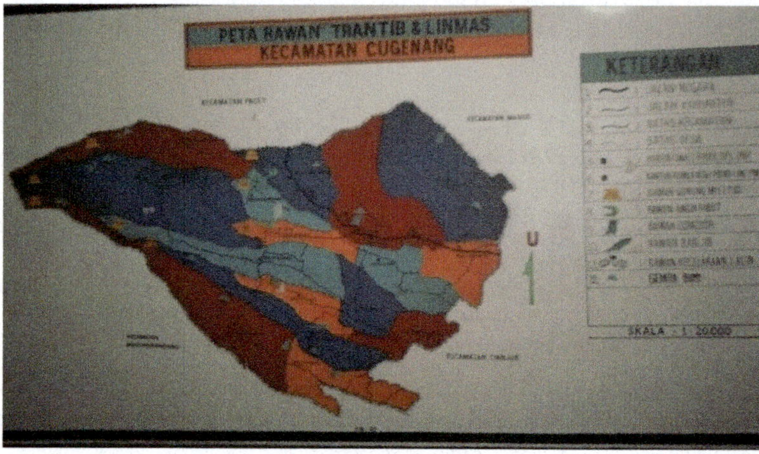

The small shops line narrow winding roads to Ciangur giving way to informal tent-like structures selling food at lower prices. These warungs are favoured by the truck drivers who transport food from Ciangur to the city and who deliver products to the supermarkets.

The ongoing competition between shops and supermarkets pose a challenge for the viability of small shops and restaurants. The number of abandoned restaurants provided evidence of financial hardship. This is in part because the new toll road to Jakarta has bypassed the area. Now people prefer to travel to Jakarta via the toll road as it takes 3 h and not 7 h as previously.

The focus group discussion was hosted by the sub district leader who invited 34 participants ranging from his wife (leader of the PKK women's group, a Headmistress of a primary school, three female staff members of the District Council Office, A Public Health Official, farmers of varying ages with experience of the challenges in the district, youth leaders, and youth trainers. Two villages were represented, one with better socio-economic conditions and one with lower life chances.

The key challenges raised in the women's discussion focused on the fact that only 25% of the village receives piped water. The Provincial Government has sent most of the water elsewhere and the village relies mostly on installing wells. This can cost up to 1500 Australian dollars for sinking a well and maintaining it.

The village primary school is reliant on water from the river and during heavy rains it is inadvisable to use it, because it becomes contaminated with rubbish and E. coli. One of the most reported diseases is diarrhoea in the area.

One of the major challenges faced by the villagers is flooding and inundation after heavy rains which also lead to landslides. In the event of a disaster they rely on help from neighbours.

Electricity is provided by the hydroelectric power and the installation was originally built by the Dutch. The average cost of electricity for higher income families is around 150 Australian Dollars per month. But most rely on small 'melon' gas cylinders that are subsidised by the government. Each small green canister is around $3 Australian dollars per week. So each family spends around $12 dollars per month on gas.

The PKK women's leader stressed that domestic violence and trafficking were two of the most important concerns they faced as a group. Although trafficking was not such a problem in their village, it was an issue in the areas where tourist visit. The problem is compounded by the fact that both women and many young people are encouraged by local pimps (paid by traffickers) to advertise the benefits of working elsewhere or entering into contract marriages for a few months. Furthermore some parents encourage their children to seek work elsewhere because the earnings from farming are less attractive than earnings from farm labour. Safe houses or government protection is available for children who feel abused by parents but it is not an option commonly followed. Many women return from working elsewhere pregnant and suffering the after effects of abuse and non-payment of wages.

The public health worker confirmed that TB was one of the highest rates of illness and we discussed the fact that sexually transmitted diseases were a risk too as a result of the rising rate of prostitution. HIV is thus a possibly latent condition linked with rising TB as is the case in South Africa where migrant labour helped to spread the disease.

The working conditions for women in local government need to take into account the fact that working women need to undertake domestic chores before arriving at work. So more flexible working hours may help to enable more women to work.

One of the senior farmers stressed that he believed that the decline in agriculture would continue unless something was done to make agriculture attractive to the younger generation. Declining markets for agricultural produce are a problem. Young people prefer to earn a living as a pedicab driver or hawker to labouring for lower income in a rice field. As a result many plots were not being planted.

This was addressed in a workshop that followed on the three sub groups who comprised the event.

UNPAD runs a training programme for all village members and places were offered to 50 people per village, namely places for 600 trainees to be subsidised by West Java Provincial Government. Each trainee would receive the equivalent of 50 dollars on completion plus a certificate. Accommodation and food would be provided during the training. The programme has successfully assisted startups and training programmes over the last 3 years at UNPAD. The aim of the training is to add value to the agricultural products.

A third group focused on the challenges facing young people. One of the key issues is the way in which the internet is misused. The potential for exposure to trafficking and pornography needs to be addressed by protecting children and young people. The said that better coordination of services is needed. Hence the software programme and its value. They also stressed that the role of the police in supporting and protecting young people needs to be explored. The lack of hospitals was highlighted. There are clinics and traditional healers who need to be encouraged to work with the biomedical system and need to be registered to ensure that they are compliant with standards.

3.5.3 Interview Schedule to Compare Life Chances Across an Urbanised, Suburban Sprawl Area and Regional Food Producing Area

The highly urbanised area of Depok City, Jakarta with a population of 1.898.537 with a high unemployment rate of 9.54 and poverty levels recorded at 2.4% possibly because of the ability to rely on the informal sector to survive. The suburban area of Jatinangor is becoming increasingly affected by loss of land as urban sprawl extends, with a population of 4407. The poverty rate is 11.85% with an unemployment rate of 6.5%. The rural area Ciangur, a food production area with a population of 570.991 and an unemployment rate of 11.43% and a poverty rate of 5.63% which indicates that having access to land helps to prevent the poverty experienced when land is lost.

Please describe your family using a map with the following symbols (I will provide the details of how to map out household relationships using standard anthropological symbols:

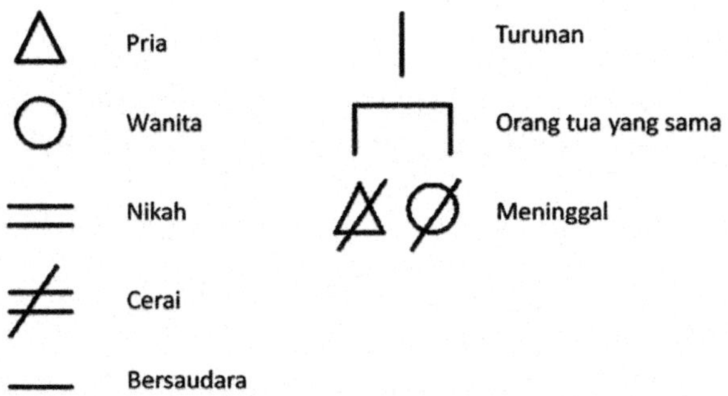

- Where do you live? Jakarta/Jatinangor/Ciangur?
- Do you/your family have a home in Jakarta/Jatinangor/Ciangur? If so explain who the house belongs to.
- If you have connections with the country do you visit? When? How often? Weekly, monthly other?
- What do you like about living in Jakarta/Jatinangor/Ciangur?
- Does living in the city make a difference in terms of socio-economic development and your own sense of wellbeing?
- The cost of living is much higher in Jakarta/Jatinangor/Ciangur. Please discuss what that means to you and your daily choices
- Do you own or aspire to own a property?
- If so where would you want to buy it? Why?
- Does living in the city make a difference in terms of socio-economic development?
- Does living in the city make a difference in terms of environmental development?
- What do you dislike about living in the Jakarta/Jatinangor/Ciangur?
- What does city life mean to your family? Status, access to better life?
- Where does the food come from that you eat each day? What food choices do you consume each day for (a) breakfast 9b) snack in morning (c) lunch (d) snack in afternoon (e) snack in evening
- Do you consider the implications of what food and beverages you drink on the environment?
- Does this knowledge influence the choices you make each day?
- Does living in Jakarta/Jatinangor/Ciangur effect wellbeing? If so how? Please tell a story about the best and worst aspects of daily life
- **What** energy choices do you make each day in the city? For instance do you drive to work or study or afterhours activities in a private vehicle—**Y/N**
- **If yes** describe the vehicle and your reasons for using it
- **If no** describe the form of public transport and your reasons for using it
- How much do you or your family/place you rent from pay for power bills each month?
- What is the source of power for (a) lighting, (b) cooking (c) cooling
- How do water costs affect you?

- How do water costs affect your family?
- How do water costs affect your neighbourhood in Jakarta?
- What company provides water to your home? What does it cost? What are the good aspects of the service if any? What are the bad aspects of the service if any?
- Do you or your family live in an area in Jakarta that is effected by flooding? If yes please discuss
- Do you or your extended family have connections with a rural area that is affected by natural disasters? Such as flooding or drought?
- How does climate change effect your society from your point of view? What indicators are relevant to your life? What indicators are relevant to your family?
- Does climate change effect wellbeing?
- How does climate change effect social wellbeing?
- How does climate change effect the economy from your point of view? What indicators are relevant to your life? What indicators are relevant to your family?
- How does climate change effect economic wellbeing?
- How does climate change effect the environment from your point of view? What indicators are relevant to your life? What indicators are relevant to your family?
- How does climate change eff*ect economic wellbeing?*

FGD in Ciangur

The FGD began with viewing publicly available material in the form of videos by Costnaza.[51] And Vandana Shiva[52] as well as a basic introduction to the nature of the Anthropocene.[53] The focus group participants were asked to consider the following questions:

- What do you think are the implications for the Public/Private/NGO sector?
- What are the challenges for food, energy and water security in Jakarta and more regional areas as the city grows in size?
- Please give examples of food security challenges
- Please give examples of energy security challenges
- Please give examples of water security challenges
- Please discuss the implications of changes and challenges posed by urbanisation for Natural capital, Built capital and Social capital

3.5.4 Questions for the Survey and Focus Group

Then we will ask participants about social, built and environmental capital in terms of 'if then scenarios' spanning business as usual, making small changes and living virtuously and well:

What do I have in my life that could help me to live ethically and well? For example: *I have* the following things in my life – understanding of human rights, respect for biodiversity, fear for the future/ hope for the future, a confidence, lack of confidence, loss of home due to natural or other disaster, no family/ community support, responsibility to care for others and very high levels of stress. *My own story is:*

What do I need in my life to help me live ethically and well? For example: *I need* in my life – a home, a sense of safety, access to public transport and hope for the future. *My own story is:* --------------

What am I prepared to do to change my life by adding new ideas or actions? For example *I will add* to my life – more community supports from a range of services and /or more community engagement to lobby for resources, more connection to nature *My own story is:*-------------------------------

What am I prepared to do to change my life by discarding some preconceptions or harmful habits? For example

- *I will discard* from my life – a sense of hopelessness, a sense of entitlement, excessive consumption. *My own story is:* -----------------------------------

  What can I learn by thinking about what has changed in my personal life, my family, community and environment? Self-reflection on *the turning points* for the better or worse – hope that consumption can be replaced with greater sense of attachment to others and the environment. *My own story is:* --
  ----
- *What can I learn by considering the barriers* that currently exist and thinking about what could be done to transform society and our relationship to the environment? *My own story is:*-----------------
  ----

[51]Notes for the Public/NGO/Private sector. Please watch the following video by Robert Costanza as a precursor for our discussions on Natural capital, Built capital, Social capital. http://blogs.worldwatch.org/sustainabilitypossible/state-of-the-world-2013/ Please watch the following video explaining the implications of the size of our carbon footprint.

[52]Vananda Shiva Growth = Poverty Published on Nov 10, 2013 http://sydneyoperahouse.com/ideas Ideas at the House: http://www.youtube.com/ideasatthehouse, http://www.youtube.com/watch?v=7M3WJQbnHKc.

[53]Anthropocene-magnitude of human footprint—implications for new forms of governance. https://www.linkedin.com/grp/post/6973569-6044442047962697729 http://www.ageoftransition.org/#!transformation/c6fi.

3.6 Policy Context and Options for Stewardship

Specifically the research responds to the policy context,[54] in order to achieve better forms of distributive governance to address the UN Sustainable Development Goals and adapted from Murray et al. (2007) who developed the so-called 'Tuvalu test', Local Agenda 21, Triple bottom line accounting (Elkington 1992, 1997) and in line with Hayden (2010):

- "United Nations Local Agenda 21 and variants that strive to scale up triple bottom line accountability not only at a local level, but whether networks of local organisations internationally could effectively the approach. (2a) Accept that with the right to own manage and use natural resources comes the duty to prevent environmental harm and to protect the rights of people. (9a) Guarantee the right to potable water, clean air, food security, uncontaminated soil, shelter, and safe sanitation, allocating the national and international resources required. (10a) Promote the equitable distribution of wealth within nations and among nations. The Aarhus convention as a means to implement (13a) uphold the right of everyone to receive clear and timely information on environmental matters and all development plans and activities which are likely to affect them or in which they have an interest
- (13b) Support local, regional and global civil society, and promote the meaningful participation of all interested individuals and organisations in decision making" (cited by Hayden in Wallace Brown and Held 2010: 369)
- "Common Future and UN Declaration of Human Rights and Earth Charter in particular—4a Recognise that the freedom and rights of each generation is qualified by the needs of future generations. 6c Ensure that decision-making addresses the cumulative, long term, indirect, long distance, and global consequences of human activities".

The decisions taken at local level on how we live have an impact on the way others live. So when personal decisions are made to use food, energy or water resources it has public implications for others as well as the environment. Thus the research enables the development of a personal score card to manage social, economic and environmental decisions. Triple Bottom Line Accounting and accountability is applied through United Nations Local Agenda 21 at the local level. This is a macro policy that has been applied by Local Governments in Australia.

[54]Anthropocentrism and humanism need to move towards respect for Biodiversity (Nagoya Summit in Japan in 2010). The purpose of this research paper is not to rehearse the same arguments about rights and responsibilities—these are taken as a given. This paper is also not about 'what is the case' or 'what ought to be the case'. Instead it takes it as given that social injustice and inequality exists and that the disappearance of biodiversity will make a difference to the eco systemic web of life and to human wellbeing. The loss of insects, such as bees along with greed and hubris will impact on food security—just as it will jeopardise seed security. Instead this research is located in the domain of how to develop a new architecture in response to Dahl's (1967) pessimism about extending the scale of democracy and governance.

The challenge is to develop indicators that are meaningful for residents, who need to understand the links across personal choices and the implications these could have for adapting to climate change and mitigating the worst affects. Furthermore it provides the means to achieve scoring personal choices to enable people to consider what they have, what they need to achieve wellbeing, what changes they are prepared to take through adding actions or adding changes in their attitudes towards others or the environment and through discarding certain taken for granted ideas.

The sticking point is that weak cosmopolitanism argues for equality of all human beings without realising that recognition of human rights, requires securing the basic conditions for life. Today the understanding of carbon-based cultures and economies is relevant to climate change. So the starting point is a recognition that economically wellbeing needs to be the outcome of economic choices—not merely productivity.[55] So how can we address the issue of rights and responsibilities at a local level? Could this be the basis for scaling up management of our Social, Economic and Environmental choices at a regional level? Florini suggests the potential of the Aarhus convention. Thus the participatory action research grasps the nettle to: 'address the challenge posed by the Earth Charter: 'Everyone shares responsibility for the present and future wellbeing of the human family and the larger living world' (cited by Hayden 2010: 68).'

The challenge is to move beyond the rhetoric of cosmopolitan citizenship and to address both justice and sovereignty. The nation state needs to be held to account by an Earth Charter that is supported by overlapping regional institutions (supported by inclusive regional research institutions, policies, the rule of law including parliaments and courts).

3.6.1 Purpose of Software

- To provide practical empirical data to help develop blue economy[56]
- To enable people to participate in ways that make a difference to their own lives

[55]These public interests include environmental considerations. Today the decision to for example to use energy selfishly and excessively has implications for other communities who may be more affected by extreme weather events as a result of our choices. For example we may choose to drive an SUV when it is unnecessary or when it is possible to walk to meet our friends. We may use water excessively for forms of agriculture that will have a downstream risk to other communities needs to be governed by laws that protect the common good. Furthermore, complex decisions need to be informed by those who will be affected by the decisions Ashby's rule of Requisite Variety (1956) to establish ways to enhance representation, accountability and engagement.

[56]Rice–fish farming in Indonesia is discussed as a means to add value to one market and to create employment opportunities for women http://www.thejakartapost.com/news/2016/11/04/climate-smart-agriculture-not-oxymoron.html.

- Create a new way of representing people's diverse needs and to enable organisations to be accountable at the local level.

3.6.2 Specifications for the Software

- Drop down menus already prompt people to think in terms of what they have, what they need, what prepared to add or discard, turning points for the better and worse and barriers to addressing the following goals:
- So the Software needs to address how to address ways to meet each of the 17 [57] goals at the individual, personal and interpersonal level. See the United Nations Sustainable Development Goals below:

Sustainable development goals
• Goal 1. End poverty in all its forms everywhere
• Goal 2. End hunger, achieve food security and improved nutrition and promote sustainable agriculture
• Goal 3. Ensure healthy lives and promote wellbeing for all at all ages
• Goal 4. Ensure inclusive and equitable quality education and promote lifelong learning opportunities for all
• Goal 5. Achieve gender equality and empower all women and girls
• Goal 6. Ensure availability and sustainable management of water and sanitation for all
• Goal 7. Ensure access to affordable, reliable, sustainable and modern energy for all
• Goal 8. Promote sustained, inclusive and sustainable economic growth, full and productive employment and decent work for all
• Goal 9. Build resilient infrastructure, promote inclusive and sustainable industrialization and foster innovation
• Goal 10. Reduce inequalitywithin and among countries
• Goal 11. Make cities and human settlements inclusive, safe, resilient and sustainable
• Goal 12. Ensure sustainable consumption and production patterns
• Goal 13. Take urgent action to combat climate change and its impacts
• Goal 14. Conserve and sustainably use the oceans, seas and marine resources for sustainable development
• Goal 15. Protect, restore and promote sustainable use of terrestrial ecosystems, sustainably manage forests, combat desertification, and halt and reverse land degradation and halt biodiversity loss
• Goal 16. Promote peaceful and inclusive societies for sustainable development, provide access to justice for all and build effective, accountable and inclusive institutions at all levels
• Goal 17. Strengthen the means of implementation and revitalise the Global Partnership for Sustainable Development

[57]http://www.un.org/sustainabledevelopment/sustainable-development-goals/

- The spread sheets have the purpose of showing cross-tabulations across the domains that are linked with socio-demographics. So age, gender are cross-tabulated with haves, needs, what people are prepared to change, etc. This part of the software is the section that develops the 'Being, Doing, Having and Interacting' Index (Human Development Index to build capabilities in own life and community life). It is a way to engage young people and those who are currently excluded. So it can be used as a way to help people by matching up services and skills to needs.
- The data can be mined or 'read' by service users and providers to explore own progress over time and the progress of groups of people within the community or region.
- Progress is defined in terms of the user and the provider perceptions of 'what works, why and how' to enable individuals to move from one point in a continuum to another.

Wellbeing—not the economic bottom line is important—social and environmental indicators of wellbeing need to support economic indicators need. Government response to human wellbeing is often based on economic development, which inadvertently increases consumerism, resulting in greater environmental degradation. Stewardship like all concepts is shaped by assumptions and values. Flannery (2012) considers that stewardship is based on awareness that the land and biologically diverse ecosystems are a cultural heritage on which the wellbeing of current and future generations depends. Anthropocentric versus non-anthropocentric stewardship needs to be carefully considered to ensure that human consumption is not at the expense of environmental considerations and future generations of life (Parker 2012). Wellbeing is a carefully assembled raft of concepts that form the basis for co-developed user-centric, multidimensional measure of wellbeing[58] that is discussed in more detail in the following chapter that

[58]The multivariate nature of the research is appropriate to user-centric design to address policy such as the UN Sustainable Development and Millennium Goals. transdisciplinary projects including an ARC linkage grant (LP0560406) on Indigenous wellbeing and addressed the complex health, housing and social inclusion factors spanning perceived social, cultural, economic and environmental dimensions The process of participation is the subject of her recent publications and other monographs, entitled '*User-centric policy design*' (McIntyre-Mills 2008), based on research conducted in South Australia and funded by an Australian Research Council Linkage Grant and '*Critical Systemic Praxis for Social and Environmental Justice*' based on research conducted in the Northern Territory and funded by Local Government (McIntyre-Mills 2003), respectively made the case that participation enhances attachment to policy ideas and narrows the gap between perceived needs and service outcomes. In '*Wall Street to Wellbeing*' (based on the Australian Local Government funded research) McIntyre discusses in detail public engagement to enable people to make better social, economic and environmental consumption choices and thus to think about wellbeing and sustainability. Specifically, this research builds on the findings of three previous projects: (1) a Participatory Action Research (PAR) project with Aboriginal Australians focused on wellbeing (ARC linkage LP0560406) and (2) a PAR project on decision-making to promote adaptation to climate change (Local Government Grant https://wellbeing.csem.flinders.edu.au/) and (3) Ethical non-anthropocentric approaches to decision-making. The proposed DP

strives to make sense of the many interrelated factors that shape wellbeing in every day decisions.

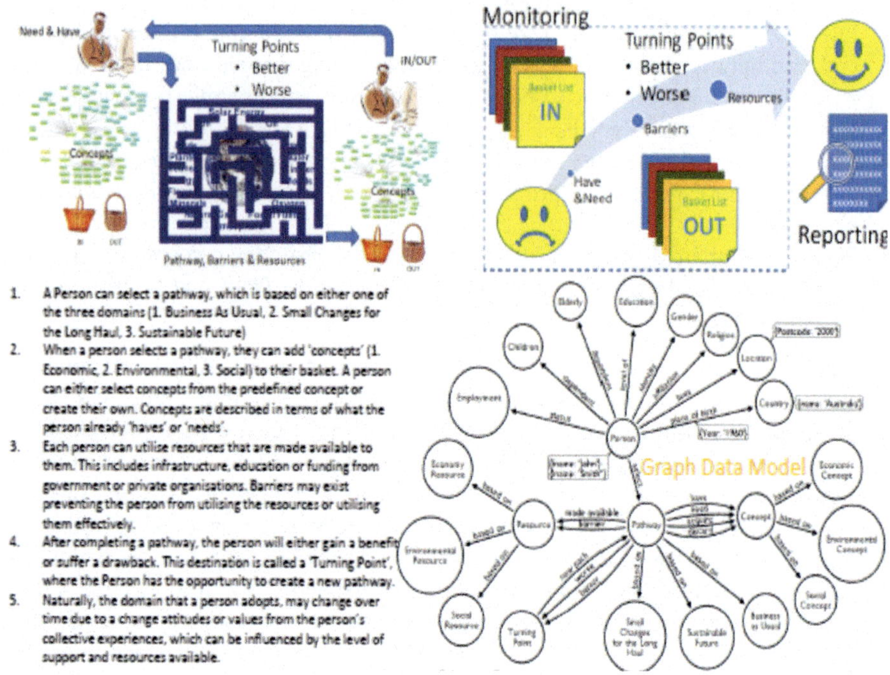

Source McIntyre-Mills and Wirawan, 2017 and see http://wirasoft.com/pathways-to-wellbeing/, http://wirasoftfoundation.org/en_GB/web/smartenergy/plenary-presentation. User name = test and password = test).

3.6.3 The Potential and Weaknesses of Digital Futures

Mobility is greater than it has ever been in the past. Powerful decision makers live lives that are mobile and information rich, because they have wide networks. The current research with a local community in South Australia indicates an increasing

(Footnote 58 continued)

builds on research monographs and articles published in highly ranked and impactful journals on consciousness, systems and behavioural science.

awareness of the need to live sustainably.[59] Living locally and eating locally could become the norm in the event of a collapse in the carbon economy and rising costs, but it will need the careful governance of resources. The project makes the case for:

- Systemic matching of governance options in context through a range of engagement processes that help people to think about the consequences of their choices for themselves, others and the environment on which they depend.
- A reframed approach to wellbeing.

Local government is able to address policy and governance based on study their perceptions and choices. By considering the implications of three scenarios the users of the interactive software programme are inspired to think about social, economic and environmental choices. The scenarios range from 'business as usual' to 'small adjustments to green economies' to 'sustainability and wellbeing'.

Contradictions need to be identified and addressed by policy (supported by legal frameworks) at the local, national and post-national regional level to achieve transformation. The challenge is to enable people to become aware of the need for a cultural shift from a large carbon footprint created by 'business as usual' and to move towards making logical steps towards sustainable living, rather than making confused gestures at best, or making contradictory decisions in personal and public lives.[60] A digital future need not be cities of skyscrapers in a Dubai-type landscape that is controlled through digital means and top down Orwellian controls.[61] It could instead be based on a form of participatory democracy and governance aimed at

[59]The mobile hunter gatherer is as normal as the localised cultivator of the earth in human history. Archaeological history and anthropology have also demonstrated that food security is not achieved only by the hunter gatherers (who forage beyond the immediate neighbourhood) but also through localised agriculture, that needs to be sustainable, or it will lead to the sort of collapse that occurred at Chichen Itza, Mexico or on Easter Island.

[60]But the point that is made in 'Identity, Democracy and Sustainability' (2011) is that a legal bipartisan agreement between the major emitters—USA, China and India is vital. Good faith needs to be demonstrated by local networks that prefigure change on the ground and that link up with global networks, such as 350.org to make a change. '350.org is an international campaign dedicated to building a movement to unite the world around solutions to the climate crisis—the solutions that science and justice demand. www.350.org/en/node/3108' Accessed 27/05/2012 other social movements such as Avaaz or open democracy are examples of other online movements that can ask for responses, such as the following: 'We urge you to save the Earth Summit and the planet by delivering a real deal to end fossil fuel subsidies'.

[61]Surveillance systems could be developed from below—to monitor Big Brother and blow the whistle on corruption—in order to avert control from above and to ensure that local areas do not lapse into so-called 'ungovernable zones' of 'warlordism' (Elliot and Urry 2010). But the paradox is that in the resource rich cities of the world mobile people spend their time rushing from appointment to another. They use mobile technology in ways that can make life frenetic. But digital technology can be used in different ways—not to constantly reschedule meetings but to enable life to be lived at a slower pace. It could be used to enable people to think about alternative ways of living and to 'join up the dots'. The containment of anxiety has been stressed as one of the main reasons for using digital technology by many very mobile workers. Containment anxiety is felt increasingly by over wrought people living very fast paced lives (Elliot 2010: 21) and so digital technology tends to be used to address isolation and rescheduling—but it can be used for

balancing individual and collective needs (McIntyre-Mills and De Vries 2011). Monitory democracy (Keane 2009; Wilson 2012) will need to support post-national governance (McIntyre-Mills 2012). The aim of the local pilot is to provide a test the concept and process prior to extending it to a sub national regional level within Indonesia and then to do a test at a post-national regional test. This was currently underway in West Java at the time this chapter was finalised.

A 'small is beautiful' (Schumacher 1973), post-carbon future in which energy alternatives are still at an early stage of redevelopment (sun, wind, and ocean powered tidal systems) may be a likely future. For this reason learning to live sustainably in smaller, slower and more localised communities could be an important way to adapt to climate change. Local community resilience could be enhanced through engaging with people to decide on how scarce resources should be used, in order to avert a breakdown in governance and democracy. Furthermore, monitory democracy and governance becomes increasingly relevant for adaptation to climate change, in order to mitigate the effects (greater extremes in temperatures, water shortages in droughts, inundations during floods or storm surges and greater risk to bush fires in summer). All these risks will become increasingly frequent. Wellbeing—not the economic bottom line—needs to be the focus of our attention for cultural transformation, according to Stiglitz et al. (2010). But this requires re-conceptualising culture.

Wellbeing rests on mindfulness of our relationships with others and with the environment, not on status or ability to live 'high carbon lifestyles' of our own choosing that are epitomised by mobility for pleasure, maintaining family connections or for work purposes (Elliot 2012). Mobility that is enforced could become greater as a result of increased regional conflict as a result of resource scarcity, disaster or climate change. Whilst there is resistance to this scenario at the moment whilst many live according to the principles of conspicuous consumption,[62] because carbon-based lives are still within the reach of the middle classes—it is likely that

(Footnote 61 continued)

transformation. This report suggests a different way of using technology, in order to support being rooted in one particular place. http://www.dubai-architecture.info/DUB-004.htm.

[62]Originally conspicuous consumption was possible only for the leisure class and now they are more widely emulated, in order to keep up appearances in capitalist society. 'Conspicuous consumption of valuable goods is a means of reputability to the gentleman of leisure. As wealth accumulates on his hands, his own unaided effort will not avail to sufficiently put his opulence in evidence by this method. The aid of friends and competitors is therefore brought in by resorting to the giving of valuable presents and expensive feasts and entertainments. Presents and feasts had probably another origin than that of naïve ostentation, but they acquired their utility for this purpose very early, and they have retained that character to the present; so that their utility in this respect has now long been the substantial ground on which these usages rest. Costly entertainments, such as the potlatch or the ball, are peculiarly adapted to serve this end. The competitor with whom the entertainer wishes to institute a comparison is, by this method, made to serve as a means to the end. He consumes vicariously for his host at the same time that he is a witness to the consumption of that excess of good things' Project Gutenberg's The Theory of the Leisure Class, by Thorstein Veblen's eBook or online at www.gutenberg.orgTitle. Accessed 25/06/2012.

this will be increasingly the preserve of the super-rich elites. The monitoring of scarce resources will become increasingly important for the democracy project, in order to ensure that cynicism and distrust does not alienate electorates. This requires many kinds of engagement—ranging from face-to-face focus groups to community agoras with an 'online' link to regional parliaments. This could enable all those who live within a region to voice their opinions. Giving 'a voice to the mob' is regarded as fraught by liberals and neoliberals—not only because majorities do not always uphold social and environmental justice, but because the current status quo could be overturned! But in a post consumerist world wealth would need to be re-evaluated and re-conceptualised, because, to draw on, adapt and apply Einstein's well-known aphorism: We cannot solve the economic and environmental problems of today with the same ideas of *property and consumption* that created the problem of an *unsustainable* way of life. But to enable freedom of thought science and democratic testing out ideas requires a democratic state—but the problem is that the nation state has not acted in the interests of freedom and democracy as Keane (2009a, b) points out.

Emotions run high as resources run out, but it is pointless to polarise emotional and rationalist responses to climate change. Our capability as human beingsto think about our thinking requires reflexivity. The more we are able to think about our emotions and the emotions of others—the more likely we are to make decisions that are rational. At times of high risk this becomes increasingly important for policy makers. The subjective, objective and intersubjective domains of knowledge are vital, and as Habermas has stressed it is the latter that require the most work in policy making. Reflexivity is both emotional and rational (see Holmes 2010).

The recognition of the symbolic meanings people attribute to ideas needs to be given attention (Collins 2008, 2010). Human agency to respond to rising temperatures cannot be avoided and will require rational responses that take into account diverse emotions—particular the outrage of the disenfranchised.

The Participatory Action Research is based on the axiom that freedom and diversity ought to be fostered to support wellbeing and creativity, but only to the extent that the rights of others and the next generation of life are respected. Many researchers suggest that we make a difference to climate change through our choices and our impact on the carbon footprint. The IPCC (Intergovernmental Panel on Climate Change) Theorem is as follows:

E (emissions) = Population × Consumption per person × Energy Efficiency × Energy Emissions (Charlton 2011).

Charlton criticises the Southern Cross-Climate coalition for arguing against coal, because he does not see any way to meet both the social and environmental justice concerns simultaneously. The GFC and the environmental melt down seem to be seen as two problems—not one. Human beings cannot achieve new growth through increased extraction of profit. Instead living elegantly and well through redistribution and simpler life styles supported by green economies, reciprocal sharing of resources and skills in green communities, supported by energy efficient systems. The options are as follows: Business as usual, Small Adjustments for the Long Haul to achieve living in ways that do not risk our way of life, and Sustainable Futures (see McIntyre-Mills and De Vries 2011 for a detailed argument and the details of the software). The approach developed in the pilot diverges from the limited calculations of economics and is instead rooted in notions of what wellbeing actually means for cultural change. The approach is also informed by sociocultural cybernetics and physics, drawing on West Churchman (1971), De Beer (1992), Shiva (2011),[63] Christakis and Bausch (2006) and Bausch and Flanagan (2011). Emeritus Professor Alexander Christakis of Global Agoras, together with Flanagan, Bausch and his team have stressed the importance of democratic engagement to enhance an understanding of climate change.

Recently Professor Wilkinson, together with Pickett (2009) linked social inclusion in more equal societies with better quality of life in their book entitled 'The Spirit Level'. Nussbaum and Glover (1995), Sen (2000) and Flannery (2010) stress that a better quality of life is linked with a demographic transition to smaller population growth rates. Furthermore, the current winner of the Sydney Peace Prize, Dr. Vandana Shiva, a physicist, links wellbeing with the protection of our food supplies; whilst Joy Murray a senior academic physicist at the University of Sydney links the size of our carbon footprint with the wellbeing of our whole region. Physicists such as Shiva, Murray and Christakis see the world in terms of its interconnections. The nonlinear approach to working systemically at multiple levels could help to mitigate climate change, avert catastrophic change and adapt. Changing ourselves to respond the environment is part of the process of adaptation. The time for working in isolation is long since over.

According to Gamble (2000), the basis for sound economic governance is stability through full employment and a stable currency that holds its value. This needs

[63]'The International Commission on the Future of Food and Agriculture was created in 2003 in Tuscany, Italy, as a result of an international gathering of leaders in the food and agriculture movement brought together by Claudio Martini, President of the Regional Government of Tuscany for the annual conference on globalisation issues. …Chaired by noted activist, Dr. Vandana Shiva, the Commission is composed of a group of leading activists, academics, scientists, politicians and farmers from North and South working toward shaping more socially and ecologically sustainable food and agriculture systems. Its mission is the promotion and protection of biodiversity, local food production and consumption, food security and food safety, and the rights of small farmers. …It has published four far-reaching Manifestos on issues of critical importance to the future of the planet: the future of food, the future of seeds, climate change and the future of food security, and the need for new knowledge systems for a healthy planet'. http://www.future-food.org/ Accessed 10 April 2012. Shiva argues that if we wish to engage with the might of WTO and multinationals who patent seeds—then a way forward can be as profound and simple as saving seeds, growing food in as many places as possible in urban areas, using solar panels, saving water, collecting rain water and using legal force to prevent the criminalisation of farmers and those who oppose the patenting of life. Multinational companies and the WTO have made it possible to patent the conditions of life thereby causing almost a quarter of a million suicides by Indian farmers who cannot afford to purchase the seeds and the pesticides that are needed (despite the claims by the manufacturers). This is not so very far removed from the Australian context as the number of suicides amongst farmers has also increased as detailed elsewhere (McIntyre-Mills 2012). Furthermore the attempt to criminalise farmers who store old varieties of seed could lead to very vulnerable monocultures of foods. The idea that the very basis for life-seeds and genes can be patented is part of the process of commodifying people, animals and the fabric of life.

to be supported by development that invests in human and physical capital. But Gamble neglects to factor in what Stiglitz has called 'stocks for future generations', or diminishing returns and the opportunity costs of destroying the means of our common survival. Most importantly he does not factor in the dimension of care and the time to build respectful relationships with one another and the environment or the tensions that exist between regional neighbours within and beyond the nation state.

Just as the Mekong River causes tensions between neighouring nation states who resent the development of hydroelectric power in the upper reaches of the Mekong, because it impacts on fishing in Laos and Cambodia, the residents of South Australia understand the tensions around water usage in the river Murray and the implications for the sale of Cubby Station, because the property dams water that could flow downstream, according to some participants in the debate.

The question remains how to balance individual and collective rights at a regional and post-national regional level. Recently Australia passed legislation to limit fishing off our shores by a super trawler. The standoff between Australia and the Norwegian trawler echoes the tensions within our region in the East and South China Sea over access to fishing and mineral rights.

China and India have economies that are growing faster than the US and they too need to protect 'stocks for the future' (Stiglitz et al. 2010).[64] This chapter makes the case that developing connections with others could provide a way forward. Decisions need to be informed by connecting many kinds of knowledge.[65]

If people make conceptual connections through thinking through scenarios it could enable those of us who are not physicists to think about the big picture. This is a simpler approach based on a humility that we return to the earth or water or air on which we rely for our wellbeing. Our wellbeing depends on the wellbeing of the planet. Our home or sense of place is thus both local and global.

---

[64]It is undeniable that the approach to development needs to respect the environment. The market failures Held (2005: 15) anticipates fall short of what has actually occurred. The problem is not only concerning externalities that are not factored into calculations of the degradation to the environment, it is a way of thinking and 'being in the world' that shifts the extraction of profit to where labour is cheaper and where governments and citizens are less likely to complain about degradation of environment. Short-term profits are made at the expense of future generations. This undermines the conditions of employment so that full time employment becomes less available. The globalised market needs post-national controls to ensure control of the movement funds (see McIntyre-Mills 2011). Enabling a transformation of identity through consciousness raising and governance is the challenge for governance which has become increasingly difficult in the wake of the convergent crises.

[65]But the problem with media control in any nation state—is that they could lead to an erosion of democracy per se. It is better to be able to develop critical systemic thinking skills so that one can read through newspapers and are able to locate the different arguments in commentary and editorials that are pro the zero-sum or containerist approaches, pro market and unaware of or denying the interconnections or the impact on wellbeing. The Mail and Guardian Newspaper editor stressed that self-regulation is vital for democracy, see: McDermott (2012) 'Clash of paradigms' as editors defend self-regulation, Feb 01 2012.

The following definition of wellbeing is developed on the basis of research firstly in the Northern Territory with Aboriginal Australians and then in South Australia in the Southern Region of Adelaide on complex needs. The relevance of this definition is that the NT and SA are, respectively, one of the hottest and driest parts of Australia and thus most affected by climate change. Furthermore sustainability thinking is now more relevant than ever for planetary wellbeing. Most importantly it is relevant, because it is likely that the sorts of challenges faced by the poorest on the planet—are the sorts of conditions that we face in the future, if we continue our addiction to consumption of resources. The definition of wellbeing used in this research spans 5 axial themes:

> 1. Home safety (personal safety, social indicators and physical indicators, physical safety of the home and environs) 2. Physical and mental health (Wilkinson and Pickett 2009) 3. Sense of purpose (indicators of satisfaction with employment), 4. Connection/belonging to healthy community and healthy environment (people and land or place), 5. Self-respect and confidence to speak out.

Then the axial themes were explored in FGD in Indonesia and in conversations linked with workshops and a rapid appraisal with an NGO in South Africa, detailed in the following chapters. The initial findings so far on the research (in progress) is that the 5 themes are relevant and the planetary passport to address the UN Sustainable Development Goals in terms of being, doing, having and interacting to achieve these themes resonate across contexts, albeit with contextual variations.[66]

Stiglitz (2010) the ex-Chief Economist and senior vice-president of the World Bank along with his colleagues has adopted similar ideas based on his experiences in response to an invitation from Sarkozy to address the poverty, and global financial issues facing Europe. It recognises the convergent social, economic and environmental challenges. Stiglitz et al. (2010: 15) use a multidimensional measure of wellbeing spanning:

> 1. Material living standards (income, consumption and wealth), 2. Health, 3. Education, 4. Personal activities including work, 5. Political voice and governance, 6. Social connections and relationships, 7. Environment (present and future conditions), 8. Insecurity, of an economy as well as a physical nature.

To sum up, according to Stiglitz et al. (2010), the essence of their findings is that wealth needs to include stocks for the future. These are social, economic and environmental. Leisure should also be given a price. If all measures consider wellbeing in terms of median households, then consumption, wealth and income could be compared with rich and poor households.[67] Together with vulnerability to

[66]These are seen as overlapping needs which need to be addressed (McIntyre-Mills and De Vries 2008: 307). Axial themes according to Max Neef (1991) also resonate with the axial themes identified in McIntyre-Mills et al. (2008, 2011, 2014).

[67]It is possible for measures of average to disguise the fact that a high income does not enable exhausted workers to enjoy the same kind of leisure that a lower income family enjoys who has access to the safety net of capital, a home (rather than a mortgage) and the security that loss of income will not lead to a loss of their home or their social networks.

job loss, many urban residents also face risks associated with fires, flood and drought.[68]

3.6.4 *Translating Thinking into Praxis: Stewardship Conditions Through Extending the Social Contract*

How can we 'curate the conditions for a thrivable planet' (International Systems Sciences 2013) by transforming democracy and governance? Human beings have the capacity for both consensus and conflict. We need to think locally, regionally and globally and to develop multilevel democracy for multilevel governance. What matters at the moment in global history is that all people feel engaged and respected and that their ideas are 'listened to'. Representation is one of the major challenges if democracy is to be enhanced and if people are to be able to understand the consequences of their decisions.

A transformation of the culture of greed and waste is needed to develop more sustainable futures. But the signing of the United Nations Declaration of Human Rights has not lead to the means to implement the norms. The non-local nature of human rights was recognised, but the political and economic interests of nation states have prevailed to prevent the realisation of multilateral changes.

It could be argued that for human rights to be implemented. It will require our realisation that our survival is interconnected with the survival of others, including non-human animals.

Some overarching regionalist approaches to ensure that differences in life chances are not translated into disrespect for others and an overweening pride that helps us to deny the rights of others. Failed states and environmental wastelands should not be ignored by democracies, but we should be mindful that in three of the world's so-called 'advanced democracies', namely America, United Kingdom and Australia the divides between rich and poor are very wide. Far wider than in Sweden or Denmark or Japan (Wilkinson and Pickett 2009). Before discussing what should be done to make our societies more equal, it is worth pointing out that focusing attention on the inequalities within them does not mean ignoring the international inequalities between rich and poor countries. The evidence strongly suggests that

---

[68]Conservative economists argue that consumption can be modified by those with an income to build up wealth. The difference in this report is that consumption refers to stocks that are usually not included in economic measures. The most important point made is that by contributing to the public good, individual households could enhance their wellbeing. In Australia households that were protected by levee banks weathered the recent floods in New South Wales. State governments that levy taxes to build resilient communities are providing for the common good. Reducing consumption in order to pay taxes to protect our homes and families makes sense as New South Wales, Victoria and Queensland continue to count the costs of flooding in Australia.

narrowing income differences within rich countries will make them more responsive to the needs of poorer countries (Wilkinson and Pickett 2009: 229).

The root cause of consumption is power without responsibility—so whoever comes to power needs to be held to account through mechanisms to develop social, economic and environmental indicators that secure the wellbeing stocks for the future.

Minority greed is at the expense of the majority and future generations of life who face hunger, thirst, disease, displacement without access across open borders and death

> The IPCC estimates that between 1.1 and 3.2 billion people could experience water scarcity by 2080….Around 655 million people in Asia lack safe drinking water, while 1.9 billion lack access to basic sanitation…. (Saul et al. 204).

The IPCC formula suggests that the privileged lives of some could lead to 'existential risk' for people and the planet (Bostrom 2011). This has implications for the way we live and the need to change our way of life through living sustainably. Representation, accountability and sustainability challenges need to be met through addressing consumption choices that are currently very unequal. The gaps between rich and poor, the powerful and the powerless have become wider and wider.

The greatest challenges are the consequences of inaction that will potentially pose an 'existential risk to humanity' (Bostrom 2011). These challenges include representation of the increasingly diverse populations within nation states along with accountability to ensure that resources (e.g. water, food, and energy) are used fairly, equitably and sustainably in local and regional biospheres.

The ways forward to govern our ecological footprint are fourfold:

1. Extend the social contract through new forms of communications for discursive democracy (Dryzek 1999, 2000, 2005, 2010) and inclusive governance within and beyond the nation states to include regional post-national constellations.
2. Declare an a priori global covenant, norms, ideals and laws to protect the planet through constitutions like Bolivia and Ecuador and ensure that these declarations are more than window dressing. This can be achieved through more engagement by civil society in monitoring the way in which resources are valued as well as the way in which they are used, exchanged or redistributed.
3. Transform a posteriori performance accounting and accountability from below through new measures that protect rights and that ensure that being socially and environmentally responsible is protected in terms of a new set of indicators, spanning doing work that supports living systems, having a social and environmental conscience (demonstrated in terms of the choices and decisions we make daily) to protect the social and environmental fabric of life and—thus protecting our ecological foot print. Global policy needs to address food, energy and water security. Policy requires balancing individual and collective interests. This requires addressing the wider regional biosphere. Wilkinson and Pickett (2009) argue that after a certain point, more consumption does not lead to more

happiness. Thus, increased consumption based on increased economic growth does not make people happier—in fact it poses 'existential risks'. 'More equal societies almost always do better' socially, economically and environmentally.
4. Public education to protect the global commons and the public good through transforming the curriculum through formal and informal means to enhance the capacity for transformation through teaching people to value living systems through protecting water, food and renewable forms of energy. These basic, systemic needs are best protected through valuing cultural diversity to the extent that the rights of sentient being and the environment are protected. The approach strives to avoid the dangers of the commodification approach in 'Mis-measuring our lives' made by Stiglitz et al. (2010) and instead makes a plea for changing the way in which some human beings live (at the expense of others and future generations of life) based on a recognition that the 'personal is indeed political'.

Transformation is about enabling people to voice their intrinsic values and to shape agendas from below it is not merely about neoliberal markets shaping and commodifying nature. Pilots to prefigure changes that provide examples of learning how to do governance differently are necessary (McIntyre-Mills et al. 2014; McIntyre-Mills 2014b), in order to maintain or reduce the size of our ecological footprint. For example: Murray's notion of a Tuvalu test for decisionmaking. Ways to reduce the size of our ecological footprint through alternative forms of governance and democracy has been demonstrated with Global Agoras team (Bausch et al. 2012). The software for the previous pilot was designed by Denise de Vries and Natasun Binchai (2014) and extended by Rudolf Wirawan (2016). Praxis takes as its starting point the issue of commodification and suggests not merely a new form of performance accountability but also a transforming what we value and how we live.

Complex challenges such as poverty, climate change and terrorism comprise many, interrelated variables that are perceived differently by those who hold different values. Religion, morality, politics and aesthetics are the values that make us human. They enable us to cooperate and to compete. They are what Churchman calls, the so-called the 'enemies within'. The notion of wicked problems is his concept (albeit popularised by Rittell and Webber). The notion of taming is not a critical heuristics concept. It is oppressive and it assumes power over others and including non sentients and nature!

Critical heuristics when applied to thinking through 'if then' policy scenarios is a precursor for decision making. This does require 'drawing a line'. But who draws it, what is included and excluded, where it is drawn and how it is drawn matters. Ethically we need to consider many ways of knowing before making a decision. This is different from the language of 'taming wicked problems' or governing from above. It is about stewardship based on appreciating many dimensions and many ways of knowing. We need to think of:

- Logical coherence of an argument, the extent to which the perceptions and lived experience and the quantitative data are available.
- Idealism in terms of what ought to be done to ensure human rights, the capabilities of all sentient beings and planetary rights. We are part of an ecosystem on which we depend.
- Engagement to address alternative arguments needs to be based on considering the consequences of decisions for human and environmental wellbeing. Discursive engagement enables people to test out ideas and to enhance their capacity to think critically and to join up the policy dots. It also enhances their capabilities to engage in the policy and politics as to who gets what, when why, how and to what effect.
- An alternative future needs to be modelled and co-determined by with and for people of good will. This is the post-nationalist agenda. It is not naïve. It can be argued to be logical and scientific if we consider the work of Buckminster Fuller on governance of Space Ship Earth and Kenneth Boulding's work on domains of knowledge and evolution from inorganic to organic life to living systems comprising human systems that can consciously evolve in directions that they can shape with a sense of purpose. The evolutionary direction is shaped by values. Hence the justification for co-determining and co-creating the direction of policy and politics (Hulme 2009, 2010; McIntyre-Mills 2014 a, b).

3.7 Systemic Governance for Stewardship

In order to protect the ecological web of life of which humanity is a strand—we need to do more than 'build stocks for the future'; we need to achieve a quantum leap towards recognising our potential role as stewards for current and future generations or destroyers of living systems. The choice is ours. It is a value based choice to become stewards for current and future generations of life. By thinking carefully about the consequences of our choices (conceptually, spatially and in temporal terms) we can arrive at a rational sense of what stewardship ought to entail as well as a deep understanding of the consequences of the growing disparity in life chances.

Surveillance 'from below' is needed to protect food, energy and water security by protecting the non-renewables and by ensuring that local people can live lives that are worth living and are capable of achieving quality of life (Nussbaum 2006).

It would be ideal to be able to protect planetary citizens in ways that respect rights and responsibilities across national boundaries through international law. But to enable the implementation of social and environmental justice we need a new form of governance and a new form of democracy.

3.7.1 Pan Opticon: Surveillance 'From Above'

The revelations of the Snowden findings have led to friction between Australia and Indonesia as well as between the US and the German chancellor Angela Merkel.[69] Although Obama quickly clarified and apologised for the invasion of civil liberties, the Australian premier Abbott refused to apologise and then sent a private letter to President Yudhoyono to explain the situation. The exposure has led to the end of military cooperation with Australia until further explanations and conversations occur. But the process of healing the rift has begun. Similarly other embassies within the region from South Korea and Singapore over media reports that their 'governments have been involved in wiretapping'.[70]

The surveillance from above by State military/industrial complex has been raised by Snowden. The use of drones that lead to civilian deaths has been raised by Etzioni. The space base in Alice Springs to launch surveillance for military intervention.[71] Arguably, surveillance 'from below' is needed to protect human security by protecting the non-renewables and by ensuring that local people can live lives that are worth living and are capable of achieving quality of life.

It would be ideal to be able to protect planetary citizens in ways that respect rights and responsibilities across national boundaries through international law. But to enable the implementation of social and environmental justice we need a new form of governance and a new form of democracy that enables participation and monitoring. But we also need free accessible ways to ensure that transparency does not become a vehicle for the panoptic on authoritarian state.

It is insufficient to argue for global covenants in overlapping regions. We need:

- A movement that ensures we live differently and that we value sustainable living.
- Overlapping republican rights not a Penal state supported by global panoptic on
- Power of the state through judiciary, police and army to subjugate and incarcerate.
- Power of surveillance the challenge is to avoid surveillance from above without monitoring from below.

Comaroff discusses 'Law, labour and the contours of contemporary capitalism' in a key note address at the '*Knowledge and value in a globalising world: disentangling dichotomies, querying unities*' (July 7th 2011). She stressed: 'Europe is going South'. It could be argued that America has also headed south as indicated by its loss of a triple AAA rating from Standard and Poor's.

---

[69]http://www.theguardian.com/world/2013/dec/04/nsa-storing-cell-phone-records-daily-snowden.

[70]SBY seems satisfied with Abbott's response Nov 27th 2013.

[71]A Wiki leak orchestrated by Assange and supported by Manning and Snowden has led to incarceration of Manning and no outcry beyond a few newspaper articles. Clearly surveillance is common.

The question is—if it can happen in Europe could it happen in any post welfare state? What hope is there to achieve social democratic aspirations in developing countries? The possibilities for 'states in crisis' will need to be achieved through exploring more options in context. I make the case in this chapter and previous publications (McIntyre-Mills 2014b) that this requires expanded pragmatism in governance and new forms of cosmopolitan politics.

Governance and democracy have to deal with three options pertaining to truth (McIntyre-Mills 2000; Crowder 2003): (a) One truth responses defended by grand narratives or conflict, (b) No truth (postmodernist) approached defended by relativism, (c) Mediated (harmonised) responses based on stewardship.

The chapter makes the case for social change through exploring interdisciplinary frameworks to address greed, the commodification of the powerless and the environment. It will be informed by debates on ethics, human reasoning capability, emotions and environmental politics and the practical implications for social democracy (Judt 2010).

Beck (2009: 96) stresses the need for sub politics 'from above' and 'below'. How to achieve this is the goal of current research on balancing individual and collective needs. The focus is on the local, so that the poor will not be silenced and so that local wisdom can be drawn upon. This enables creative responsive transformation and the emergence of new relationships.

> At the beginning of the twenty-first century, we see modern society with new eyes, and this birth of a 'cosmopolitan vision' (Beck 2006) is among the unexpected phenomena out of which a still indeterminate world of risk society is emerging. Henceforth, there are no merely local occurrences. All genuine threats have become global threats. The situation of every nation, every people, every religion, every class and every individual is also the result and cause of the human situation. The key point is that henceforth concern about the whole has become a task. It is not optional but the human condition (Beck 2009: 19)

The Intergovernmental Panel on Climate Change (IPCC) formula suggests that the privileged lives of some could lead to 'existential risk' for people and the planet (Bostrom 2011). This has implications for the way we live and the need to change our way of life through living sustainably. The notion of the ecological footprint has been shaped largely by theorists who work critically and systemically across a number of disciplines. Systemic relationships are relevant to our understanding of ecological footprints, as is the development of an understanding of the interdependency of all living systems explained in the speech to colonists by Chief Seattle (1854). Several versions exist but the key points are: 'Every part of the earth is sacred… the earth is our mother…the rivers are our brothers, they quench our thirst'.[72]

The link between the policy choices that human beings make and capitalist economics (based on profit) needs to be appreciated if the notion of the human ecological footprint is to be understood. Boulding (1966) contributed to the

[72]http://www.bing.com/images/search?q=Native+American+Speeches+Chief+Seattle&FORM=RESTAB#view=detail&id=3E8E2FE587AF7B9FD38D954D496969D1F93B8428&selected.

discussion on economics and ecological footprints by focusing on the need to cooperate on 'space ship earth' and the need to avoid so-called 'cowboy economics' based on competition between 'us versus them' without understand the systemic ethical implications. Boulding's '*Skeleton of Science*' (1956) idea is that evolution through culture and society is vital for transformation towards a more sustainable society. Bateson's (1972) notion of the '*Ecology of Mind*' is relevant in this regard as he stresses the hybrid relationshipbetween people's thinking and choices that shape the environment and, this, in turn, shapes us.

Gro Brundtland, a medical practitioner, realised that personal health could only be secured by peace and that human security could in turn only become a reality through protecting 'Our Common Future' (Brundtland Report 1987). This landmark report helped to lay the groundwork for an appreciation of our interconnectedness, the limits of growth and it provided the basis of the sustainable development movement. Meadows and Randers (1992) in 'Beyond the Limits' set the basis for the Club of Rome approach, an approach which criticised the way in which natural resources were being exploited to support an unsustainable way of life at the expense of future generations. Productivity and profits often come at the cost of living systems without an understanding of our hybrid relationship. The cybernetics contribution by Beer (1992) and Von Foerster (1995) explained the role of people in shaping the environment and our dependency on it. Thus Elkington (1994, 1997) argued that social, economic and environmental dimensions need to be factored into the economic bottom line. This provided the first steps towards applying a new form of accounting and accountability to managing EF. The work of Howard and Elizabeth Odum (2000) on modelling for all scales demonstrated that urban designs are unsustainable and that human security is jeopardised by the current consumption of energy resources. Rees and Wakernagel (2008) explain that the ecological footprint can be understood as 'carrying capacity as maximum load'.

The greatest challenges are the consequences of inaction that will potentially pose an 'existential risk to humanity' (Bostrom 2011). These challenges include representation of the increasingly diverse populations within nation states along with accountability to ensure that resources (e.g. water, food, and energy) are used fairly, equitably and sustainably in local and regional biospheres. But addressing the EF has implications for policy and politics, because the gaps between rich and poor, the powerful and the powerless have become wider and wider (Wilkinson and Pickett 2009).

The Intergovernmental Panel on Climate Change formula addresses the implications of polarising 'people versus the planet' (Charlton 2011). Representation and accountability rest on showing the links across excessive consumption, unfair distribution and harm (Saul et al. 2012: 167).

Table 3.4 Rethinking the social contract approach

Legal category	Protection	Limitations	Comments and adaptations
Refugee	Protects a person from return to a place of persecution	Prosecution is unlikely to encompass victims of carbon pollution, unless discriminatory	The problem is that asylum seekers need to be processed to be recognised as refugees. At this stage they are even more vulnerable
Complementary protection	Protects a person from arbitrary deprivation of life or cruel, inhuman or degrading treatment	It is unclear whether climate – related harms would be recognised as such	Current laws and protocols are outdated
IDP	Rights based framework to protect all persons displaced within a country	Guiding principles are not binding and cannot be enforced against neglectful government	A cosmopolitan approach to governance would render support of displaced people a constitutional requirement, based on an extension of the social contract.
Litigation	Potentially offers a binding judicial decision with compensation or other remedy	Foreign governments cannot be sued in local courts; international courts may not have jurisdiction	A new architecture to enable the implementation of social and environmental justice requires management of size of carbon footprint, fines for infringements and incarceration of perpetrators deemed to be undermining the fabric of life for current and future generations

Source Adapted from Saul et al. (2012: 169 in McIntyre-Mills 2014: 132–133)

3.7.2 *Social Justice for Non-citizens and Environmental Protection of the Planet*

The global figure for displaced people passed 50 million for first time since the Second World War.[73] The needs of the displaced are not addressed through the current architectures of democracy and governance. The protection of food, energy and water supplies beyond the boundaries of the nation state is inadequate resulting in the displacement of people from climate related socio, economic and environmental disasters. The protection for climate refuges is inadequate (Table 3.4):

The treatment of asylum seekers has deteriorated as both sides of Australian politics play a role in supporting border protection.[74] The notion of tow backs has been stressed recently (Oct 2014) by Labour supporters on ABC news as a potential game changer, but it is not supported in Indonesia.[75] The need to reframe the debate has been stressed by the greens in Australia some independents and academics.[76]

---

[73]http://www.theguardian.com/world/video/2014/jun/20/global-refugee-figure-passes-50m-un-guterres-video. "According to the United Nations High Commission for Refugees, Antonio Guterres (15 January, 2014): 'Within a few years, Syria has gone from being the world's second largest refugee hosting to becoming its fastest refugee producing country…. UNHCR has registered far over 2.3 million Syrians as refugees in the region, and governments estimate the total number of those who fled at over 3 million…. …. Few refugee influxes have generated this profound an impact on their host countries, with dramatic demographic, economic and social consequences. The pressure is felt heavily in all areas of daily life, as budget deficits are increasing, growth suffers and jobs, salaries and price levels are affected across the region, leaving local families struggling to make ends meet'. The discourse above remains rooted in the notion that bounded nation states are inevitable. The argument developed in this paper is that the nation state needs to take as a given that the cosmopolitan human rights come first and that nation states serve regional needs. They need to be seen as part of an overlapping region with responsibility at a regional level. The costs of climate change will place heavy burden on the region". Benhabib and Archibugi (2008) and Benhabib (2007) stress that we need a form of federalist republicanism. But how do we move from the containerist approach that undermines justice through limiting compassion to citizens at the expense of human rights? Bohman (2005p 102, 111) stresses Arendt's concept of 'the capacity' as a way to 'begin a democratic dialogue': "For democracy to promote justice, it must already be just. Call this the democratic circle. While it can never be said to disappear, the circle can become virtuous through what I refer to as the "democratic minimum": the achievement of a democratic arrangement sufficient for citizens to exercise their creative powers to reshape democracy according to the demands of justice—that is the capacity to initiate democratic deliberation" (2014, p 111).

[74]The most powerless and voiceless are those who do not have the vote. They are young people, children, asylum seekers, the dis Abled, sentient beings who are commodified and traded and of course the planet which has no rights at all. So the focus needs to be on the inadequacies of the nation state and the social contract.

[75]http://www.abc.net.au/news/2014-10-27/kingsbury-turning-back-the-boats/5844926.

[76]http://insidestory.org.au/beyond-deterrence-reframing-the-asylum-seeker-debate. According to a report by McNevin, A. Mares, P. Mitric, D. Neumann, K. & Taylor, S: 'The situation in camps in Papua New Guinea, Nauru and Christmas Island is a mix of explosive volatility and self-destructive despair. The lives of men, women and children are being destroyed. Asylum seekers in the community on bridging visas live in an agony of uncertainty that prevents them from

An Independent candidate has suggested that it is an issue for the International Criminal Court.[77]

The social contract is increasingly inadequate to protect citizens of developed nations let alone those without citizenship rights or those living in nation states that do not respect social and environmental justice. Increasingly Muslims internationally feel marginalised. This is because they have lower levels of employment and in Australia marginalisation according to Akbarzadeh (2013) is indicated by lower home ownership and being more than 3 times as likely as other Australians to be unemployed.

3.7.3 Australia Needs to Re-consider Its Relationships with Others, the Environment and the Wider Region

Currently the role of the state in Western so-called democracies has been to act as protector of residual welfare rights in terms of the social contract applied to citizens within the boundaries of a state or federation. Those outside these constructed

(Footnote 76 continued)

rebuilding their lives. Temporary protection—as currently conceived by the federal government—will only prolong this agony. The policy of using the navy to return asylum seekers to Indonesia creates great stress and anxiety among defence force personnel and adds considerable diplomatic risk to Australia's delicate relations with a close and important neighbor…'.

[77]"https://www.icrc.org/applic/ihl/ihl.nsf/INTRO/585?OpenDocument accessed 29/10/2014 Recently Federal independent MP Andrew Wilkie has stressed that he: 'Wants the International Criminal Court (ICC) to investigate whether the Abbott Government has committed crimes against asylum seekers'. Mr. Wilkie has formally asked the ICC prosecuting authority to investigate whether the treatment of asylum seekers contravenes international conventions. He has named Prime Minister Tony Abbott and each member of the federal Cabinet. Mr. Wilkie said the Government's actions were criminal and the ICC had jurisdiction to investigate. 'The fact is it is not illegal to come to Australia and claim asylum, and we have a fundamental obligation to hear those claims and to give those people refuge if those claims are accurate', he said. 'But instead the practice of forcibly sending people to other countries, for example, Nauru, Papua New Guinea and back to Sri Lanka—that is a crime against humanity in accordance with article seven of the Rome statute'. According to the International Committee for the Red Cross: 'The United Nations has been considering the establishment of a permanent international criminal court since its creation. After years of negotiations, a Diplomatic Conference was held from 15 June to 17 July 1998 in Rome which finalised and adopted the Statute for the International Criminal Court (ICC). The Statute was finally adopted by a vote where 120 were in favour, 7 against and 21 abstained'."

containers do not receive protection in states that are becoming increasingly like fortresses.[78]

National security in many democracies is at the expense of human security which is a planetary issue. The top down surveillance of citizens and non-citizens to protect the boundaries of the state are in place with the support of businesses, as revealed by phone companies and Google. The surveillance or whistle blowing 'from below' needs to be supported by an extended form of Aarhus Convention supported by digital democracy and governance.

The idea of externalising (read buck passing) the processing of asylum seekers or rendering suspects for questioning elsewhere or revoking citizenship or at worst including them on elimination lists—called euphemistically a 'disposition matrix' for drone attacks on enemies of the state (Etzioni 2010). These are just a few examples. It would be best if a planetary passport were regarded as an essential right for all. This will become increasingly important as competition for resources increases with climate change as food, energy and water resources become scarce.

Uniting against climate change and a common threat is a rational response. It is also essential for social and environmental justice. The current forms of democracy and governance that operate within the boundaries of the nation state do not support

---

[78]"http://www.theguardian.com/commentisfree/2014/may/12/new-zealand-refuses-climate-change-refugees-mass-action-is-now-needed. Accessed 27/05/14 'New Zealand's court of appeal has refused refugee status to a family from Kiribati, a Pacific island which is quickly sinking beneath the sea'. 'An abandoned house that is affected by sea water during high tides stands next to a small lagoon near the village of Tangintebu'.....The idealist in me hoped the court of appeal would create an option for that mass migration. The realists on the court wouldn't have a bar of that. Teitiota was not a refugee under 1A (2) of the refugee convention. The court conceded Teitiota was a refugee under a 'sociological definition', but not the legal definition. The court approached the case as a threshold question: was 'there a real chance of the refugee claimant being persecuted 'and, if so, was "there a convention reason for that persecution"? Teitiota couldn't satisfy the test'. The court took the orthodox position that persecution is primarily political and internal. There had to be a 'violation' of human rights and a 'failure' of state protection. Teitiota's claim turned that definition on its head. In essence, his claim held that the violation and failure was on the part of the international community. The persecution was external, not internal, and environmental, not political. The decision reveals—in all its misery—the protection deficit in international law. A judicial decision is an uncodified statement of power relations. Never could there be a more unequal power relationshipthan here: on one side, the I-Kiribati No refugees please, we're New Zealanders. That's the message from New Zealand's court of appeal. In a decision released last week the court endorsed earlier rulings that Ioana Teitiota—a Kiribati national—is not a climate change refugee. Teitiota, his wife and their children will be deported to Kiribati, where the court believes they can 'resume their prior subsistence life with dignity'. ... The IPCC projects that the Pacific Ocean will swallow most Kiribati by the end of this century. Life in Kiribati is life against the clock. There's little dignity in that. But before Kiribati sinks beneath the sea, ocean creep will make the islands uninhabitable. Sea level rise—coupled with more intense storm cycles—will contaminate the water table, pollute the small pockets of agricultural land and destroy homes and businesses. Tarawa, the main atoll, is a tiny sand strip some 6 miles^2 in size. There is, quite literally, no escaping the misery climate change will cause. The conditions for social unrest are manifest and the I-Kiribati know it. Density means disease and unrest spread quickly. There are 50,000 people tightly packed on Tarawa. The president, Anote Tong, is exploring options for mass migration'."

the natural environmental biosphere on which we support because the approach is one of extraction of profit or shifting the blame or responsibility in a zero sum approach that belies physics and a quantum future that is interconnected (Turok 2012).

The most important challenges are understanding the interconnected social, economic and environmental problems pose an 'existential risk' to life as we know it. Food, energy and water supplies can only be protected when we develop an approach to representation; accountability and sustainability that span national boundaries and that recognise our interconnectedness.

The environment of the problem and the range of inter related ways in which the challenges can be addressed requires not merely trans disciplinary and cross cultural capability when designing policy responses, but the will to grasp the nettle as to why some live at the expense of the majority in this generation and the next.

Lowering consumption and enhancing fair distribution of resources is vital. The Millennium Goals can be described as little more than window dressing, because they do not engage or provide a voice to those whose needs are greatest. The United Nations Sustainability Development Goals could be regarded as a means to translate these goals into practice through using social, economic and environmental indicators that are generated by local residents within a local context that reflects their cultural needs and engages them in considering the consequences of their choices within limits set by means of government. A multilevel approach to governance could thus be developed.

3.7.4 Valuing Self, Others, the Environment and Future Generations of Life: Implications for Socio-economic and Environmental Governance and Democracy

The historical decision taken in Paris to limit emissions below 2° and to enable all 200 participating nations to agree is perhaps the most heartening decision in 2015, alongside the role of the German PM's example of agreeing to take refugees from Syria, even if border controls were later tightened. It was an example of moral leadership. The bombing of Syria by France and now UK in response to the atrocities in Paris provide low points in 2015 and 2016 as does the ongoing bombing by Russia. The impact of drought has been cited by some as one of the reasons for the war in Syria which had experienced one of the worst droughts in a decade which resulted in migrations to the cities, others disagree (Pearce 2015: 26). But it is undeniable that as resources decrease, conflict in congested ghettoes and camps will result in conflict and that the conflict in Syria and the Sudan was also linked with the political issues and a government system that lacked transparency and fairness.

3.7.4.1 Competition for Scarce Resources

Whilst Australia makes trade deals with Japan others within the region, such as China see the arrangements with concern.[79] Sri Lanka and Australia also have an uneasy relationshipthrough turning back refugees.[80] The principle of non-refoulement states:

> A refugee's right to be protected against forcible return, or refoulement, is set out in the 1951 Convention relating to the Status of Refugees: 'No Contracting State shall expel or return ('refouler') a refugee in any manner whatsoever to the frontiers of territories where his life or freedom would be threatened on account of his race, religion, nationality, membership of a particular social or political opinion' (Article 33(1) which states in section 9:
>
> It is widely accepted that the prohibition of forcible return is part of customary international law. This means that even States that are not party to the 1951 Refugee Convention must respect the principle of non-refoulement.[9] Therefore, States are obligated under the Convention and under customary international law to respect the principle of non-refoulement. 'If and when this principle is threatened, UNHCR can respond by intervening with relevant authorities, and if it deems necessary, will inform the public'.[81]

Similarly Australia's strong relationship with regard to Indonesia has been placed under strain through the asylum seeker issue and concerns about the way in which the law was applied to young drug smugglers. In this context the role of non-state actors becomes increasingly important. According to Priyambudi (2010: 130) the role of one to one friendship and non-state actors plays an important role:

> For Indonesia, Australia is always important … However, it must be said that the public interest in Indonesia about Australia is not greater than its interest in the United States, China and the Islamic world. The international events that occur in the US have dominated the Indonesia media (especially since Barack Obama became President). China is also increasingly important and is regarded by many Indonesians as the rising superpower in Asia. Events in the Islamic world also receive a lot of attention domestically because many Indonesians have a strong sense of solidarity with other Muslims. Despite this, Australia has gradually captured the attention of the Indonesian people with more interaction and regular contact between government officials and ordinary people. I have argued that in strengthening the bilateral relationshipit is important to acknowledge the role of parliament and non-state actors in Indonesia and that we need to consider how both Indonesia and Australia are now democracies.
>
> As the state relations ebb and flow as a result of politics as well as the personalities of leaders, it will be increasingly important for non-state actors to foster relationships through

---

[79]Tehan, D 2014 Financial Review Monday 7 July New Deal pivots us back to Japan, p. 46. 'The first bilateral visit to Australia by a Japanese prime minister since 2002 …Jap[an has been Australia's most reliable and significant partner since the two countries put tensions aside in 1957 and signed an early trade agreement. So when Prime Minister Tony Abbott and Shinzo Abe sign a new economic partnership (EPA) agreement on Tuesday they will update one of the most productive relationships…'

[80]McKenzie Murray, M. 2014 'Refugee Policy at Sea', July 6–11 Saturday Paper, p. 4.

[81]http://en.wikipedia.org/wiki/Convention_Relating_to_the_Status_of_Refugees#cite_note-UNHCR_Guide-9 and http://www.unhcr.org/protect/PROTECTION/3b66c2aa10.pdf.

ongoing dialogue in education, professional, business, tourism and alumni friendship and cultural networks that can help to support the need for democratic engagement for all sectors of the community.

The attitude towards multiculturalism has been threatened in the wake of the Paris massacre. The conservative parties in France and Britain have reacted and ensuring that the removal of civil rights does not undermine democracy:

Getting the balance right between fighting the external dimensions of Isis (its self-proclaimed caliphate in Syria-Iraq) and countering its internal dimension in Europe (its terror cells and ideology) will be essential. (Nougayrede 2015)[82]

In Australia this message is equally relevant to restore support for multiculturalism and to prevent a replay of the Cronulla Riots on the anniversary (Dec 13th) and to be measured in responses that do not undermine civil liberties. The recognition of Jamal Rifi[83] as Australian of the Year was recognised for his stand against violence an important step in the right direction to address the need to recognise the role of diverse Australians at a time of increased numbers (approximately 90) of people leaving Australia to join Islamic State fighters and 30 women have travelled to Syria wish to become brides of the cause. It is not surprising that young people are radicalised if they feel that the world is becoming an increasingly unjust place. Many who join do so for a range of reasons some may be irrational but most are not—but because they feel that the world's values are out of step with their own. This is indeed cause for concern.

Dr. Rifi has become the spokesperson for the Muslim community. In the wake of the Sydney Siege in Martin Place, he said:

These people are a disgrace …We shouldn't allow them to tear the public and social cohesions of our community and society. "We are as horrified as anyone else, if not more, by these events. We feel very strongly against their action that is condemned in the strongest possible words.[84]

The treatment of asylum seekers has deteriorated as both sides of Australian politics play a role in supporting border protection.[85] The notion of tow backs has been stressed recently (Oct 2014) by Labour supporters on ABC News as a potential game changer, but it is not supported in Indonesia.[86] The need to reframe the debate has been stressed by the greens in Australia some independents and academics.

[82]Nougayrede (2015). 'Britain and France can take lead on EU security' *The Guardian* 11.12.15, p. 19.

[83]Our shared values must unite us all The Australian, Jan 24.25 2015.

[84]http://www.couriermail.com.au/news/martin-place-siege-muslim-leader-jamal-rifi-says-gunmen-are-a-disgrace-to-islam/story-e6freon6-1227156638520?nk=769378d2ffad063ae0791898f24e12af.

[85]http://www.themonthly.com.au/issue/2013/september/1377957600/christos-tsiolkas/why-australia-hates-asylum-seekers.

[86]http://www.abc.net.au/news/2014-10-27/kingsbury-turning-back-the-boats/5844926.

3.7.5 Penal States, Protection for the Displaced, Disappearing States and Obsolete State and Property Law

Justice Julian Burnside stresses that temporary 3 year visas are inadequate to protect asylum seekers:

> TPVs were first suggested by Pauline Hanson in 1996. Her proposal was criticized by Philip Ruddock as 'unconscionable'. Three years later, TPVs were introduced by the Howard Government; Ruddock was immigration minister. TPVs suffer from a couple of serious vices: not least that they are likely to cause more deaths at sea—something both major parties claim to be concerned about. TPVs provide asylum seekers with three years' protection only, and they deny the visa holder the right to be reunited with their family. If one member of a family makes it to Australia and satisfies the authorities of their status as refugees, then it is likely that other members of their immediate family are also refugees.... TPV was the major cause of the SIEVX disaster. On 19 October 2001, a boat carrying 400 asylum seekers sank in its attempt to reach Australia. 353 people drowned. Most of them were women and children coming to Australia to be reunited with the men in their family who were already in Australia on TPVs. The people who drowned were true refugees, doing what human instincts dictate: getting the family back together. The Rudd government abolished TPVs in 2008. If the Coalition is elected to government in September and follows suit, it will be a convincing demonstration that they do not care whether people drown trying to reach Australia. It will highlight that their real purpose is to stop people asking for our help.
>
> It is a dismal prospect, considering that allied troops are pulling out of Afghanistan later this year. When they leave, there will be a bloodbath as the Taliban declare open season on Hazaras. Some of them will only have enough resources to head towards Australia, as it is more expensive to get to Europe or America. The Coalition promises to add more misery to their lives....Is this really the Australian way?[87]

Malcolm Fraser on Coalition asylum plans: no limits to the inhumanity, Former Liberal Prime Minister Malcolm Fraser speaks to Guardian Australia about a 'breach of common decency'[88]:

Off shore processing in states whose economy is reliant on Aid could lead to the development of oppressive camps along the lines of convict colonies. The propensity for riots has already been demonstrated along with the reaction of increased control and increased distancing of people as 'other'. Off shore processing of asylum seekers in Nauru and PNG by Australia[89] has been implemented

---

[87]http://www.theguardian.com/commentisfree/2013/aug/16/temporary-protection-visas-refugees.

[88]theguardian.com, Friday 16 August 2013 13.38 AEST http://www.theguardian.com/world/2013/aug/16/malcolm-fraser-coalition-asylum-policy 19/08/2013.

[89]http://www.theaustralian.com.au/national-affairs/immigration/asylum-seeker-bluff-set-to-be-exposed-by-the-numbers/story-fn9hml gu-1226686524164 downloaded 28/07/2013.

http://www.theguardian.com/world/2013/aug/15/nauru-transfers-in-doubt-after-fire downloaded 19/08 2013.

http://www.theguardian.com/world/2013/jul/19/kevin-rudd-labor-asylum-border-protection 19/08/2013.

by the Liberal government as a so-called election promise to control the boundaries of Australia and to control who has the right to enter:

> Australia is preparing to move the first cohort of refugees on Manus Island into temporary accommodation on the island, but detainees are resisting, saying they fear being killed if moved into the community.
>
> In a transcript of a speech intended for detainees at the Australian-run detention centre – a copy of which has been obtained by Guardian Australia – an official from the PNG immigration department said those recognised as refugees would be forced from detention. 'Your choices are simple: make Papua New Guinea your home, or leave'.
>
> 'But detainees said they have been told they will be attacked and killed if made to live in the community'.[90]
>
> They want the United Nations to intervene.[91]

3.8 Policy and Governance Praxis: A Way Forward for Collective Co-determinism

The role of law ought to be to ensure fair distribution of rights and responsibilities to mitigate the causes of climate change. Climate change refugees and climate change displacement and migration' are raised by Saul et al. (2012).

But unfortunately Saul et al. (2012: 61), fall into the trap of following Hardin's '*Tragedy of the Commons*'. They do not realise that Hardin's approach is in line with John Locke's defence of property and the enclosure movements of the seventeenth Century in Britain. Hardin's core argument is that in the absence of private property there would be lawlessness. Hardin's classic argument for nationalism and private property is developed as follows:

> Picture a pasture open to all. It is expected that each herdsman will try to keep as many cattle as possible on the commons. Such an arrangement may work reasonably satisfactorily for centuries because tribal wars, poaching, and disease keep the numbers of both man and beast well below the carrying capacity of the land. Finally, however, comes the day of reckoning, that is the day when the long-desired goal of social stability becomes a reality. At this point, the 1 logic of the commons remorselessly generates tragedy. (Hardin 1968: 1243–1248, cited by Shiva 2002: 26)[92]

---

[90]The Guardian 6th January http://www.theguardian.com/australia-news/2015/jan/06/manus-detainees-recognised-as-refugees-will-be-forced-into-png-community Accessed 8th January.

[91]http://www.smh.com.au/federal-politics/political-news/manus-island-protest-escalates-up-to-700-detainees-on-hunger-strike-20150118-12sszu.html.

[92]"Each of 10 people own one 1000 lb cow and all 10 cows graze in a common area. If an additional cow, is added then all the cows would eat less grass and they would weigh 900 lbs. But I do not care because I have an additional cow and so have 900 lbs × 2 rather than just 1 cow of 1000 lbs. The health of all the cows is threatened if I continue to introduce more cows, but in the short and medium term I will get richer and more powerful, until the number of cows outstrips

Shiva argues that this is incorrect. The greatest misuse of natural resources has been by nations and corporations and not by local communities that have acted as water democracies, for example in Bali, Indonesia and in Gujurat, India water was sacred and carefully managed.

> Community rights are necessary for both ecology and democracy. Bureaucratic control by distant and external agencies and market control by commercial interests and corporations create disincentives for conservation… (Shiva 2002: 30–31).

By treating water as a commodity it is traded and people have to pay for an essential resource. We regard air as free to breathe, water is another basic right.[93] Perspectives are out of step with global dimensions and dynamics of the challenge.

The so-called liberal approach was emphasised at the Society for Alternative Economics (SASE conference, 2013) but with an emphasis of avoiding grand narratives and the need to learn by doing and to work on finding solutions in context. Patrick la Gales emphasised possibilities of change and the need to avoid top down intentions. He stressed the need to learn from diverse case studies internationally.

Adelman's (2013) plenary address at the 'States in Crisis Conference', stressed the importance of the work of Albert Hirschman and the panel applied the themes of 'Exit, loyalty and voice' to a discussion on 'states in crisis'. But the level of analysis needs to be extended to address the convergent challenges (social, economic and environmental) that do not allow exit from the problem which is in global. The panel and key note speakers stressed that they chose the themes of possibility, rather than the bleakness of the work of Hannah Arendt who (also a refugee like Hirschman) stressed the banality of evil and in her work on 'Crisis of the Republic' stressed that the problems are pervasive and just as the slaughter of Jews was not the fault of a single monster, Eichmann, it was the entire machinery of the state.[94] Le Gales, did however refer to the work of John Scott, who stresses the need to avoid 'Seeing like a state' and to provide space for diversity because mono culture leads to the end of creativity.

But I think the message and bleakness of Hannah Arendt's work on 'the banality of evil' provides the other side of the coin. We are all vulnerable to evil and we are all dependent on one another to do the 'right thing'. Recognition of our vulnerability needs to underpin our working towards being resilient. Judith Butler holds the

(Footnote 92 continued)

the amount of grass for them to eat- then all the cows fall ill and die. Then the rich and the poor cattle owners will suffer as their animals die (adapted from Lloyd 1833 by Hardin 1968)".

[93]The terrifying notion of having to work to pay for air was explored in a recent Hollywood film.

[94]Similarly today we face issues in Australia, namely the return of Tamils to Singhalese persecutors in Sri Lanka by the Abbott government. This principle of refoulement has been ignored until an appeal was made. This principle was raised in the WW2 as a means of protecting Jews fleeing persecution in Europe. Julian Burnside has also raised the possibility of piracy—the principle of rescuing people at sea and then failing to protect them.

'Hannah Arendt Professorial Chair' at Berkley. She stresses this very point about our vulnerability.

We need to refocus on the question: How should we live? This paper addresses democracy, governance and ethics. It is taken for granted these days that the silo approach to the nation state is acceptable. But given the entangled nature of globalisation and the need to protect the global commons we can no longer address socio-economic and environmental challenges within bounded areas.

In times of crisis—Hirschman (1970) suggested three options—loyalty, voice or exit. Although it was possible for him to apply all three options to great acclaim in his life time—to exit totalitarian states and to demonstrate alternative ways of doing things, times have changed. His work could be characterised as striving to reveal 'in the small new ways of seeing the whole' (Adelman 2013: 9). He contributed to reframing the way in which areas of concern were perceived. But the challenges we face today cannot be addressed by working within the boundaries of a nation state or leaving a nation state that has become totalitarian.

The work of Hannah Arendt emphasises both potential and risks. It holds resonance today, because she talks about the banality of evil which seems perhaps even more relevant today than it did when she wrote about Eichmann's trial. As a Jew, she stressed that the man Eichmann was part of a monstrous system that appeared quite banal as it was widespread and part of a culture and a bureaucracy that was taken for granted.

Today we need to be thinking critically about taken for granted structures. We need to take the liberative potential of small scale projects (as suggested by Hirschman) and try to think of their potential as a way to do things differently. Hannah Arendt stressed the banality of evil and the need to consider the broad context and the structures that lead people to make unethical choices.

The work of Hall (1992, 1996, 1997), a cultural theorist discussed the diversity within the modern state and stressed the need to consider the way resources are distributed and consumed within nation states by asking: What is the social, cultural, economic and environmental context? Who gets What, When, Why and So what? His contribution to praxis is important as he stressed that the notion of national identity today requires a long narrative response to explain each of our origins. Where we live, where our parents live, what they did for a living and where they came from and why. When we think of cultural identity we also think of food, hospitality and fostering friendships through sharing.

Where our food comes from is now a global concern and we also need to think of our responsibility when consuming cheap food that is supplied by food chains that make a profit as a result of the lack of rights of the food producers, for example:

Slavery in Thailand provides cheap prawns in their supply chains, Cumberbatch. 2014:

We were kept in the wind and the rain

We were treated like animals

But we are not animals we are human beings.[95]

If we accept a strong version of cosmopolitanism along the lines suggested by Nussbaum (2006) in 'frontiers of justice', then we must accept that immigration is tied to the notion of separate nation states that are no longer relevant. In an era of social, economic and environmental convergence of risk we need to accept the possibility of new forms of governance to protect those who are not protected by the nation state, because they are not citizens. The work of Ann Florini provides possibilities as does the work of Danielle Archibugi.

We need to address the challenges of governance by working across disciplines. Gibbons et al. (1994) argue that the ability to work across boundaries is vital for 'The new production of knowledge' and vital for '*the dynamics of research*' *to address current complex challenges.* A core capability will be to transcend disciplinary boundaries, synthesise diverse perspectives and critically analyse the role of science in public policy.

These perspectives on the capabilities required to effectively contribute to the exploration and resolution of significant social and policy issues. The need to develop a support for human capabilities is vital. Australian politicians and academics are becoming more competitive and dualistic.

Boundaries are sociological constructions. In biological systems the most basic boundary is the membrane of a cell which opens and closes according to context.

If we consider and adapt a hierarchy of systems (developed by Boulding 1956) biological systems appear above inorganic systems because they are open and responsive to their environment and able to adapt and learn as self-evolving systems (see McIntyre-Mills 2014 for more details).

The so-called 'tragedy of the commons' argument developed by Hardin (1968) is a construction informed by simplistic thinking. Ironically, it has become an archetype for bounded thinking. Ironically and tragically it is often used as the starting point for environmental thinking, but in fact it was an argument developed by Locke in support of the enclosure movement and private property in Britain. The argument being that contained areas of land are cared for better than the areas of land that are held in common and shared. The example of common grazing land is shared. But ironically it is the privatisation and commodification of land and natural resources that has led to environmental pollution and degradation.

The lack of trust between nations has evoked the rhetoric of nationalism and state protectionism has led to 'seeing like a state' (Scott 1999), which in turn leads to the argument for competition and the zero sum approach which argues that one nation can profit at the expense of others. The organisational philosophy and

---

[95]'A 6-month Guardian multimedia investigation has, for the first time, tracked how some of the world's big-supermarkets relying on slave labour to put cheap prawns on their shelves. Slavery is back and here's the proof. Narrated by Benedict Cumberbatch' http://www.theguardian.com/global-development/video/2014/jun/10/slavery-supermarket-supply-trail-prawns-video.

governance arguments that flow from this philosophy are based on bounded compartments that can profit at the expense of others. The issue of population growth and demographic transition could be addressed by empowering the most marginalised members of the world's population, namely women.

> The world's population is estimated to grow from 6.8 billion to 9 billion over the next 40 years, placing even greater pressure on agricultural production. As one scientist observes: we are going to have to produce as much food in the next 50 years as was produced in the past 5,000 years. Food scarcity may be aggravated by increasingly high living standards in developing countries, which brings changes to consumer diets among the new middle classes—particularly more demand for meat, thus diverting grain supplies from staple foods to livestock feed— placing further pressure on food prices. (Saul 2012: 197)

Fair competition does not exist and added to this it is used to subsidise unviable and unsustainable businesses

> To put it most starkly, an average European cow receives more per day in European Union subsidies (US 2.20) than the daily earnings of 1.2 billion people living in poverty. (Saul et al. 2012: 201)

Arendt (1972) stresses that the abolitionists (even Jefferson) did not think that slaves should be afforded space within the nation state. The social contract did not extend rights to slaves! So the nation state and the social contract have always had within it a very limited sense of rights and responsibilities. Today we face the challenge of political systems that are not open to very different ideas.

The elephant in the room is that we need educational opportunities to support a cultural and demographic transition to lower birth rates. By respecting diversity and freedom to the extent that diversity and freedom does not undermine the rights of others, perhaps democratic governments could legalise diverse relationships in marriage and pass laws to support the quality of life of sentient beings. Veganism and vegetarianism could be seen as culturally appropriate as could driving solar cars or using solar public transport. Reframing the nature of representation, and accountability is vital for a culture of post-national resilience. Central to cultural transformation is our ability to identify with others and to develop a sense of solidarity and compassion for sentient beings. The culture of identifying the worth of human—animal relationships (Sharpe 2005) and animal rights cannot be dismissed—after all we are human animals. If the rights of sentient beings are acknowledged (Kirby 2011; Nussbaum 2006; White 2002) it could lead to a complete redesign of ethics, democracy and governance buttressed through international law and post-national legislation. Maguire (1996: 186 cites the Voice of the Earth by Roszak 1993: 16):

> … the endangered species, the imperiled biosphere, cannot speak for themselves. We must be their voice….It is the brave beginning of a project that both the person and the planet require.

Justice Michael Kirby has stressed the need to respect diversity and has recently written a paper that advocates for protecting animals[96]:

> I learned of the horrors that can attend animal slaughter, the suffering of chickens in corporatized chicken farms; the cruel treatment of poultry kept substantially immobile in tiny spaces in defiance of the strong social inclinations of their species, the sow stalls in which pigs injected with hormones, are retained immobile so as to fulfill a human purpose of expansion to provide pork for consumption….and unable to fulfill the intended life experience of their kind…The very thought of eating meat offends me.

Kirby asks the question, what would Ghandi have done? What was his praxis, because Ghandi did not merely philosophise, he attempted to translate thinking into practice, but he did suggest that as human beings we could evolve beyond the meat eating stage? Kirby (2013: 31–34) reflects that this is unlikely in the short term, but concludes that at the very least:

> reducing the corporatized slaughter of animals is a subject for increasing reflection. In meat exporting countries, the very least we must insist on, and legislate for, are strict conditions in which we farm and slaughter our meat…

Unfortunately the right to monitor the conditions of farm animals is being restricted (with the complicity of the organisations that are concerned about the impact of calls for animal rights on the profits of agribusiness).[97]

Social and environmental justice including the right to a decent quality of life for sentient beings needs to be addressed through social movements that balance both human rights and sustainable living—that are buttressed by international law that is in turn supported by regional courts.

The recognition of the EU (despite its many failings) with a Nobel Peace Prize is indicative of the potential of the EU. In a confederation citizens do not have a direct say in international affairs and an example of confederalist model is the EU. Archibugi (2010) stresses this is different from the 'more rigid constitutional structure' of the United States or Australia, for example. Florini (2003) in her book 'the coming democracy' outlines an argument for the potential of the European Union (and other con federations) of scaling up the Aarhus convention is to enable

---

[96]Kirby (2011) 'Animals deserver our protection'. Weekend Australian October 1–2 p. 22.

[97]http://www.abc.net.au/radionational/programs/lawreport/animal-rights-activists-26-the-law/5430446 'Attempts to undertake monitoring from below are being subject to gag orders, for example:

The use of surveillance from below is now being restricted in the interests of protecting the business of farmers. Ag-gag laws are a response to a wave of undercover investigations by animal protection groups that have shown horrific animal cruelty on factory farms and slaughterhouses. And industry, in response, rather than addressing those abuses or calling for reform, is trying to make them illegal. So these ag-gag bills make it a criminal offence to photograph or videotape animal cruelty on factory farms. These have been passed in a few states already. We are seeing more states considering them this year, they are expanding to some other industries as well'.

all citizens who are members of the EU to have a say, in order to address both environmental concerns and social concerns.

This could provide the architecture for a balanced approach to involvement by members of a federalist union that respects the identity of sovereign nations and their citizens—to the extent that their freedom and diversity does not undermine the rights of others. This ought to include young people, future generations, the voiceless, disabled and those who are not protected by citizenship rights. The rights of the voiceless need protection by advocates. Neglect or abuse of the elderly or those with disabilities needs to be given much more attention than is currently possible on the basis of a social contract that protects those who can vote and who can express their rights vocally.

The globalised market needs post-national controls to ensure control of the movement funds (see McIntyre-Mills 2011). Enabling the transformation of identity through governance and public education is the challenge for ethical systemic governance. This has become increasingly difficult as governments focus on political survival, rather than addressing the social, economic and environmental crises. Morality becomes politically disposable, because of a lack of understanding of our interconnected existence.[98]

Strong cosmopolitans recognise that the need to contain capitalism. Our so-called containment anxiety, linked with our identity and need to define ourselves, cannot be addressed by living a schizophrenic existence in which we see ourselves as benefitting at the expense of other nation states. We already have regional conflicts fueled by energy shortages and competition for the last of the non-renewables. Satellite monitoring from above seems to be more of a priority at the moment by USA than space travel. Digital communications are already widely used, but their potential is under tapped as stressed in this chapter and other research on wellbeing, representation, accountability and sustainability (McIntyre-Mills and de Vries 2011, 2012).

Stiglitz (2010) the ex-head of the World Bank along with his colleagues has adopted similar ideas based on his experiences in response to an invitation from Sarkozy to address the poverty, and global financial issues facing Europe.[99] It recognises the systemic social, economic and environmental challenges.

---

[98]The market failures Held (2005: 15) anticipates fall short of what has actually occurred. The problem is not only concerning externalities that are not factored into calculations of the degradation to the environment, it is a way of thinking and 'being in the world' that shifts the extraction of profit to where labour is cheaper and where governments and citizens are less likely to complain about degradation of environment. Short-term profits are made at the expense of future generations. This undermines the conditions of employment so that full time employment becomes less available.

[99]These are as follows: '1. Material living standards (income, consumption and wealth), 2. Health, 3. Education, 4. Personal activities including work, 5. Political voice and governance, 6. Social connections and relationships, 7. Environment (present and future conditions), 8. Insecurity, of an economy as well as a physical nature. Leisure should also be valued'. According to Stiglitz et al. the essence of their findings is that wealth needs to include stocks for the future—these are social,

A sustainable local community is determined by a sustainable region in which food, energy and water supplies are considered as major determinants for wellbeing.[100] No community can be expected to transform from a high carbon life style (or aspiring to this life style) without feeling part of the design process and owning the decisions as to how resources should be used. But if young people can be enabled to tap into 'eco-facturing' in a 'cascade economy' (Pauli 2010) based on finding sources of abundance in nature and in the misdirected systems (Ackoff and Pourdehnad 2001) created by the current economy, then new possibilities can be created. Pauli (2010: 79) citing Peter Drucker stresses that: 'the needs of the poor are opportunities waiting for entrepreneurs.' Furthermore the contributions made by those who live simply and well need to be demonstrated and recognised in their planetary passport.

If we are prepared to recognise opportunity, the potential for resilience and also our mutual vulnerability, it provides a basis for stewardship. We are all reliant on others and need to be able to depend on our connections with others in a cascade economy. What if we could recognise our vulnerability and what if we could foster a sense of caring for others that recognises our humanity and our mutuality?[101]

3.9 Accountability for the Loss of Food and Loss of Place Could be the Next Step in Law

Biodiversity is threatened by markets that control life, for example the Trade Related Aspects of Intellectual Property Rights Agreement of WTO (TRIPS) enables the corporatisation and commodification of life:

(Footnote 99 continued)

economic and environmental. Together with vulnerability to job loss, many people across the world also face risks associated with fires, flood and drought.

[100]Stiglitz et al. (2010: 15) use a multidimensional measure of wellbeing. Arguments for and against the national, transnational and supra national organisations have been raised by idealists, pragmatists and realists. A way to bridge the divides is through expanding pragmatism to consider the consequences for current and future generations through addressing wellbeing and developing governance systems to ensure that stocks for the future are developed through limiting the way in which production processes and distribution practices benefit the minority at the expense of the majority.

[101]Young people (Osler and Starkeyb 2005); the disabled, asylum seekers and sentient beings (Nussbaum 2006) along with future generations live 'precarious lives' (Butler 2005). Those perceived as different are not protected (Young 1990). The ability to show compassion underpins cosmopolitanism (Butler 2011). Her work stresses 'the need to rethink *the human* as a site of interdependency'. Butler stresses that humanity needs to be able to ask for assistance and we need to be able to anticipate that we will be heard and that people will respond with compassion as well as a sense of opportunity. Unless this is possible it leads to a life that can be unbearable. Do we wish to live in a world where we do want to help one another and in which we deny the pain of sentient beings? The ability to show compassion underpins cosmopolitanism (Butler 2011).

> Parties may exclude from patentability plants and animals other than micro-organisms, and essentially biological processes for the production of plants and animals other than non-biological and microbiological processes. However, parties shall provide for the protection of plant varieties wither by patents or by an effective sui generis system or by any combination thereof. This provision shall be reviewed four years after the entry into force of the Agreement establishing the WTO. (Article 27.5.3 (b) cited by Shiva 2010: xviii).

Policy decisions flow beyond spatial boundaries and affect future generations. The refugees from Tunisia and Libya are given the boot on Lampedusa as Italy strives to keep control of the refugees. The attitude towards borders appears to be hardening as the economic climate worsens.[102]

By treating water as a commodity it is traded and people have to pay for an essential resource. We regard air as free to breathe, water is another basic right.[103]

The argument developed in this chapter is that a new architecture for governance and law needs to find a way to arbitrate across conceptual and spatial boundaries in order to protect the global commons and to ensure fair distribution and redistribution of resources.

The current structures are unsustainable. The penal state in last throes of disappearing state climate change refugees leave and are locked up.

How will the law protect those who lose their territories and become stateless? Can they continue to be punished for disasters that could in fact be linked with the toxic life styles and the extreme levels of consumption by the 1%?

Currently there seems to be a strong return to sovereignty and a return to competition for scarce resources. The current election in Australia is being run along the lines of 'whatever it takes'. Both parties seem to try to outdo the other on border protection. The systemic effects of carbon emissions could be addressed by means of the Tuvalu tests and other forms of management and governance.

But according to Saul et al. (2012: 166): states can only be held to account (at present) if they have signed up to the Kyoto Treaty in 1990 or the United Nations framework convention on Climate Change (UNFCCC):

> …[I]t is very difficult to establish legal liability for historical emissions (say before the United Nations Framework Convention on Climate Change (UNFCCC) or Kyoto Protocol in the 1990) which have predominantly contributed to climate change, because such emissions by companies or governments at that time were not unlawful. The case for liability only becomes stronger following the adoption of emissions reductions targets since the 1990s, from which point excessive emissions might well entail legal liability. (op cit.)

The need for fine grained accountability will become increasingly important. As does the need to continuously re-frame international law to address current policy concerns. At a recent conference on creating the future in Melbourne, November 2012, Professor Susan Greenfield summed up the issue of the plasticity of the brain, meaning that the brain makes neural connections, the more connections, the more

---

[102] http://www.watoday.com.au/opinion/politics/we-can-do-better-than-italys-warehousing-on-lampedusa-20091014-gwh7.html.

[103] The terrifying notion of having to work to pay for air was explored in a recent Hollywood film.

mindful we become. The making of connections is personal. It is the result of our personal experiences, so even two identical twins will have the potential to develop different connections as result of their experiences. To what extent can the use of conceptual tools enable us to extend our sense of the world and to enable us to see ourselves as members of a wider group to whom we are responsible and with whom we have solidarity? Can leadership that fosters 'neurogenesis' or making of connections make a difference—or is it all about power and the will to power?

Although activities like playing the piano will help to develop new neural connections, even thinking about playing can develop these connections. So the idea behind this research was that perhaps by getting people to think differently about the way in which they live, and to consider, what if I were to do things differently? Perhaps it will enable the participants who do these mental exercises, to transform the way in which they live? I did this research first with people with complex health, housing and social inclusion needs. Many were addressing unemployment and had a range of associated needs and health related problems—including recovering from drug and other alcohol related illnesses. Addiction to a way of life can be both a cause and the effect of many other problems, as Beer (1974) stressed in his research. This is why the approach to research needs to be one of engagement and being the change.

> 'Systemic racism': discriminatory patterns and practices which involve more than the action of a few individual attitudes, but rather the systemic practices of racism built into society's major institutions (Feagin 2006).

The pain of rejection based on categories such as age, gender, culture including religion, ideology or physical or mental dis-Ability also trigger anger. So this appears to be a long leap from neural research. But by connecting the dots it becomes apparent that it is not![104]

The challenge to move beyond the rhetoric of cosmopolitan citizenship and to address both justice and sovereignty. The nation state needs to be held to account by an Earth Charter that is supported by overlapping regional institutions (supported by inclusive regional research institutions, policies, the rule of law including parliaments and courts).

The GFC and the environmental melt down seem to be seen as two problems—not one. Human beings cannot achieve new growth through increased extraction of profit. Instead living elegantly and well through re-distribution and simpler life styles supported by green economies, reciprocal sharing of resources and skills in green communities, supported by energy efficient systems.

---

[104]Florini (2003) suggests the potential of the Aarhus convention could be scaled up to enhance accountability. This book argues that social, economic and environmental accounting could be assisted by means of participation to make this so-called triple bottom line accountability viable. Thus the participatory action research grasps the nettle to: 'address the challenge posed by the Earth Charter: "Everyone shares responsibility for the present and future wellbeing of the human family and the larger living world"' (cited by Hayden 2010: 368).

Ideally, a universal respect for social and environmental justice could enable subsidiarity as a means to support freedom to the extent that it does not undermine the quality of life of this generation or the next.

Designs need to be supported by constitutions, based on a priori *norms*, and consequentialist or a posteriori *approaches*, based on testing out ideas within context and with future generations in mind.

Global axioms to protect future generations and the poor of this generation need to guide legal constructs and legal decisions at the local level.

The principle of subsidiarity could rest comfortably with global axioms—provided that the caveat is made that—*people at the local level can be free and diverse to the extent that their freedom and diversity does not undermine the rights of others or future generations of life*.

This challenge of balancing individual and collective social, cultural, political and economic concerns needs to be buttressed by regional parliaments and courts that focus on social and environmental justice at a post-national level.

The desire of the rich to consume and the desire of the majority of the poor to leave the ghetto are expressed through emulating the rich.

On the one hand, socio-demographic research shows that debt and bankruptcy in developed nations—such as the USA and nations within the EU—are driven by advertising, easy credit and the desire to 'keep up appearances' by emulating the standards of the very rich. (Frank 2007; Wilkinson and Pickett 2009).

Making and invoking treaties and conventions that control commodification needs to be a priority and needs to be ongoing if justice is to be maintained. The approach could be buttressed through federations supported by, for example the Lisbon Treaty (Horvath and Odor 2010) which requires that social, economic and environmental legal considerations be met. Unfortunately the structural and process mechanisms of the EU are not able to manage the distribution of power or funding (Rhodes 1997). Clearly if the EU is to survive it needs to be supported by a means to balance individual and national interests with the collective good of the union. More and more members of the union are disenchanted for a number of reasons.

How can we increase engagement in the environment that balances individual and collective wellbeing and simultaneously protects the environment when we know little about this nexus? Stewardship for future generations underpin the philosophy of Aboriginal custodians of the land and their dreaming sites. However, there is evidence that many non-Aboriginal urban citizens wish to spend more time living slower lives, walking wherever possible, riding bikes, growing local food, recycling and reusing and consuming less, instead of living stressful, competitive lives that save time, but waste resources as they are reliant on fast food in 'throw away' containers, fast travel and a 'time is money attitude' (McIntyre-Mills et al. 2012, 2013, 2014 and forthcoming). These findings clearly link environmental health and human wellbeing and raise the question of what can we learn from mobile and egalitarian, place–attached people (Vaske and Kobrin 2001) whose history is recorded in the landscape? (Guddemi 2006; Rose 2004).

3.9.1 *The Social Contract Is Eroded to Prop Up the Failing Economy*

Currently the social contract is being eroded, in order to prop up austerity measures in a vain attempt to restore the capitalist economic order that depends on profits at the expense of the environment and the social fabric.

Systemic needs and the social contract
The move has been from away from systemic needs towards less enlightened thinking, less innovation and more neoliberalism, increased control and more authoritarianism.

In order to re-frame the current markets we need to value flows of energy by supporting relationships that foster socio-environmental wellbeing. This requires time and a change of pace. The response to community interests requires engagement facilitated by the policy detailed by the Aarhus convention that gives rights to people to access information and to be heard, it shifts a responsibility back to people who have always valued the global commons,unlike the argument developed by Hardin (Shiva 2002: 28):

> Corporations and laws such as TRIPS and penalties for polluters or trading permits for polluters has shifted the sacred integrity of the relationship between human beings as stewards, to commodifying the commons as a resource, ironically resource actual means regenerate and the extraction of profit runs the risk of bringing existential risk closer to a reality…as stressed by IPCC reports that show the 6 degree Celsius rise in temperature is inevitable, given the move away from reducing emissions as governments focus on short term austerity measures to try to restart an economy that depends on extraction of time and surplus value.

The challenge is to address the current erosion of the social contract. But the current mainstream labour agendas for social democracy can achieve little unless nation states develop regional shard goals.

The problem is that regional initiatives like the EU are under threat because they are based on the wrong kind of economics. States will remain in crisis until a new way of valuing social and environmental dimensions is achieved, rather than emphasising profit at the expense of the environment. This can only occur through new scaled up forms of accounting and accountability that are driven by monitory democracy and a scaled up form of governance. This is what we are trying to achieve through our research.

The post-national agenda does not need to undermine strong states; it could build on and extend a social contract, in the way suggested by Habermas. It could provide a way forward to address challenges of food, energy and water security. The UN is not the organisation for republican government according to Habermas (2008: 340) Instead need regional governments that recognise that overdoing concerns of managing the global commons require specific interventions to balance individual and collective needs. This is what the governance architecture that I am trialling attempts to do. It is not based on top down invention; instead it is asking people to reflect on their own lives and to try to make changes through self-regulation, in order that their rights and their responsibilities are secured to the event that they do not undermine the rights of others. Geertz (2000: 65) was only half right when he stressed the importance of relativism, in which he suggests: 'Indeed if one cannot face difference one should stay at home'. But now home is diverse and nation states are becoming increasingly diverse. So the challenge is how to balance individual and collective needs.

3.9.2 Intimidation and Silence of the Media: Maintaining the Status Quo and Business as Usual

Whistleblowing in the public sector is protected by legislation (that differs across the states in Australia) and that is regarded with coolness and some ambivalence according to Uhr (2005) in his work 'Terms of Trust'. The motivations of the whistleblowers is often questioned along the lines of personality differences, conflicts of interest or simple revenge. But nevertheless according to the *House of Representatives Report* (2012): roughly one in three employees in the workplace has experienced some form of bullying or poor behaviour and one in two has witnessed it'. Bullying is a synecdoche of the challenges facing public administration and policymaking. The 'global panopticon' is associated with the unrestricted power of surveillance 'from above'. A case is made for co-determination through architectures of cosmopolitan governance guided by:

1. The principal of subsidiarity and
2. The axiom that democracy and sustainability need to be supported to the extent that they do not undermine the rights of others or future generations of life.

Not only quality of life but the very existence of some disappearing states' (such as Tuvalu) are at risk.

By closing our boundaries to the displaced fleeing social and environmental disaster or locating them on islands in the Pacific, Australia (along with other nations) is contributing to creating so-called 'penal states'.

Facing up to an Unequal World has praxis implications for sociologists. The current treatment of asylum seekers and refugees by Australia has been raised by Justice Julian Burnside and by the independent Andrew Wilkie. The tow back policy is at risk of becoming a bipartisan position which entrenches cruelty by both major parties.

> The International Criminal Court (ICC) in The Hague is currently considering a submission calling for an investigation into Australia's treatment of asylum seekers. The submission was officially accepted by the ICC on May 19, 2014, it names senior members of government.
>
> Kon Karapanagiotidis told *No Fibs* in a statement today[105]:
>
> Under the Abbott Government we have seen for the first time the average time in detention surpass 300 days. The loss of access to legal representation while in detention, a spike in

[105]http://www.asrc.org.au OAM (@Kon__K) from the Asylum Seeker Resource Centre.

reports of self-harm and increased intimidation and pressure to depart Australia while in detention.[106]

The previous Abbot government and the current government address asylum seekers through silent control and a lack of engagement with the media. Listening and dialogue are impossible when the approach is to silence discussion and block access to information. Over reliance on the good name achieved through good neigbourliness during previous disasters and not responding to ongoing flows of people from natural and social disasters. The global refugee figure has exceeded '50 million for first time since second world war', according to Guterres (2014)[107] and yet the needs of the displaced are not addressed through the current architectures of democracy. Instead, protecting boundaries has led to penal states being constructed off shore to act as processing stations for asylum seekers from vulnerable or disappearing states affected by social and natural disasters. The surveillance and drone approach to USA and the fortress Europe approach are allied with the approach to national security followed by Australia on displaced people asylum seekers.[108]

This section makes a plea for a different form of governance that weighs up the social, economic and environmental indicators of wellbeing, in order to enable

---

[106]'No Fibs has now seen the full submission, however, after advice from experienced investigators, we decided against publishing the document. The advice was publication could harm any possible future investigation. We have published the executive summary …The submission was prepared by lawyer and migration agent. The submission calls on ICC prosecutors to use Article 17(2) of the Rome Statute in relation to all onshore and offshore processing. It alleges the Australian government is committing atrocities that are in breach of Article 7 of the convention…'.

[107]http://www.theguardian.com/world/video/2014/jun/20/global-refugee-figure-passes-50m-un-guterres-video.

[108]http://www.theguardian.com/world/2014/jun/05/the-life-and-awful-death-of-a-tamil-asylum-seeker-in-australia. Tamils who escaped Sri Lanka sought refuge in India. Some are en route to Australia as reported by the ABC (2/July 2014). The government refuses to acknowledge this. Complete silence is the approach taken. The last election was won on the basis of 'stop the boats'. It is a perceived badge of pride that since the election the boats have indeed 'stopped'. The suicides and attempted self-immolation reported recently in the Guardian by Tamils fearing being returned are indicative of the desperation of asylum seekers who are not given refuge. The riots on Manus Island PNG Off shore processing in states whose economy is reliant on Aid could lead to the development of oppressive camps along the lines of convict colonies. The propensity for riots has already been demonstrated along with the reaction of increased control and increased distancing of people as 'other'. Off shore processing of asylum seekers in Nauru and PNG by Australia have led to deaths that could have been prevented. The whistle blower was a labour party candidate. The Commission has raised concern about third country processing in Nauru and Manus Island. The Commission is of the view that people who arrive here seeking asylum should have their claims assessed in Australia by the Australian Government, and we should provide protection to those found to be refugees. See Commission's submission to the Parliamentary Joint Committee on Human Rights examination of the Migration (Regional Processing) package of legislation: http://www.humanrights.gov.au/submissions/examination-migration-regional-processing-package-legislation. 'The Commission has also recently made a submission to the Senate Legal and Constitutional Affairs Committee Inquiry into the incident at the Manus Island Detention Centre from 16 February to 18 February 2014: http://www.humanrights.gov.au/submissions/inquiry-incident-manus-island-detention-centre-16-february-18-february-2014 ….'

equitable distribution of resources and to ensure that some are not living at the expense of others and future generations of life. It moves beyond a critique to suggesting an alternative form of governance and democracy that spans communities of interest at a planetary level.

3.9.3 Held's Global Covenant

Held's proposal[109] provides a macro level approach to protecting the global commons based on social democratic control to protect the common good and the global commons. ***The will*** to make a difference is the challenge (Held 2005: 33–34):

> We may lack the will but it cannot be said that we lack the means. …What do we require to make a substantial difference to the basic wellbeing of the world's poorest? Again the statistics are available…Required would be 6 billion per annum on basic education, 9 billion per annum for water and sanitation, 12 billion per annum for the reproductive health of women, 13 billion per annum for basic health nutrition…These figures are substantial, but when judged against the major consumption expenditure in the US and EU they are not beyond our reach. Moreover if all the OECD agricultural subsidies were removed and spent on the world's poorest peoples this would release some 300 billion per annum ….In addition a small shift between military and aid budgets- 900 billion and 50 billion a year globally would make a marked difference to the human security agenda….

[109]'Cosmopolitan democracy and governance requires a multilevel and multiagency approach within and across overlapping regions'. We need to work within a common framework for co-determination that operationalizes a form of Global Covenant that protects social and environmental justice through monitoring social, economic and environmental indicators of wellbeing in a responsive and contextual manner. Ensuring that extreme luxury is fined/taxed at levels that not only acknowledge the costs to the environment and future generation, but criminalise through supporting whistle blowers through international law and systemic governance to ensure social and environmental justice. The social changes that are easy, such as the low hanging fruit of recycling, repacking and reusing along with promoting vegetarianism, respect for diversity and approving marriage equality for all needs to be considered in nation states that are able to legalise these policies without too much cultural opposition. This would help to buy time for more wide-ranging efforts to lower our carbon footprint. The argument that is developed is that protection of the global commons and human security needs to be protected through an extended International Court of Justice, in other words monitoring from above as well as by monitoring 'from below' by means of a scaled up version of the Aarhus Convention and Tuvalu test to ensure that people do not live at the expense of others or future generations of life. Current institutional and constitutional responses are inadequate to address the needs of people living in disappearing and unlivable states. The current approach to processing of asylum seekers who wish to be recognised as refugees because they are in need of protection is raised as a relevant international concern that needs to be addressed through a new approach to co-determination of protection of food, energy and water resources locally, nationally and in post-national states.

> The Sydney Peace Prize winner Vandana Shiva (2002) sums up the challenge for this century as one of preventing the commodification of life and stresses the need to learn from nature and across disciplinary specialisations.
>
> The ability to see the big picture should be encouraged ranging from social thinkers to physicists:
>
> all the indications are that the universe is at its simplest at the smallest and the largest scales……One can draw further parallels with the selfish, individualistic behaviours that are often the root cause of our environmental and financial crises. 'Within physics I see the idea of a "multiverse" as a similarly fragmented perspective, representing a loss of confidence in the prospects for basic science. Yet I believe all of these crises will ultimately be helpful if they force us, like the quantum physicist, to remake our world in more basic and far sighted ways….If we can only link our intelligence to our hearts, the doors are wide open to a brighter future, to a more unified planet…to quantum technologies that extend our perception…. (Turok 2012: 256–257)

Multinational companies and the WTO have made it possible to patent the conditions of life thereby causing almost a quarter of a million suicides by Indian farmers who cannot afford to purchase the seeds and the pesticides that are needed (despite the claims by the manufacturers). Furthermore the attempt to criminalise farmers who store old varieties of seed could lead to very vulnerable monocultures of foods. The idea that the very basis for life-seeds and genes can be patented is part of the process of commodifying people, animals and the fabric of life.

The open democracy forum was the starting point for research in response to David Held's proposal that the core challenge of the day is to address the vast differences in standard of living between rich and poor through reforming world trade and using the market to address global warming by means of a global covenant that prevents unilateral interventions. In response to his critics he claims that the resources exist for a reformed United Nations to make a difference, if there was a will. The UN according to Scruton (2005: 47–48) is corrupt, the bureaucrats are overpaid and he argues that national sovereignty is the only way to make a difference through controlling corruption.

The market failures Held anticipated fall short of what has actually occurred. The problem is not only concerning externalities that are not factored into calculations of the degradation to the environment, it also involves shifting the extraction of profit to where labour is cheaper and where governments and citizens are less likely to complain about degradation of environment and short-term profits at the expense of future generations. This undermines—not only the conditions of employment so that full time employment becomes less available—but also the very fabric of life on which all economic development depends. This is aided by the lack of control of the market and the movement funds (see McIntyre-Mills 2011, ISA paper).

3.9.4 Local Agenda 21 and Aarhus Convention for Scaling Up Principle of Subsidiarity and Testing Out Ashby's Rule of Requisite Variety

The application of policies such as the UN Local Agenda 21 and the Aarhus convention provide potential for a scaled up approach to managing resources through considering what people have, what they need and what they are prepared to add or discard from their lives in order to use resources carefully and to feed forward resources for future sharing. This works along the lines of the veil of ignorance approach underpinning the social contact but it is expanded to the entire human population on the planet. As we do not know where the next disaster will be located within the region, human security and hospitality is extended across the region. Resources need to be managed carefully so that resource rich areas share with those in need. Wellbeing stocks need to be stored but shared beyond contained areas or limited interest groups. They need to be shared according to a system of accountability that measures resources in terms of a planetary passport and score card. This has implications for complex decisionmaking within and across regions.

People who are displaced need to be able to track and link with other people in their local area. The software could provide a means to do this by enabling displaced people to continue to link with people in their geographical space and with those with whom they connect beyond their immediate community. In other worlds it provides a means of tracking resource use and networks of people. Ensuring that extreme luxury is taxed at levels that acknowledge the costs to the environment and future generation and act as a deterrence against profligate consumption. Transparency and tracking could help to support international law and systemic governance to ensure social and environmental justice so that taxation is linked with a passport of consumption and assets.

The social changes that are easy, such as the low hanging fruit of recycling, repacking and reusing along with promoting vegetarianism, respect for diversity for all needs to be considered in nation states that are able to legalise these policies without too much cultural opposition. This would help to buy time for more wide-ranging efforts to lower our carbon footprint.

The argument that is developed is that protection of the global commons and human security needs to be protected through an extended International Court of Justice, in other words monitoring from above as well as by monitoring 'from below' by means of a scaled up version of the Aarhus Convention and Tuvalu test to ensure that people do not live at the expense of others or future generations of life.

Current institutional and constitutional responses are inadequate to address the needs of people living in disappearing and unlivable states. The current approach to processing of asylum seekers who wish to be recognised as refugees because they are in need of protection is raised as a relevant international concern that needs to be addressed through a new approach to co-determination of protection of food, energy and water resources locally, nationally and in post-national states.

3.10 From Stocks to Flows as a Synecdoche for Planetary Democracy and Distributive Governance

The water management approach is a synecdoche for the challenges for managing the global commons. The laws for water management highlight the challenges, but also the way forward for water management. I will discuss the four theories of water rights raised by Shiva (2002: 77):

1. The first lawis the Harmon Doctrine which protects the sovereignty of states and is derived from British private property law. Sharing of water for downstream users is only moderated by the suggestion that good neighbourliness is preferable.
2. The second approach is the natural water flow theory based on protecting the rights of up and down stream riparian users. The reintroduction of the hydro-electric power agenda on dams is relevant. The implications of the World Banks attitudes towards linking with market partners to provide water is increasingly problematic as the commons is threatened by the increased emphasis on the economic bottomland that creates scarcity and a market for selling water.
3. The third is the Helsinki principle[110] based on the equitable apportionment of water, but as Shiva (2002: 78) stresses: 'equitable utility is defined as the maximum benefit accruing to all the riparian states, in light of their differing economic and social needs'.
4. Community of interest is closely related to the Helsinki Rules on the Uses of the Waters of International Rivers, 'Each basin state is entitled, within its territory, to a reasonable and equitable share in the beneficial uses of the waters of an international drainage basin'. (Shiva 2002: 79)

The complexity of the use of flows, not stocks also needs to take into account community interests or values that are central to a sense of wellbeing Shiva (2002: 80). Hence the need to critically analyse the work by Stiglitz.

Raising the educational capabilities of marginalised women would be a step in the right direction and not disempowering them through multinational companies. Shiva on enhancing the strategic needs of women, not just their basic needs and giving women a voice in decisions at the domestic level, community level, government and the market. The knowledge that women farmers and as nurturers have developed over millennia we have swept aside as seeds are patented. TRIPS or trade related intellectual property rights have to be contested by empowering networks of local people to hold corporations and their political supporters to account. Ann Florini has stressed in a series of books the importance of empowering people from below. Demographic transition through enhancing capabilities and empowering the strategic rights of women. Through empowering women and accepting

[110]http://w0ww.mpil.de/shared/data/pdf/pdf/8helsinki_rules_on_the_waters_of_international_rivers_ila.pdf downloaded 19 May.

diversity within human relationships we could make some steps towards achieving a demographic transition.

I am also inspired by the work of Haraway, because she stresses that women must not be subject to the design of others, they need to be designers of their own future. It is for this reason that we have worked on technology for monitory democracy and governance to try to give a voice to people and to empower them at the local level. The liberation potential of the Aarhus convention to enable local people to have access to information and the right to speak out and to be heard on environmental issues provide a means to scale up the principle of subsidiarity. If this is combined with a revised Triple Bottom Line Approach that develops social, economic and environmental wellbeing indicators to ensure quality of life within and across national boundaries, then policy could move closer towards addressing social and environmental justice for current and future generations of life.

Development need not modernise the slow, locally based lives of people defined as poor. Poverty is not having time and social relationships. It is not knowing your neighbour and distrusting the quality of the food that we eat. It is competing for a step up the run of status, instead of trying to live lives that support a sense of well being.

It is a well-known historical fact that competition for scarce resources should not be underestimated. The current structure of the global economies supported by competitive nation states pose a challenge to food security and to peace, because the carbon economy continues to grow.[111] Le monde[112] reported that China has questioned Japanese rule of Okinawa which is a USA Military base and: 'the calls over Okinawa may well revive tensions between Beijing and Tokyo, already embroiled in a territorial dispute over the Senkaku Islands, an archipelago in the south China Sea known by China as the Diaoyu islands'.

3.10.1 Water Management and Global Commons Management

Cosmopolitan politics requires a multilevel and multiagency approach within and across overlapping regions. De Gales key note speech stressed the importance of avoiding blue print approaches and working in a responsive and contextual manner, but without agreement on basic principles and shared values, the freedom to work creatively in a free society will be curtailed. The challenge for Europe is to balance this diversity locally. The regional spaces used in regional meetings by world

[111]Carrington, D. 'Global carbon dioxide in atmosphere passes milestone'. Guardian Weekly. 17.05.13, p. 13.

The United Nations Summit in 2015 is the deadline for a 'binding international treaty to curb emissions'.

[112]Le Monde China urged to 'reconsider Japanese role of Okinawa? Cited by Guardian Weekly International News. 17.05.13, p. 10.

leaders, such as the G8 or G20 whilst relevant for those at the meetings do not enable local people to engage in dialogue which is so important to enable a change in thinking and practice. Electorates need to be in touch with the governments that serve them. Unless a seamless set of conversations from local to regional to global occurs through face to face and on line dialogues, that an foster a sense of solidarity.

The solidarity of African unions, European Union and Asian groupings need to achieve cross cutting discussions that focus on food, energy and water.

A case in point is the ongoing tension within the region over fishing and resources in contested territories within the region. These are points of tension that test the nerve and power of USA, China, Vietnam, Japan and Korea. http://www.bbc.co.uk/news/world-europe-22744728.

The UN is not the organisation for republican government according to Habermas (2005: 340). Instead regional governments need to recognise that managing the global commons requires specific interventions to balance individual and collective needs. This is what the governance architecture that I am trialling attempts to do. It is not based on top down invention; instead it prefigures a way to enable people to reflect on their own lives and to try to make changes through self-regulation, in order that interests are not achieved at the expense of others.[113]

Food and energy are the bases for human wellbeing and security. Current approaches to international relations and war (Etzioni 2004, 2011) are unsustainable. As the experience of suffering becomes more widespread the notion of class will become central once again, because 'the poorest people in the world will be the hardest hit' (Beck 2009: 37).

Water management as a synecdoche for new architectures and processes for governance and democracy and it has implications of building *stocks for the future* (Stiglitz et al. 2010) and re-framing not only economics but our relationships and understanding the systemic feedback of poverty and carbon emissions based on Max-Neef Human Sustainability Index applied to valuing, measuring and managing environmental and human assets. The index that supports the engagement process to help people think about their wellbeing is called the *Being, having, doing and*

[113]The praxis on which my research is based distances itself from the game theory approach based on the zero sum approach. Game theory is promoted through the sovereign states. But if sovereignty was softened by a recognition of overlapping regional interests and a recognition of reframed relationships across state market and civil society, through an economics that stewards resources for future generations, then this expanded form of pragmatism, based on testing out the social, economic and environmental implications of choices for this generation and the next, could become the new basis for governance, democracy and ethics that are required to move the vision from us/them to a recognition that the effects of profit are the erosion of both the environment and social relationships. The austerity measures to prop up business as usual, based on saving on welfare and saving on environmental protection will lead to a descent into chaos. We need to understand that by identifying with others and with biodiversity (Lever-Tracey 2011)—we extend the circle of those with whom we feel solidarity and compassion and redefine our identity without hubris—then we may be able to transform the culture (Hulme 2010, 2011) which shapes the planet on which we depend. This requires some space for diversity and freedom and a 'sense of irony' (Rorty 1989).

interacting Index. It enables people to think about the implications of what we have, what we need, what we are prepared to add or discard from our lives and the turning points for the better or worse in social, economic and environmental terms related to the UN Sustainable Development Goals.

These indicators are assessed by the interactive, web based Resilience Score card. The research explores whether engagement in thinking through 'if then scenarios' could aid discursive dialogue within communities of practice to support post materialism and wellbeing.

Chapter 4
Political Freedom Versus Democracy

4.1 Introduction

In South Africa and in particular in the Western Cape hunger is a daily challenge. Although the Africa Recovery report is dated as 2004 it is worth citing as it is even more applicable in 2015 as the current drought in South Africa impacts crops and food security. I start by citing this report to demonstrate that the issue raised in the current report by Oxfam on hunger stresses that the situation is worsening not improving due to drought and the weakness of agricultural workers.[1]

> A third successive year of drought has devastated the harvest in Southern Africa and will leave millions of people at risk of hunger over the coming year, the World Food Programme (WFP) predicts. Two of the six countries most affected, Lesotho and Swaziland, have already declared national emergencies, while estimates of the number of people in need of food aid in Zimbabwe alone exceed 7.5 million – nearly 60% of that country's total population.... The causes - drought, the late arrival of the seasonal rains and a severe shortage of seed, fertilizer and other inputs - have been exacerbated by the HIV/AIDS pandemic. The disease has devastated the agricultural workforce, weakened community coping mechanisms and left tens of thousands of families too poor to purchase food supplies.
>
> South Africa is supposedly a 'food-secure' nation, producing enough calories to adequately feed every one of its 53 million citizens. The reality is that, despite some progress since the birth of democracy in the country in 1994, one in four people currently suffers hunger on a regular basis and more than half of the population live in such precarious circumstances that

[1] "...A reduced harvest in drought-stricken South Africa, the region's breadbasket, is expected to worsen the funding crunch by driving up the price of the staple food, maize, and forcing WFP to rely on costly imports" Africa Recovery, Vol. 18, No 1 (April 2004), Watch page, Southern Africa: Another year of hunger Website: www.africarecovery.org Contact us by email: africa_recovery@un.org.

J. McIntyre-Mills, *Planetary Passport*, Contemporary Systems Thinking,
DOI 10.1007/978-3-319-58011-1_4

they are at risk of going hungry. The South African National Health and Nutrition Examination Survey (SANHANES) revealed in 2013 that 26% of the population was actually facing hunger and a further 28% were at risk[2]

I begin the chapter with this example because it underlines the concerns associated with lack of quality of life and the lack of life chances that will face those living in poverty in urban areas. The need to build capacity is vital and it is a starting point for Neil Turok[3] who said that instead of beginning with a sense of hopelessness, he believes in creating opportunities.

In South Africa I spent time with staff at the University of South Africa where they stressed that unemployment and appropriate educational pathways that recognise the skills of people who have no formal education (but plenty of life skills in survival) are urgently in need of recognition as a starting point for entry to further education. They stressed that they did not think about environmental issues per se as a primary concern, but were aware that hunger was an issue for many who are unemployed in the cities. This point was also by the UCT food security network on the need to address the challenges in cities that could become food deserts and the need for both urban and rural agricultural development opportunities.

The Food Security Network[4] has concluded that the reliance on growing food in urban areas whilst a step in the right direction is not enough to address the needs of an increasingly urbanized population in South Africa. Instead the emphasis needs to be on supporting more agricultural development and protecting arable land. The UCT food security network has published papers on the need to address the challenges in cities that could become food deserts. Competition for resources has led to a proliferation of gangs who traffic crack and maintain power through intimidation of local residents. In Masiphumelele the local residents have taken the law into their own hands, because the extent of the problem has not been effectively prevented or policed, despite the efforts of a nearby NGO and by the efforts to develop community policing.

The challenge of meeting the needs of the hungry are visible on a daily basis as people pick through bins and as the struggle for survival is reflected in the levels of crime. Forty-two murders occur a day in South Africa and this has spiked in the past year to 49 per day.[5] According to Smith (2015):

[2]Hidden Hunger in South Africa 2015 report www.oxfam.org/grow cites Community Childhood Hunger Identification Project (CCHIP). Hunger (food insecurity) was assessed by means of the Community Childhood Hunger Identification Project (CCHIP).

[3]http://m.youtube.com/watch?v=5nN8ep2mJpU. He has set up a school to teach mathematics and physics for the brightest African students in Cape Town. He hopes to create opportunities for the next Einstein to be from Africa. But between reaching out to the best and brightest and those without hope, there remains an ever widening gap created by the current social, economic and environmental system.

[4]See Crush and Fayne 2010, Battersby, and Crush 2014, Frayne, B and McCordic, C. and Shilomboleni 2014.

[5]http://www.theguardian.com/world/2015/sep/29/south-africa-a-country-at-war-as-rate-soars-to-nearly-49-a-day.

> The stark figures threaten to undo work done in recent years to overturn South Africa's reputation for violent crime that deterred foreign tourists and drove middle class emigration. They also left politicians and academics trying to understand why the country is bucking global trends, which have seen homicides drop to record lows in cities such as London and New York. South Africa's murder rate of 33 per 100,000 is more than five times higher than the 2013 worldwide average of 6.2 per 100,000. The murder rate among young black men in some communities is 300 to 400 per 100,000.
>
> But the answer, some believe, is unusually straightforward:
>
> "The single most important reason is because we have not cracked inequality as a problem," said Eusebius McKaiser, an author and newspaper columnist. "We've known for years there is a reliable correlation between gratuitously violent crime and inequality. ... South Africa... is among the most unequal countries in the world... Despite the advent of democracy, a 2011 census found that the average annual income of white households was six times higher than that of black households. Corporate boardrooms and rich lists remain dominated by white faces. Ostentatious wealth and brutal poverty often sit cheek-by-jowl in the geography of what has been called "post-apartheid apartheid".[6]

Leadership on saying no to violence is a focus in South Africa through reducing inequality not merely policing the effects. In Australia one in six women has been affected by violence and this has resulted in a safety package to help women stop the violence. But the approach has been criticised because more needs to be done to

[6]http://www.theguardian.com/world/2015/oct/01/south-africa-violent-crime-murders-increase-inequality.

prevent physical violence by perpetrators[7] in ways that recognise the complexity of social, cultural, economic factors as well as the challenges posed by gender relations. It is undeniable that *both* men and women can be perpetrators of violence, even if the latter are more vulnerable.

The anniversary of Marikane[8] was on 15th August, 2014. This anniversary provides a point at which we need to pause to consider the state of democracy and governance in South Africa. According to Xalabile (2014):

> [D]uring a recent march to the Ntabankulu municipal offices to protest against inadequate service delivery officials told protesters that 'marches don't help anyone because when you march another Marikane will happen.

His report reflects that although 'South Africa is the beacon of hope for the world' and that 'truth and reconciliation are possible' and that 'the spirit of Ubuntu' or being a person through others can 'triumph over Apartheid' it also 'marks a point' where South Africa needs to 'learn from the risks' through openness and transparency.[9]

It is estimated that 65% of the population live in cities. South Africa is becoming increasingly urbanised[10] and this poses a threat for food security (Battersby and Crush 2014; Crush and Fayne 2010; Frayne et al. 2014). The future for South Africa lies in developing learning organisations and learning communities where

---

[7]"http://www.malcolmturnbull.com.au/media/release-womens-safety-package-to-stoptheviolence: Immediate practical actions to keep women safe include: "$12 million to trial with states the use of innovative technology to keep women safe (such as GPS trackers for perpetrators), with funding to be matched by states and territories. $5 million for safer technology, including working with telecommunications companies to distribute safe phones to women, and with the eSafety Commissioner to develop a resource package about online safety for women, including for women from CALD communities. $17 million to keep women safe in their homes by expanding successful initiatives like the Safer in the Home programme to install CCTV cameras and other safety equipment, and a grant to the Salvation Army to work with security experts to conduct risk assessments on victim's homes, help change their locks and scan for bugs. $5 million to expand 1800RESPECT, the national telephone and online counselling and information service, to ensure more women can get support. $2 million increased funding for Men's Line for tools and resources to support perpetrators not to reoffend. Up to $15 million to enable police in Qld to better respond to domestic violence in remote communities and for measures that reduce reoffending by Indigenous perpetrators. $3.6 million for the Cross Border Domestic Violence Intelligence Desk to share information on victims and perpetrators who move around the cross border region of WA, SA and the NT."

[8]An uprising in South Africa by mine workers concerning their pay and conditions which resulted in the death of 44 mineworkers on 15th August 2012.

[9]http://marikana.mg.co.za/#families_xalabile Mail and Guardian accessed 13/08/2014.

[10]Urbanization poses a systemic threat to quality of life and has implications for policy. Food deserts are the likely scenario if more emphasis is not placed on balance, greening cities and supporting small farmers. It has been wrongly assumed that growth in the economy will sustain a growing population. During my sabbatical I attended workshops or held conversations with colleagues at Living Hope, Embrace Dignity and Africa Tikkum in Cape Town. The concerns they raised were for food security and how it relates to educational and employment opportunities for all especially young people.

people feel that they have a voice and a right to a voice and where they do not only participate in vertical democracy through elections that vote in elites who lose connections with the people they are supposed to represent. So the future lies in participatory democracy and more direct interventions:

> Participation is the cardinal principle of democracy—not only because of its intrinsic value, but also because it increases the political efficacy of citizens by giving them direct training in the policies and tools of governance. Almost 200 years ago, John Stuart Mill suggested that this kind of democratic training is best obtainable at the local level, where citizens can make decisions about issues they can immediately relate to, and then generalize that knowledge to the broader, national political system. (Mangcu 2014: 31).

What is needed in South Africa is a greater emphasis on core principles and engagement with the people in discursive democracy, if not direct voting as suggested by Mangcu (op. cit).

4.2 Representation and Accountability in South Africa: Freedom Cannot Exist at the Expense of Democracy or Vice Versa

The current anniversary of Marikane provides a point at which we need to pause to consider the state of democracy and governance in South Africa (Mail and Guardian report).[11] On the one hand, South Africa is the beacon of hope for the world that 'truth and reconciliation' are possible and that the spirit of 'ubuntu' or 'being a person through others' can triumph over Apartheid. On the other hand, the reality is that Marikane marks a point where South Africa can either learn from the risks associated with power and remedy the situation through an open and transparent 'calling to account' of all those directly and indirectly to blame. This involves acknowledging the so-called 'banality of evil' (Arendt 1963) and that wide ranging challenges that need to be addressed. Systemic problems will continue to lead to social, economic and environmental injustice unless unemployment and poverty are recognised as priorities that need to be addressed through capacity building, pathways to training and employment supported across the public, private and non-government sectors.

The prospect of 65% of South African's living in cities has implications for food, energy and water security. This requires addressing sustainability at the local level in municipalities. For this to occur people need to participate in local municipalities. South Africa is food insecure. South Africa imports more food than it exports, which effectively mean (sic) that with the food produced in South Africa only, the nation cannot feed itself.

[11]http://marikana.mg.co.za/#families_xalabile Mail and Guardian accessed 13/08/2014.

A recently published Human Sciences Research Council Report shows that only:

> 45.6% of the population are food secure, 28.3% are at risk of hunger and 26.0% experienced hunger (were food insecure). The largest percentage of participants who experienced hunger (food insecurity) was in urban informal (32.4%) and rural formal (37.0%) localities (HSRC, 2013)…

Firstly, decision makers ought to be above the taint of corruption, collusion and nepotism. The most obvious being that board members of companies should not also be able to use political power to foster their interests that can be seen to be at the expense of others. Secondly, people feel that they have a right to a voice and that it will be heard. For democracy to exist there needs to be opposition parties and civil society groups who believe that people can voice protest in a frank and fearless manner. But these protests should also respect the axiom that we can be free and diverse to the extent that our freedom and diversity does not undermine the rights of others in this generation or the next.

This means that education for the future and opportunities for the young are important. It also means that instead of placing growth only in cities, we need to develop an insight into the importance of protecting agricultural land (SAPA, see: unicef| johannesburg http://www.news24.com/Tags/Places/johannesburg).[12]

The prospect of 65% of South African's living in cities has implications for food, energy and water security. This requires addressing sustainability at the local level in municipalities. For this to occur people need to have a voice and feel that they have a right to voice their ideas openly. To cite the Mail and Guardian report.[13] In an interview with Peter Alexander (8 January 2013) in 'International Socialism' he stressed that nationalism was emphasised as an option by young people who saw no future for themselves and that unless an effort was made to address the current levels of unemployment and poverty that this could be regarded as an option by this section of the electorate. In terms of the size of the massacre he stressed that only the Soweto massacre in 1976 is comparable, namely 34. At that time the murders

[12]"According to Unicef's "Generation 2030 Africa" report, next year, out of South Africa's projected population of 53 million people, 18 million of those would be under the age of 18. This is reported in the following article on 24.com/Web/News24/."

"Over one third of South Africa's population is expected to be under the age of 18" in 2015, according to a United Nations Children's Fund (Unicef) report released on Tuesday. "South Africa was also expected to have 65% of its population living in urban areas next year, the ninth highest level in Africa. According to the report, in 2050 around 41% of all births world-wide would take place in Africa, while in the same year 25 people out of every 100 would be African. This was against the expected figures in 2015, where Africans would make up 16 people out of every 100 around the world. In 2015, 40% of Africa's population was expected to be living in cities, versus over 50% in 2050" News24/SouthAfrica/Articles.

[13]http://marikana.mg.co.za/#families_xalabile Mail and Guardian accessed 13/08/2014.

the previous week brought it to 44.[14] But it is fair to say that who the perpetrators are is unclear[15]:

> According to News 24 "F34 were killed in a clash, over 70 wounded and over 250 arrested …and in the preceding week 10 people, including two police men and two Longman security guards were arrested.[16]

As yet those responsible for the massacre have not been brought to justice. The events of Marikane have been likened with the massacres of the past, such as Sharpville, The Marikane Commission of Inquiry (chaired by retired Judge Farlam and held in the Tshwane Council) has for answers.[17]

The issue raised by socialist political commenters[18] is that business interests were at stake because Ramaphosa was a board member on the mine where strikes were occurring. Ramaphosa contended that in fact losses were already extreme and that he had no intention of trying to stem losses by controlling the strikes. The issue was to control the violence.

The root of the strikes however needs to be considered as being far deeper than a single incident of unrest. The garnishing of wages by workers who were in debt was another factor that has already been raised as relevant to the issue as detailed in McIntyre-Mills 2014). Mineworkers wanted higher wages because they could not make ends meet and many were severely in debt to hire purchase companies that sold items to the workers and ensured that payment was deducted directly from

---

[14]The Marikane Anniversary of the massacre of 44 mineworkers on 15th August 2012 is on 15th August 2014.

[15]This is the second anniversary. http://www.socialistworld.net/print/6442, the committee of workers international CWI argues (see Socialist news http://www.socialistworld.net/print/6442: "On August 16, the anniversary of the police massacre of striking workers at the Lonmin platinum mine in Marikana, over 5000 mineworkers gathered on 'the mountain' where their comrades had been shot down one year earlier. Democratic Socialist Movement (CWI South Africa) reports that the massacre was the most significant political event of the post-Apartheid period. "The hardest thing is that I don't know why I survived… when nothing has changed", said one of the survivors who was one in the group that was mowed down in front of TV cameras. He was shot with 14 bullets and lay in a coma for two months…. Family members of the murdered workers spoke of their loss and pain. The Lonmin workers' strike leaders spoke of the struggle they still face as the company tries to take back the unprecedented wage increases it conceded a month after the massacre.". … Not only were the NUM and all other ANC-aligned structures absent from the commemoration, but, against a wave of sympathy across the country for the plight of the workers and indignation at the refusal to fund the workers legal representatives, the ANC head office was compelled to condemn the ANC North West for their insensitive remarks. The NUM itself has just lost a long battle it fought by means of fraud, intimidation and court action to retain its recognition…"

[16]http://www.news24.com/SouthAfrica/News/Ramaphosa-is-a-murderer-protesters-20140812.

[17]http://www.news24.com/SouthAfrica/News/Ramaphosa-is-a-murderer-protesters-20140812.

[18]www.socialistworld.net, 23/08/2013 website of the committee for a workers' international, CWI, South Africa, Marikana: workers remember the massacre a year ago. www.socialistworld.net, 23/08/2013 website of the committee for a workers' international, CWI.

their wages. In some instances workers were incorrectly charged as they did not have the necessary literacy or numeracy skills to ensure a fair transaction. In terms of the perceptions of people who wish to protest over injustice, the sense is that marches may lead to similar shootings:

> Comparing the Marikana massacre with killings under apartheid, Sibusiso says: "I didn't think it would happen again, it surprised everyone. We are free, we didn't think it would happen again — for the police to shoot workers like that". http://marikana.mg.co.za/#families_xalabile
>
> And now Marikana is being used against everyone else, Lanford says. During a recent march to the Ntabankulu municipal offices to protest against inadequate service delivery, he says, officials told protesters that "marches don't help anyone because when you march another Marikana will happen". http://marikana.mg.co.za/#families_xalabile.

According to Molefe (2014) despite Zuma's political survival he will need to spend time addressing:

> A yawning gap between rich and poor, stratospheric levels of unemployment and millions working for less than what's required to provide for their basic needs. ....

Price (2012) stresses that youth unemployment in South Africa's 'ticking bomb.'[19]

4.3 Food Insecurity: Reflection on Current Areas of Concern in South Africa

Democracy is in need of constant regeneration. We need to remember that American democracy is based on a compromise to slave owners who were allowed to give count slaves as 3/5ths of one non slave or free man.[20]

Current attacks by the police are understood as the result of racial profiling that shapes their decisions in the split second to act out violence because of guilt and fear.

If democracy is about representation and accountability, then people need to have a voice, but the silence in the media and the silencing of those who are other is not only problematic, but historically familiar, if we consider the history of democracy. The conservative narrative of the presidential candidate Donald Trump contrasts with the hopeful narrative expressed in the 2016 state of the Union Address by Obama, who emphasized the need to get beyond the old approach to democracy based on points scoring and the need for bipartisan support on tax reform so that the 1% do not continue to benefit at the expense of the rest. But the response after the address was another statement from Republicans of the need to

---

[19]Price, C 2012 "http://za.effectivemeasure.net/em_image" alt = """ style ="position: absolute; left: -5px;".

[20]http://en.wikipedia.org/wiki/United_States_Constitution accessed 11 Feb 2015.

spend tax dollars carefully and for government not to waste it. This is a fair point—but the wastage has occurred by banks that used the bailout to support business as usual. It is worth noting the following themes:

That the United States ought to rise above petty differences to address big issues—to provide a future worth living for young people, to become more sensitive to gender, to avoid deporting young people, to prevent racism and to cherish civil liberties is indeed worth highlighting as a way to promote social justice. The focus of the rhetoric by Obama was on stocks for the future through addressing climate change and the challenge now is to address this in practice as a model for America to implement as a democracy that walks the talk.

The long walk to freedom in South Africa continues as is evident in Zuma's State of the Nation 2015 on the 11 February which is also the anniversary of Mandela's release from prison after 25 years.

How much has changed in South Africa? It is also the anniversary of District 6, the day that residents were forcibly removed from their homes.[21] They have not been able to return and a Technicon is being built on the site where they commemorate their removal each year. Some of the residents are in their 80's now and one said: "when will I get my house?"

The State of the Union address for 2012 and 2013 by Zuma[22] were not very well received because the challenges of poverty[23] and food insecurity have not been adequately addressed. The violence towards women and children did not receive adequate attention and the massacre at Marikane continues to cast a shadow over the presidency. But the focus of attention remains the way in which corruption at the highest levels (police and government) remains unchecked. The spending of about 23 million on the renovations of a rural home, Nkandla, has become a byword for the problems of the presidency. Speculation was rife at the time as to whether the speech will go ahead or whether it will be disrupted by the Economic Freedom Front, led by Maleme of the Economic freedom Front or 'Red Berets'. At that time the media speculated as to whether energy outages, 'known as planned and unplanned load shedding', could lead to the inability to broadcast the speech. The speech became secondary to the disruption caused by the attempt to jam the signals of journalists attending the speech and the filibustering in parliament by members of the EFF asking when the funds for Nkandla would be repaid. In fact Manansala had specified that only some of the funds needed to be repaid as they did not fall within the ambit for security upgrade of the residence. But EFF requested that the questions they had poses should be answered in the speech.

---

[21]http://www.iol.co.za/capetimes/emotional-commemoration-of-district-six-1.1816782#.VNyo_mIaySM http://www.iol.co.za/news/south-africa/western-cape/former-district-six-residents-honour-the-past-1.1817068#.VNyo2mIaySM.

[22]http://africacheck.org/reports/a-first-look-at-president-jacob-zumas-2014-state-of-the-nation-address/.

[23]http://africacheck.org/reports/deputy-presidents-speech-downplayed-the-levels-of-child-hunger-in-south-africa/.

Some of the members were ejected after the Speaker had warned several times. The Democratic Alliance requested that the signal be reinstated before they continue and that the speaker informs the house whether the security detail were in fact members of the police force. The threat to freedom of information was central to the analysis of the State of the Nation and the content of the speech on the need to address food security, energy and employment opportunities and gender issue although on track became largely secondary.

I would like to segue with a discussion of life chances in South Africa and how the current neo-liberal economy provides few options to the majority.

> The government will have to rethink its anti-poverty initiatives after … a new report showed that more people in South Africa are poor…. A rebasing of three national poverty indicators shows that more people are poor than previously thought. The proportion of the population who were deemed to be living in poverty. The reason is that the upper-bound poverty line, which measures the income needed for essential items after meeting their basic food needs, was recalculated from R620 a month to R799 per month… airtime, transport and energy costs, as well as changes in diet that are a result of increasing urbanization…. The rebased poverty line is 26.70 per person a day …
>
> The proportion of South Africans living in extreme poverty rose 1.5% points – from 20.2% of the population to 21.7%…. A substantial increase in social support grants is deemed overdue… likely to be resisted by business and conservative lobbies ….”[24]
>
> The fact that food deserts have not been explicitly identified or discussed in relation to African cities does not, of course, mean that they do not exist. African cities contain many poor neighbourhoods whose residents are far more food insecure and malnourished than their counterparts in the UK or North America. A baseline food security survey in 11 Southern African cities by the African Food Security Urban Network (AFSUN) for example found that 57% of households in poor neighbourhoods were severely food insecure and only 17% were completely food secure (Battersby and Crush 2012). On average food purchases made up to 50% of household expenditures. Dietary diversity was very low in all cities with heavy reliance on starch staples. …Supermarkets undermine local business (Ligthelm 2008).

Urban populations tend to rely on informal sector and small business if they survive, an indicator of food security is the level of employment and extent that they have support networks.

A current talk back program hosted a young woman who has written a best seller on how to become financially independent by managing money effectively. She stressed that she came from a family where her mother had survived by working hard as an office worker to assist educate her family of 3 after her husband died. She said that she remembered her mother stressing: work hard and then all will be well. But she became indebted, not realizing that money management was needed to avoid offers debt. Her own mother's greatest concern was that she would 'outlive her nest egg'. This working class attitude that hard work is enough does not equip people to avoid the temptations of borrowing. The concerns of those who are

---

[24]Musgrave, A. 2015 'More people living below breadline, says Stats SA' The Star, p. 4.

unemployed are to survive. Those without family to support them live on the street or beg at traffic lights.

> As the ruling ANC prepares for its big policy conference in June, it faces another challenge —a new alliance of radical trade unionists and grass roots activists…(Orderson 2015)[25] This is a socialist group founded in December 2014 comprising 400 people from 71 labour, religious and activist organizations:
>
> Karl Cloete, deputy general secretary of the National Union of Metal workers of South Africa (NUMSA) …: The UF is the birth of the new left, a movement that is anti-neoliberal and would certainly bring together sectors of society. This is a far reaching shift on the political landscape of South Africa. The report explains that NUMSA clashed with and was expelled from the Congress of South African Trade Unions (COSATU) in November 2014.

The split is understandable given the links between ANC and COSATU and the concerns that the interests of the very poor have not been addressed. But COSATU stresses that it supports the SACP and that splitting the trade union movement is not the best option for addressing the needs of the very poor.

The slogan of this new group is 'enough is enough'. It is a response to the lack of accountability and the poor level of service delivery to people in municipalities.

According to Prior (2015) democracy has declined as the neoliberal markets have begun to control the state.

Prior cites the article by Thabo Mbeki (2014) in 'The Thinker's' first quarter edition in which he critiques the extent to which democracy has been achieved in African states across Africa:

> State systems have been reduced to a patrimony of a predatory elite, controlled by its self-serving professional political class…thus…the putative democratic states become a social institution which serves the interests of a rent seeking elite whose goals amount to no more than preserving its political power and using its political power to extract rent which ensures its enrichment….
>
> Prior stresses that the problems of self-serving politicians is faced globally. The issue is that neo-liberalism does not sit well with democratic principles.

Mmusinyane (2015) cites Dickens from the Tale of Two Cities which sums up the difference in life chances between people living in South Africa:

> It was the best of times, it was the worst of times, it was the age of wisdom, it was the age of foolishness, it was the epoch of belief, it was the epoch of incredulity, it was the season of light, it was the season of darkness, it was the spring of hope, it was the winder of despair, we had everything before us…we had nothing before us[26]

In his article: 'Nurturing thinking and responsible citizens' Mmusinyane (2015) makes a case for more public or civic education to teach people about democracy. But the anger of the masses and the protests about services and the xenophobia against migrant workers and migrant shop keepers is not a matter of civic education,

---

[25]Orderson, C 2015 'The birth of the new left'. The Africa Report, Feb, 34–36.

[26]Mmusinyane, L 2015 'Nurturing thinking and responsible citizens' The Thinker volume 63 pg. 67.

it is about addressing rising levels of poverty. The education needed is for those elected to represent the masses to do so in a transparent and accountable manner. The comments made by the Minister for Small Business about the xenophobic attacks has led to complaints to the Human Rights Commissioner (2015)[27] blaming those whose entrepreneurial activities are successful does not provide an appropriate example for others in South Africa to follow.

The excesses of Nkandla combined with the problem of maintaining ESCOM provide an obvious example of the points made above by Mbeki. The economic fall out of the ESCOM crisis is likely to lead to innovation in developing the green economy, but in the short term disinvestment is likely to harm the economy.[28]

In South Africa the gap between rich and poor has widened. Basic needs (let alone strategic needs, meaning being given a voice) are not met. In this context it becomes increasingly important to find ways to move forward in a constructive and engaging way that gives opportunities to those who are (as yet) unheard. A credible option is 'engaging rivals to overcoming crises"[29] Frans Cronje identifies the following anomalies:

> 15,000 riot police employed by the state in the period 1995 to 2001
>
> About 7000 in 2002 and 2006 dropping to around 2500 between 2006 and 2010, 4500 in 2014
>
> South Africa has 18 million registered tax payers, only 5 million are liable to submit returns, and only 10 percent earn more than 700,000 per year.
>
> With only 5.5% of black households spending more than 10,000 per month, the risk of middle class size is overstated.
>
> …And young people are a key factor—their living standards are much higher than their parents 'they are active on social media, but if they do not have a job and see no realistic prospect of entering the economy, that's a problem….
>
> Approval of government has dropped from 72% in 2000 to 54% in 2014.

4.3.1 Vignette: Food Insecurity and the 'Bin Pickers' in Suburban Cape Town

On the mountainside of Fish Hoek the bins are put out weekly on the kerbside. The local residents have a Neighbourhood Watch system. Those who harvest food from the bins are called "the bin pickers". They are regarded with suspicion and

---

[27]http://www.news24.com/SouthAfrica/News/Complaint-lodged-over-ministers-foreigner-comment-20150203.

[28]http://www.iol.co.za/the-star/mugabe-the-ghost-of-africa-s-past-1.1812439#.VNeP12IayK0
http://justiceafrica.org/?p=1752 http://www.news24.com/SouthAfrica/News/SA-looks-to-Zuma-to-shed-light-on-Eskom-crisis-20150208.

[29]Morris, M 'Engaging rivals key to overcoming crises: with weak leadership, a worrying economic outlook and declining public trust, the time is ripe for constructive engagement, liberal analyst Frans Cronje tells Michael Morris' Feb 14th 2015.

annoyance, rather than with concern.[30] The complaint is that the bins are overturned and rubbish flies around in this windy spot. Residents are requested not to put their bins out until the night before they are collected by the Neighbourhood Watch.

The careful sorting of rubbish (in a bid to address climate change) has made it slightly easier for the homeless, destitute or those hiking past in search of jobs to find something to sustain themselves. On the occasions people are encountered as I went for an evening walk on the night before rubbish collection, a donation of food or money was greeted with surprised gratitude by these efficient, resilient survivors.

> Bottles are placed in separate transparent bags. Organic waste is placed separately.
>
> Why not place the left overs on tables rather than in bins? It would be more welcoming.
>
> The response was: We do not want to attract them.
> But they are walking past anyway, was my response.
> But old habits die hard and the destitute continue to grow in numbers.
> Apartheid is no longer merely race based, it is class based.

Mangcu (2014) stresses that the challenge facing South Africa is to remember that Freedom and Democracy need to be worked at. He draws on Isiah Berlin's "Two Concepts of Liberty" and stresses that fighting for freedom successfully does not make one a democratic.

4.4 A Case Study of Systemic, Integrated Care

Food security and human security are important issues that need to be addressed. Living Hope focuses on addressing the heavy burden on the urban poor. Living Hope has stepped up to the challenge of meeting the immediate and longer term

---

[30]In South Africa I spent time with staff at the University of South Africa where they stressed that unemployment and appropriate educational pathways that recognise the skills of people who have no formal education (but plenty of life skills in survival) are urgently in need of recognition as a starting point for entry to further education. They stressed that they did not think about environmental issues per se as a primary concern, but were aware that hunger was an issue for many who are unemployed in the cities. This point was echoed during my time in Cape Town, where for example the UCT food security network has published papers on the need to address the challenges in cities that could become food deserts. The challenge of meeting the needs of the hungry are visible on a daily basis as a person pick through bins and as the struggle for survival are reflected in the levels of crime. Forty-two murders occur a day in South Africa. I was spent time during my sabbatical at an NGO that provides support called: 'Living Hope' for those who live in 7 areas along the coast. One of these is Ocean View, where the twenty-five thousand people live in poverty and face a range of challenges that are reflected in the morbidity and mortality rates, meaning the illnesses of poverty such as TB which is endemic and linked with being HIV positive. The vulnerability of children was recognised by this NGO and provides a hot meal and sandwich with fruit to over 7500 children. This service does not go far enough and they are concerned that with rising numbers of people coming to the cities the challenge will become even greater. Remember this is just one part of the Western Cape and the problem is repeated over and over again across South Africa.

needs of people without access to food, energy and water and in many instances without hope. Living Hope is a faith based organisation that:

> undertakes community development inter-alia through educational, social and health related programs and plays a vital role in the prevention, care, treatment and support of people infected and affected by HIV and AIDS and other chronic illnesses.[31]

In 2001 Living Hope was founded in response to a story told to a Baptist Minister by a health care nurse, who said 44% of the people who came to her clinic were HIV positive and at risk. In fact the recorded statistic for that clinic was 16%. The statistic motivated a response to start a hospice. But no one in the community wished to admit to the stigma of HIV or AIDS. Denial, stigma and secrecy were the standard cultural responses to the illness at that time.

Through setting up community outreach projects in 7 communities including Capri, Ocean View, Masiphumulele, Redhill, Capricorn, Overcome Heights, Kraaifontein, Mfuleni, Fish Hoek the community began to hear about the NGO. There are also bases in East London and Zimbabwe. It began as the mission of one church and now employs over 226 staff including volunteers from the local Cape Town communities and young people from Netherlands, Australia and America. One of these areas is Ocean View, where the twenty-five thousand people living in poverty face a range of challenges that are reflected in the morbidity and mortality rates. TB is endemic and linked with being HIV positive.

The following section addresses a response to the systemic needs of those living in poverty. It outlines the role of an NGO in meeting not only basic but also strategic needs for development in one of the most socio-economically vulnerable communities in the Cape Province of Southern Africa. The quality of life of these communities would be negatively affected if this NGO did not continue to operate in the area. The volume provides an overview of the life chances in these communities and illustrates the way in which the NGO provides systemic integrated care that provides opportunities for developing their opportunities to achieve the social democratic outcomes that are promised in a New South Africa.

A systemic approach draws on the lived experience of both service users and providers in an effort to pool social, economic and environmental resources, create bonds of friendship and to bridge differences through establishing learning communities that foster wellbeing.

Living Hope has stepped up to the challenge of meeting the immediate and longer term needs of people without access to food, energy and water and in many instances without hope. Living Hope is a faith based organisation that:

> undertakes community development inter-alia through educational, social and health related programmes and plays a vital role in the prevention, care, treatment and support of people infected and affected by HIV and AIDS and other chronic illnesses.[32]

---

[31]Health District & Sub Health District, Living Hope falls under the Cape Metro Health District and specifically the Southern Sub-Health district.

[32]Health District & Sub Health District, Living Hope falls under the Cape Metro Health District and specifically the Southern Sub-Health district.

4.4.1 *Background*

In 2001 the NGO was founded in response to a story told to a Baptist Minister by a health care nurse, who said 44% of the people who came to her clinic were HIV positive and at risk. In fact the recorded statistic for that clinic was 16%. The statistic motivated a response to start a hospice. But no one in the community wished to admit to the stigma of HIV or AIDS. Denial, stigma and secrecy were the standard cultural responses to the illness at that time.

Through setting up community outreach projects in 7 communities including Capri, Ocean View, Masiphumulele, Redhill, Capricorn, Overcome Heights, Kraaifontein, Mfuleni, Fish Hoek the community began to hear about the NGO. There are also bases in East London and Zimbabwe. It began as the mission of one church and now employs over 226 staff including volunteers from the local Cape Town communities and young people from Netherlands, Australia and America.

4.4.2 *Rapid Appraisal Approach*

This a rapid appraisal approach drawing on the work of Robert Chambers of the Institute of Development Studies aimed to gain insights based on working with the participants during a week. It comprised participating at meetings and a training session with youth leaders, reviewing secondary data, a observing, conversations with staff and secondary data made available to funders and the public. Meetings with the manager, the director of operations, the promotions manager and the finance manager gave me access to published materials on the NGO. The programs that are available are:

4.4.2.1 Living Right Prevention and Life Skills

Living right prevention and life skills "reaches out to more than 6000 children every year through our various programs by teaching them the skills to make healthy choices for their lives. We also offer mobile, health screening services with HIV/TB testing and counselling in local clinics" The projects provide for the support of mothers, babies, school aged children, teenagers and the daily needs of young women, relationship counselling and leisure activities that help to develop social capital in a safe, caring environment.

I was also invited to join one of the training sessions for over 20 youth leaders who run the outreach programs in 7 communities. They span all cultures and language groups in South Africa and also include young volunteers from America and Australia. The aim of the Friday training session was to develop a curriculum for a 5 day holiday program that is attended by over 150 children per day in each community. The week day programs are attended by young people with the younger brothers and sisters (some as young as a year or so) and the aftercare program attracts an age group up to about 16 years of age. They are taught a range of skills and given a meal; they are also supported to undertake the school projects by proving materials that may not be available at home. Basic resources such as sanitary pads are provided to girls so that they are able to continue at school, rather than missing a whole week of school as they did previously. This is another way in which educational outcomes can be enhanced alongside improved nutrition. For some children the meal they receive from this organisation is the only meal they have each day.

Some of the young leaders are studying to become teachers, social workers counsellors, agricultural workers, or setting up their own businesses. They are encouraged to aim high and to have faith. The development of links within and beyond the NGO are vital for building self-esteem and extending their horizons.

The self-respect and self-belief that is taught to these young children and teens is faith based, with an emphasis on setting boundaries and abstaining from behaviour that will place them at risk. Although abstinence is emphasised, rather than safe sex, the argument put forward by the manager is that this is the approach she would have used for her own family and that is an appropriate message for young teens and children. Beliefs and values shape behaviour, but at a practical level so does belonging to a safe network where food, friends and fun can be found, rather than exploitation by drug traffickers and the gangs that are linked with them. A faith based abstinence approach, however needs to be taught alongside providing access to public health educators and referrals to organisations that undertake counselling directed to gender mainstreaming and safe sex messages that are age appropriate.

4.4.2.2 Creating Opportunities and Caring for People

Living Hope creates opportunities and cares for people and Living way aims to teach people the skills needed to earn their way out of poverty. Training is offered in entrepreneurship, worker readiness and agriculture as well as running a weekly job shop.

> Participation is the cardinal principle of democracy—not only because of its intrinsic value, but also because it increases the political efficacy of citizens by giving them direct training in the policies and tools of governance. Almost 200 years ago, John Stuart Mill suggested that this kind of democratic training is best obtainable at the local level, where citizens can make decisions about issues they can immediately relate to, and then generalize that knowledge to the broader, national political system (Mangcu 2014: 31).

The NGO plays a role in developing opportunities for young people to have a say in their communities and to make a difference. I was privileged to have the opportunity to attend one of the training sessions for young people. The aim was to help them to learn an active, energising curriculum that they would teach during the school holidays to the young people who usually attend their after school care. The aims were to encourage them to bring and their younger brothers and sisters with them and to provide a safe, hopeful environment for them each day of the holidays. This is a time when they are vulnerable to the gangs operating in the communities in which they live. By training young leaders and by providing opportunities for young people to volunteer their time and to learn to become role models the NGO helps to develop an understanding of citizenship rights and responsibilities. They also provide the opportunity for young people to attend Job placement programs. For example, D stressed that her daughter was becoming depressed because she had worked hard to complete her matriculation and had been employed at the 'Food Lovers Market'. She had taken up the offer to undertake a short course for vocational training and had left her job. On her return after 6 weeks she was told that her job had been filled and that she would have to apply for a new one. After 6 months she had become depressed and her mother who works as a domestic worker was worried about leaving her at home alone in the Ocean View house. By attending the NGO and volunteering she was able to connect with young people and a network who supported her. The so-called social capital achieved through connecting with others, being provided with opportunities to extend networks through job placement facilitators is important.

The NGO comprises sections for caring for the sick, feeding the hungry, giving hope through preventative care and community outreach, developing life skills, developing skills to enable people to acquire and keep a job. The referral centres linked with the day care programs are staffed by trained counsellors who liaise within the organisation and beyond to address the complex needs of the service users.

Food security and the opportunity to develop employment opportunities is another important strength of this NGO that makes an effort to address this current risk by developing local agricultural opportunities. Poly tunnels have been set up to grow tomatoes of high quality that are marketed commercially to the 'Food Lovers Market' nearby. The photographs show tomato production.

This is a better option than just teaching people to make crafts that people do not need to buy every day. Food is an essential item and the NGO teaches how to grow food in the same complex as a project for training young leaders and carers to support the community and a fully staffed health care centre. The NGO also provides training to those who wish to learn to become entrepreneurs and gives assistance to job seekers and those who need help in managing debts.

4.4.2.3 Analysis

The following analysis of internal and external strengths, weaknesses, opportunities and threats is based on a rapid appraisal and iterative face to face and emailed conversations with the stakeholders. Without the contributions made by this NGO each of the above communities would lose a systemic service and in many instances those without employment and a source of income would be destitute. The services supplied to school children are vital to promote the wellbeing of young people and to prevent their becoming prey to drug networks that operate in each community. This can be attributed to the committed staff and an energetic volunteer corps.

To sum up, the NGO provides a safe environment and alternative social networks as a balance to the violence associated with the gangs and party culture. It provides opportunities to people with a range of needs and employs staff who have lived with a range of dis-abilities.

By providing structured activities and pathways to hope and wellbeing, this NGO has made a real difference to the quality of life and the quality of palliative care. In particular it provides an important intervention for the very young and young teenagers. The organisation also provides an opportunity for the service providers who benefit from the opportunities that participation provides them. For example: A young man from US said he did not know what to do with his life and felt that his career was giving him no fulfilment. He has found that fulfilment through working as an administrator at the NGO.

The systemic approach to meeting their complex needs is innovative and aimed at giving hope to all those who wish to follow the mission of this NGO. For example: a young women told me how she had no confidence when she started at attending classes at the NGO and that now she was studying to become a primary school teacher, because she had 'found direction'.

4.5 Conclusion

We are all reliant on others and need to be able to depend on our connections with others.[33] A sustainable local community is determined by a sustainable region in which food, energy and water supplies are considered as major determinants for wellbeing. A management approach that relies on a broad raft of organizations spanning sectors could help to address some of the identified needs. The challenge of addressing complex needs requires network approaches that can help to work across many different service users and providers to address the identified needs:

The opportunity to work closely with local levels of government needs to be explored. This could lead to enhancing service delivery. By learning from the strengths and experience provided by NGO sector, municipalities could strengthen and maximise their capacity to deliver services for a growing population. Those aged above 16 and above also need broader programs that support the needs of women and men through gender sensitive responsive programs that are geared to enhancing quality of life and leading to a demographic transition of lower birth rates and lower mortality and morbidity rates. Nussbaum and Sen have argued that development opportunities that address the strategic and basic needs of women improve quality of life. Capacity building through empowering women to have a voice and ensuring their literacy, numeracy and opportunity to meet their strategic needs (1995) is vital. Gender mainstreaming is based on enhancing relationships between men and women based on respect within the household and across society (Kabeer in March et al. 1999).

Thus local government needs to do more than deliver basic needs. They need to help NGOs to deliver services and to provide people with opportunities to develop

[33]Collaboration is a core method for development in public, private and NGO sectors (Graham and Barter 1999).

leadership roles within society. The more they participate, the more they learn to practice their rights and their responsibilities.

> Municipalities must make sure that people in their areas have at least the basic services they need. There are a large number of services that they provide, the most important of which are: water supply, sewage collection and disposal, refuse removal, electricity and gas supply, municipal health services, municipal roads and storm water drainage, street lighting and municipal parks and recreation.... These services have a direct and immediate effect on the quality of the lives of the people in that community. For example, if the water that is provided is of a poor quality or refuse is not collected regularly, it will contribute to the creation of unhealthy and unsafe living environments. Poor services can also make it difficult to attract business or industry to an area and will limit job opportunities for residents. Basic services needed in rural areas may be different from those that urban communities need. ...Local government [municipality] is the sphere of government closest to the people; they are elected by citizens to represent them and are responsible to ensure that services are delivered to the community. One way in which municipalities can do this is to provide the service themselves through the use of their own resources – finance, equipment and employees. A municipality may also outsource the provision of a service. In other words, it may choose to hire someone else to deliver the service but it remains the responsibility of the municipality to choose the service provider and to make sure that they deliver the service properly.[34]

The future for South Africa is in developing learning organisations and learning communities where people feel that they have a voice and a right to a voice.

[34]Municipal service Delivery http://www.etu.org.za/toolbox/docs/localgov/munservice.html [22/08/2014 12:37:07 AM].

Chapter 5
Cross-Cultural Learning Community Challenges and Opportunities

From Clash of Cultures to Co-creation and Co-determination

5.1 Leadership for Transformation Requires Gender Mainstreaming and Capacity Building

Designing public education requires enabling participants to address representation, accountability and sustainability. West Churchman's Design of Inquiring Systems Approach (1979) helps to think about our thinking and our relationships. It spans self, other and the environment. It is based on questioning boundaries of inclusion and exclusion, it examines the so-called 'enemies within' (religion, mortality, politics and aesthetics) and it considers the consequences of our choices. It helps to address a core issue such as food security, in order to unpack it and to demonstrate that it is a systemic issue that poses both challenges and opportunities. We need to work with many forms of knowledge and across conceptual and spatial boundaries to address the big issues of the day, namely poverty and climate change within increasingly densely populated cities.

Developing learning organisations and learning communities could help to create bridges across diversity and strengthen bonds of friendship through enabling people to feel that they have a voice and a right to a voice. Participation in vertical democracy through elections, whilst important needs to be extended to include other forms of participation when discursive discussion and direct votes on policy ideas could help to foster connections and a shared ideology for a sustainable socially and environmentally just future. Opportunities exist to include creating a better balance between rural and urban areas, greening cities (Dryzek 1999, 2000, 2010) through the creation of urban agriculture in open spaces, on top of attached to and within buildings, co-determination in regions—new architecture for governance democracy and ethics—learning mindfulness to support communities of practice and learning communities. These aspects will be discussed below.

J. McIntyre-Mills, *Planetary Passport*, Contemporary Systems Thinking,
DOI 10.1007/978-3-319-58011-1_5

Polly Higgins suggests that a new planetary law should be passed to protect the viability of the planet and that current systems of law are inadequate to protect people and the planet.[1] Stiglitz has made the same point in relation to economics, whilst the need to reframe boundaries across a range of disciplines has made similar points as I stressed in a recent entry in Sociopedia called 'Reconsidering boundaries' (McIntyre-Mills 2014). Whilst the proposal was made by Higgins and others to the UN, Bolivia has succeeded in leading the way by passing a law to recognise the need to protect the earth:

> Bolivia enshrines natural world's rights with equal status for Mother Earth: Law of Mother Earth expected to prompt radical new conservation and social measures in South American nation.[2]:
>
> "It establishes a new relationship between man and nature, the harmony of which must be preserved as a guarantee of its regeneration." The law, which is part of a complete restructuring of the Bolivian legal system following a change of constitution in 2009, has been heavily influenced by a resurgent indigenous Andean spiritual world view which places the environment and the earth deity known as the Pachamama at the centre of all life. Humans are considered equal to all other entities.[3]

This could help to ensure that elected leaders remain more connected to the people they are supposed to represent. So the future lies in participatory democracy and more direct interventions. What is needed is a greater emphasis on core principles and engagement with the people in discursive democracy.

The most powerless and voiceless are those who do not have the vote. They are young people, children, asylum seekers, the disabled, sentient beings who are commodified and traded and of course the planet which has no rights at all. So the focus needs to be on the inadequacies of the nation state and the social contract.

- Working across boundaries through co-creating a community of practice to address a shared concern.
- Developing a greater number of connections enhances consciousness (Greenfield 2000, 2003, 2008) who argues that the more we are able to think about our thinking. This can help to create closer bonds with others to foster links and to bridge differences. The approach is as relevant to education as it is to community development and in the process helps to enhance representation and accountability through exploring ideas and engaging in dialogue.
- Testing the implications of decisions with others so as to balance individual needs and collective needs.
- Enable residents to make decisions from below to inform policy at the regional level and to facilitate monitory democracy and governance. How can we achieve

[1]See https://www.youtube.com/watch?v=QPUmN88htCo and Earth democracy http://m.youtube.com/watch?v=Bq0lrbznsjc.

[2]The Guardian, Sunday 10 April 2011 18.17 BST http://www.theguardian.com/environment/2011/apr/10/bolivia-enshrines-natural-worlds-rights.

[3]http://therightsofnature.org/related-books/http://www.ipsnews.net/2011/05/global-campaign-to-bestow-legal-rights-on-mother-earth/.

> cultural transformation on consumption patterns? In an era of social, economic and environmental convergence of risk, we need to accept the possibility of new forms of governance to protect those who are not protected by the nation state, because they are not citizens. The work of Ann Florini (2003) provides possibilities as does the work of Danielle Archibugi. The challenges of governance need to be addressed by working across disciplines. Gibbons et al. (1994) argue that the ability to work across boundaries is vital for 'The new production of knowledge' and vital for 'the dynamics of research' to address current complex challenges. The so-called 'tragedy of the commons' argument developed by Hardin (1968) is a construction informed by simplistic thinking. Ironically, it is often used as the starting point for environmental thinking, but in fact it originated as an argument developed by Locke in support of the enclosure movement and private property in Britain. The argument being that contained areas of land are cared for better than the areas of land that are held in common and shared. The example of common grazing land is shared.

But ironically it is the privatisation and commodification of land and natural resources that has led to environmental pollution and degradation. The lack of trust between nations has evoked the rhetoric of nationalism and state protectionism. This has led to 'seeing like a state' (Scott 1998), which in turn leads to the argument for competition and the zero-sum approach which argues that one nation can profit at the expense of others. The organisational philosophy and governance arguments that flow from this philosophy are based on limited disciplinary paradigms that can profit at the expense of others.

Ideally, a universal respect for social and environmental justice could enable subsidiarity as a means to support freedom (Follesdale 2006; Poe 2010) to the extent that it does not undermine the quality of life of this generation or the next. Dualist thinking pervades our consciousness and reflected in socially unjust and environmentally unsustainable designs for society. Designs need to be supported by constitutions, based on a priori *norms*, and consequentialist or a posteriori approaches, based on testing out ideas within context and with future generations in mind. Global axioms to protect future generations and the poor of this generation need to guide legal constructs and legal decisions at the local level. The principle of subsidiarity could rest comfortably with global axioms—provided that the caveat is made that people at the local level can be free and diverse to the extent that their freedom and diversity does not undermine the rights of others or future generations of life. This challenge of balancing individual and collective social, cultural, political and economic concerns needs to be buttressed by regional parliaments and courts that focus on social and environmental justice at a post-national level.

A Community of Practice needs to develop curricula that enable students to identify with others across the differences and to recognise our interconnectedness and to reframe the nation state. The Springer Encyclopedia of Food and Agricultural Ethics (McIntyre-Mills 2013) provides a brief outline of the landscape of this complex field, defined by intersections spanning social, cultural, political, economic and environmental contributions from the social and natural sciences.

A case is made for those who are not protected by the social contract, including young people, non-citizens, the disabled, sentient beings and the environment on which we all depend. The contribution highlights the empirical contradictions and theoretical tensions that have implications for social and environmental justice.

I was an invited plenary speaker at the Annual International Conference of Islamic Studies in Balikapan in December 2014, which was on Multiculturalism as it relates to complex wicked problems such as poverty and climate change. Wicked problems (comprising many diverse and interrelated variables) are perceived differently by different stakeholders with different values. The AICIS conference endorsed traditional culture, ethnicity and diversity.

Prof. Dede Rosyade emphasised the importance of realising the potential for Indonesian Islam to be a model for world Islam. The ideal would foster the following: openness and tolerance through supporting democratic engagement to test out ideas and to enable ongoing renewal. But the conservative turn was emphasised by key note speaker, Van Bruinessen (2014) who stressed the potential either for Indonesian Islam to set an example of openness and respect for diversity or to move in the direction of fundamentalism. At the ACIS conference (2014) I stressed the importance of valuing the contribution of agricultural communities and stressed the importance of co-determining development options with food producers instead of focusing on neoliberal paradigms that impose market controls that pressurised food producers to become increasingly subject to new technologies and corporate requirements that do not value local knowledge or local biodiversity. I also discussed the importance of moving beyond extreme forms of cultural relativism and I explored a middle ground based on co-creating meanings within context. Each discourse needs to be unpacked in terms of policy discourses and the implications for social and environmental justice (McIntyre-Mills 2000). I made the case for food security and human security, because the costs of climate change will place a heavy burden on the urban poor. Valuing the role of agricultural communities and encouraging them to stay on the land drew rounds of applause. Whilst members of our panel stressed that sovereignty should be given to some primates and that forests and natural habitat should be protected in the interests of living a good life that demonstrated piety and a respect for all life. Others stressed that learning should be encouraged by achieving the good life in urban areas, rather than right livelihood approaches and that education should enable people to progress, without unpacking what a city would like without an agricultural base.

I discussed creating a learning community which has resulted in setting up an Indonesian research consortium this year spanning UIN, UI, Padjadjaran Universitas Nasional supported by Flinders Partner Wirasoft to help undertake participatory education research on issues of food, energy and water security.

5.1.1 *Creating a Learning Organisation and Learning Community Approach*

Thinking needs to be translated into co-learning and co-determination. With growing urbanisation Indonesia faces challenges such as ensuring that the future economy responds to the need for highly skilled and democratically engaged citizens with a focus on living well and sustainably in city environments that will need to be supported by sustainable agricultural regions. The basis for a democratic and sustainable region rests on regional networks to support human security resting on meeting the quality of life concerns of the current and future generations through securing sustainable cities within sustainable agricultural land. This rests on developing appropriate education and participatory democracy and governance.

I was invited to Annual Islamic Studies Conference as a result of supervising a Ph.D. thesis by Dr. Muhammad Adib who compared secular and non-secular curricula at the Universities of Indonesia and the National Islamic University, respectively. This led to my appointment as an Adjunct Professor in the education faculty and my delivery of a plenary speech[4] at ACIS and later at the National Islamic University. The focus of my presentation was on being mindful of the implications of urbanisation for quality of life and human security and education that enables co-determination of our future. It draws on a forthcoming book called 'Knowing our place and appreciating our interconnectedness' and current publications called 'Systemic Ethics' and 'Transformation from Wall Street to Wellbeing' and 'Reconsidering Boundaries'. It has been wrongly assumed that growth in the economy will sustain a growing population. We need to understand that the current way of life is unsustainable and that we need to rethink many of the dimensions of modern culture. Culture after all is simply a way of life and a response to the challenges that we face as human beings. The way we think shapes who we are and the sense we make of our daily experiences.[5]

The notion of multicultural education based on co-creation through dialogue[6] is explored in this paper. In this paper I explored the potential for multicultural

---

[4]The paper discussed recent publications that consider whether a change in the architecture of democracy and governance could balance individual and collective needs more effectively through participation, guided by the principle of subsidiarity and the axiom that people ought to be encouraged to be free and diverse to the extent that their freedom and diversity does not undermine the rights of others.

[5]Neuroscientists such as Cliff Sarron and Al Kasniak presented at the 'Mindfulness Matters' Conference. Kazniak (2014) stressed in his paper stress that mindfulness research has shown how thinking affects the material body that we inhabit and the way we think shapes the body and the environment on which we depend. The environment affects the body and mind. For example, the researches by Professors Sarron and Kasniak have found that people who exercise in a natural environment have higher levels of concentration than people who exercise in a built-up environment.

[6]In 'Transformation from Walls Street to Wellbeing', and 'Systemic Ethics' I made the case for co-creation and co-determination within regional biospheres that respect diversity and freedom based on stewardship for current and future generations.

education and critique the hypothesis by Huntington (1993: 22) in 'the clash of civilizations?', namely the argument that a clash is inevitable becomes a self-fulfilling prophesy (Bottici and Challand 2006). Nation states are indeed no longer the only dimension of democracy and governance. We also need to consider cultural identity which is many layered—as stressed by critical theorists such Hall and Du Gay (1996) on representation and identity in which they stress that people have layered identities. Hall reflects (see Akomfrah 2013) on his own history. His parents lived in Jamaica. His parents were of European Jewish and African (slave) backgrounds. As a scholarship holder at Oxford he was part of an elite academic environment but not of the establishment. He identified with the rights of immigrants of workers and of those who were without a voice within the nation state. Nation states are far too containerist in their approach to identity. These days the landscape of democracy has changed as stressed by Habermas and Derrida in 'Philosophy in a time of terror' in which they stress the need to respect diversity and to extend hospitality to others, not merely tolerance of diversity. Today we have 50 million refugees. We will need to think more along the lines of regional governance and regional cooperation.

The argument I develop is that Shiva critiques Hardin's 'Tragedy of the Commons' approach. It applied John Locke's defence of property and the enclosure movements of the seventeenth century in Britain. Hardin's core argument is that in the absence of private property there would be lawlessness. Hardin's classic argument supports nationalism and private property. I argue that we need to develop the notion of co-creation and co-determination through education and dialogue that creates unity of purpose and respect for diverse contributions of workers in the city and in agriculture[7] as stressed by the new incoming president Jokowi.

> "We also want to be present among independent nations … to be a creative nation," he said.
>
> "We have to work together to revive Indonesia as a maritime country. Ocean, seas, straits, and bays are the future of our civilization.[8]"

The AICIS conference endorsed traditional culture, ethnicity and diversity under the philosophy of Pancasila. It invited speakers to talk on the potential of multiculturalism and the conservative turn that is evident as Islam is asked to choose between openness to others and embracing freedom and diversity to the extent that freedom and diversity does not undermine the rights of others or current and future generations of life.

Prof. Dede emphasised the importance of realising the potential for Indonesian Islam to be a model for world Islam as is open and tolerant and it embraces democratic engagement to test out ideas and to enable on going renewal.

[7]"Unity and mutual assistance is a requirement to be a big nation, we will never be big if we're divided; we will never be truly free without hard work". http://www.abc.net.au/news/2014-10-20/joko-widodo-sworn-in-as-indonesias-seventh-president/5827050. Accessed 21/10/2014.

[8]http://blogs.wsj.com/searealtime/2014/10/20/as-it-happens-indonesia-inaugurates-joko-widodo-as-president/. Accessed 21/10/2014.

My appointment as a non-Islamic woman and the only one at the conference and the UIN was indicative of this openness.

> "the ire of the Wahabis was directed not primarily against outsiders but against those whom they saw as betraying and degrading Islam from within: on the one hand those who attempted any kind of modernizing reform; on the other- and this was the more immediate target – those whom the Wahabis saw as corrupting and debasing the true Islamic heritage of the Prophet and his companions. They were of course strongly opposed to any school or version of Islam, whether Sunni or Shi ite, other than their own. They were particularly opposed to Sufism, condemning not only its mysticism and tolerance but also what they saw as the pagan cults associated with it…." (Lewis 2003: 105)

The AICIS conference endorsed traditional culture, ethnicity and diversity under the philosophy of Pancasila.[9] It invited speakers to talk on the potential of multiculturalism and the conservative turn that is evident as Islam is asked to choose between openness to others and embracing freedom and diversity to the extent that freedom and diversity does not undermine the rights of others or current and future generations of life. The conference comprised liberals who support the philosophy of Gus Dur/Abdurrahman Wahid.

According to Rosyada (2014) the Ministry of Religious Affairs administers only about 24%, of madrasah, while the rest are students of private schools. Thus the role of the government in relation to the provision of non-secular education very low compared with private-owned institutions. The Ministry through the Director General of Islamic education strives to address the curriculum and to develop the capacity of teachers as mandated in the constitution.[10] But the conservative turn was emphasised by key note speaker, Van Bruinessen (2014), who stressed the potential for Indonesian Islam to set an example of openness and respect for diversity or to move in the direction of fundamentalism:

> By 2005 it appeared that a conservative turn had taken place in mainstream Islam, and that the modernist and liberal views that had until recently found relatively broad support within Muhammadiyah and NU were increasingly rejected. Both organizations held their five yearly congresses in 2004, and on both occasions the boards were purged of leaders considered as 'liberals', including persons who had rendered great service to their organizations. Many ulama and other Muslim leaders appear preoccupied with the struggle against 'deviant' sects and ideas. (Van Bruinessen 2011).

---

[9]Cited by Kevin Gogg and Peter Carey (2014) Pancasila (Five Principles)

- Belief in One Unitary God (Ketuhanan Yang Maha Esa)
- A just and civilised humanitarianism (Kemanusiaan yang adil dan beradab)
- The unity of Indonesia (Persatuan Indonesia)
- Democracy led by principled wisdom and consultation/representation (Kerakyatan yang dipimpin oleh hikmat kebijaksanaan dalam permusyawaratan/perwakilan)
- Social justice for all the people of Indonesia (Keadilan sosial bagi seluruh rakyat Indonesia)
- Unity in Diversity

[10]According to Rosyada's notes for his speech : The Ministry needs to Control school curriculum, school text book and the capacity of the teacher, because Indonesia still on a serious efforts to develop moral capacity of the students for being a pious professional, plurals, inclusive and behaving as a multicultural society, as was mandated in the constitution 1945".

Van Bruinessen discussed the fatwas declared against 'secularism, pluralism and religious liberalism' as well as interfaith prayer meetings for 'well-being and peace'. In his paper, delivered at the conference he poses the question as to whether the increase in democracy has allowed this rise in fundamentalism, or whether it is the result of increased links with Saudi Arabia and the return of Saudi graduates to Indonesia. The group is Jaringan Islam Liberal and linked with the philosophy of Gus Dur or Abdurrahman Wahid.

Participation at the ACIS conference, Kalimantan, 2014

The focus of my paper delivered at the Annual Islamic Studies (ACIS) conference in December 2014 in East Kalimantan was 'From Clash of civilizations to Co-creation and Co-determination'. I stressed that the opportunity exists to do the following:

- Teach learners to think systemically and to join up the dots so that they understand the policy implications of their choices.
- Help them to realise that decisions have consequences in the short, medium and long term.
- Teach them systemic ethics in both secular and non-secular places of learning.
- Help places of learning at all levels to connect with communities.
- Create a learning organisation and learning community approach.

- Thinking needs to be translated into co-learning and co-determination. In South Africa and Indonesia, for example at least 65 and 75% of the population, respectively, will be living in urban areas. Urbanisation poses a systemic threat to the survival and sustainability of culture as we know it today. What is the point of raising these concerns? The challenges pose opportunities to provide education and leadership. What is the way forward?

What follows is a suggestion based in part on the work and suggestions I have gained through working with Ph.D. students and academics in Indonesia and South Africa doing research on a range of policy concerns. I would like to discuss the implications of increased urbanisation on quality of life and the implications for policy. Food deserts are the likely scenario if more emphasis is not placed on balance, greening cities and supporting small farmers. This chapter advocates for developing curriculum. I will compare South Africa and Indonesian challenges for education and sustainable living. I make the case for:

Processes for engaging and holding people to account through a virtual environment prior to piloting FGD in Jating or a suburban area of West Java and Alice Springe in the NT (both face challenges associated with food, water and energy costs) as proposed in a current grant application. The research builds on the depth of the C1's connections in these contexts, for instance,

- Contributing to the development of an MOU between West Java and South Australian government.
- Doing collaborative research as Ph.D. supervisor and collaborating with staff at the University of Indonesia, Universitas National, Universitas Padjadjaran, National Islamic University, West Java and the SA Local Government Association (in particular in Alice Springs and Adelaide).
- Undertaking participatory action research in South Australia and Northern Territory (as evidenced by the incorporation of her research findings into Tangentyere Council Protocols and being cited at the Liquor Commission Hearing in NT).
- Facilitating a leadership workshop at Flinders University together with delegates from West Java Provincial Government on the public education training and leadership. This was in line with the Paris Development Agenda (1997) that underlines the notion that development should be based on collaborative principles. The participants in this workshop stressed the importance of the need to develop leadership opportunities for women and young people. The challenges they face in West Java are associated with the impact of urbanisation as people face congested cities, lowering living standards and high risks for women and children who are unable to express their strategic rights to a voice within households, communities local government and national-level organisations.

5.1.2 A Learning Organisation Within a Learning Community and Post-National Region

The reason for my interest in this is that with growing urbanisation there will be a need for actively engaged young people to address democracy, climate change, food security and sustainable living in urban and regional environments. This has implications for the big issues of the day, namely representation, accountability and sustainability. Learning to live sustainably and well requires knowledge spanning social, cultural, economic and environmental issues. This requires social engagement and the fostering of partnerships across the state, market and civil society to address well-being and sustainable living. Professor Pratikno (2014) stressed at the presentation for the inaugural meeting of the Flinders Culture House (Rumah Budaya) initiative that Indonesia and Australia need to foster regional partnerships through building on friendship networks and soft diplomacy. In 2014 Professor Victoria Farrar Myer (Fulbright Distinguished Chair) echoed this plea for greater emphasis on rebuilding relationships and emphasised the need to foster collective determinism through public engagement processes McRae (2014) of the Lowy Institute also echoes this concern:

> Finally, people-to-people ties remain a persistent challenge. Abbott's announcement of a new Australian Centre for Indonesian Studies is a positive initiative, as is the Coalition's signature New Colombo Plan. Neither though is a replacement to committing the resources required to promote Indonesian literacy in Australia. This requires both maintaining teaching infrastructure for Indonesian language in Australia, as well as promoting career options that would encourage Australians to attain Indonesia literacy.[11] In South Australia and Australia education and development needs to support language skills and an understanding of the challenges of living in one of the driest continents. The recent IPCC reports stressed that South Australia will be affected by climate change and that rainfall levels will decrease. In terms of public policy, Australia has announced the importance of 'studying Indonesian and fostering regional ties'. The following statistics were presented by Praktino and Akbar Susamto (2014) at a seminar at Flinders University stressed that the size of the Indonesian economy would increase from the 16th largest to the 7th largest economy in the world. That it would grow from 45 to 145 million members of the so-called 'consuming class'. 55 million skilled workers in the Indonesian economy to 135 million workers in the Indonesian economy who generate 0.5 trillion Australian dollars and this will grow to 1.8 trillion in consumer services, agriculture, fisheries and education.

Philosophy and learning from one another through open engagement remains even more important 'in times of terror' to cite the book written by Borradori (2008) based on a conversation with Habermas and Derrida in the wake of 9/11. Dialogue and engagement remains even more important today in the wake of the Paris massacre, based on the axiom that democracy needs to foster freedom to the

[11] http://www.abc.net.au/news/2013-10-02/mcrae-challenges-ahead-for-the-indonesia-relationship/4993720.

extent that it does not undermine the rights and responsibilities for current and future generations of life.

What is a learning organisation within a learning community and post-national region?

It is an organisation that is open to the social, economic and environmental context and responds to the needs of the people it purports to serve. It helps to develop opportunities to test out ideas with others. It considers the consequences for self, others and the environment by teaching the ability to think about the consequences of decision-making through applying critical heuristics (see Systemic Ethics, McIntyre-Mills 2014).

How can we engage in dialogue that supports learning?

Churchman (1979) addressed the following ways of knowing:

- Logic based on considering the links across premises that support and substantiate a conclusion. Logic refers to the way in which arguments are structured and the relationship across ideas that build from premises to a logically supported conclusion. Argument needs to be developed so that the logical flow is continuous and so that it supports the conclusion. The flow of ideas upstream will affect the pool of experience and the conclusion downstream.
- Empiricism refers to both qualitative and quantitative dimensions of knowledge that reflect the meanings of diverse stakeholders by faithfully representing their points of view. Quantitative data are based on collecting data to represent patterns and weightings of the number of times particular ideas have been expressed, for example. Empiricism (qualitative, quantitative data from human beings and learning from the patterns revealed by nature, the environment and other species)—animals know in advance of human beings when a tsunami is approaching because of their superior senses, animals do not (usually) destroy the environment on which they rely.
- Idealism is the domain of knowledge that refers to what ought to be done in ethical terms to meet ideals or essential living standards or quality of life, whilst expanding pragmatism ensures that the consequences of decisions for others and the environment are considered in the short, medium and long term through testing out ideas with people within and beyond the learning environment of universities and schools to create learning communities and institutions 'without walls'.
- Dialectic based on the capability to explore thesis, antithesis and to make syntheses. This is only possible when respectful communication is the basis for the conversation. The dialectic refers to the ability to engage in co-created knowledge that is intersubjective. The process used in this workshop modelled this approach to begin to 'unfold' the ideas of all the stakeholders and to sweep

in the social, cultural, political, economic and environmental dimensions that affect the framing of an area of concern. We explored what needs to be included or excluded when considering unemployment in South Africa and education to address the needs of people of all ages. The potential of the young to learn from the old and vice versa through appreciation of lived experience, innovation, art, narrative and design needs to be further explored through participatory action learning and action research projects. Nussbaum (2011) has stressed the capabilities approach as a way to enable people to meet strategic needs through empowering them at home, at school, at work, at play and giving people a voice. State, market and civil society need to work together to enable and empower workers and learners through co-creating partnerships.

- Considering the consequences for self, other and future generations of life is expanded (not narrow) pragmatism. These ways of knowing include the subjective (personal perceptions, values, emotions and intuition); objective (lived experiences and empirical knowledge) and intersubjective dimensions (understanding of the meaning of Ubuntu and its relevance for developing social capital), which were highlighted originally by Habermas. Churchman's five domains span: logic, empiricism, dialectic, idealism and pragmatism. This is an incomplete list, however, that needs to be extended to include.
- Non-anthropocentric knowing drawing on the environment as well as knowing through all our senses—including empathy and intuition along with:
- Respect for 'know how' developed through empirical trial and error.
- Creativity in creating sustainable relationships with the land and
- Caring for future generations of life.
- 'Phronesis' or Aristotelian wisdom to match the right knowledge in context so as to empower learners through providing them with appropriate pathways to learning.

In the social context learning to use a design based on Ulrich's 12 basic questions can be helpful as a way to consider where and when to draw the policy line based on considering the consequences of our choices in the short, medium and long term. The axiom is that we can be free and diverse in our praxis (thinking and practice) to the extent that it does not undermine the rights of others.

West Churchman's critical and systemic Meta approach to working with, rather than within the boundaries of a single paradigm inspires my work. I draw on several approaches that locate Nussbaum's 10 Central Capabilities—relevant to all sentient beings (Nussbaum 2011: 33–34)[12]—within the ethical literature, in order to address life, health and ability to live a life worth living.

---

[12]"Source: Nussbaum, M. C., 2011. Creating Capabilities: The Human Development Approach. London: Harvard University Press:

1. Life. Being able to the end of a human life of normal length; not dying prematurely, or before one's life is so reduced as to be not worth living.
2. Bodily health. Being able to have good health, including reproductive health; to be adequately nourished; to have adequate shelter.
3. Bodily integrity. Being able to move freely from place to place; to be secure against violent assault, including sexual assault and domestic violence; having opportunities for sexual satisfaction and for choice in matters of reproduction.
4. Senses, imagination and thought. Being able to use the senses, to imagine, think and reason—and to do these things in a "truly human" way, a way informed and cultivated by an adequate education, including, but by no means limited to, literacy and basic mathematical and scientific training. Being able to use imagination and thought in connection with experiencing and producing works and events of one's own choice, religious, literary, musical and so forth. Being able to use one's mind in ways protected by guarantees of freedom of expression with respect to both political and artistic speech, and freedom of religious exercise. Being able to have pleasurable experiences, and to avoid non beneficial pain.
5. Emotions. Being able to have attachments to things and people outside ourselves; to love those who love and care for us, to grieve at their absence; in general, to love to grieve, to experience longing, gratitude and justified anger. Not having one's emotional development blighted by fear and anxiety. (Supporting this capability means supporting forms of human association that can be shown to be crucial in their development.)
6. Practical reason. Being able to form a conception of the good and to engage in critical reflection about the planning of one's life. (This entails protection for the liberty of conscience and religious observance.)
7. Affiliation.

   (a) Being able to live with and toward others, to recognise and show concern for other human beings, to engage in various forms of social interaction; to be able to imagine the situation of another. (Protecting this capability means protecting institutions that constitute and nourish such forms of affiliation, and also protecting the freedom of assembly and political speech.)
   (b) Having the social bases of self-respect and non-humiliation; being able to be treated as a dignified being whose worth is equal to that of others. This entails provisions of non-discrimination on the basis of race, sex, sexual orientation, religion, caste, ethnicity, national origin.

8. Other species. Being able to live with concern for and in relation to animals, plants and the world of nature.
9. Play. Being able to laugh, to play, to enjoy recreational activities.
10. Control over one's environment.

    (a) Political. Being able to participate effectively in political choices that govern one's life; having the right of political participation, protection of free speech and association.
    (b) Material. Being able to hold property (both land and movable goods); having the right to seek employment on an equal basis with others; having freedom from unwanted search and seizure. In work, being able to work as a human being, exercising practical reason and entering into meaningful relationships of mutual recognition with other workers."

Nussbaum on a priori approaches	Haraway and Singer on a posteriori approaches	Arendt and Butler on interconnectedness	Deborah Bird Rose, Vandana Shiva
Nussbaum's work can be criticised because she ignores the role of the market in privileging the property rights of the few at the expense of the majority[a]	Haraway shows that we are already Cyborgs with connections to enhance capabilities and animal connections to sustain life through medical research. Haraway's work criticises privileging the human. Both Singer and Nussbaum also strive to avoid speciesism but their ethical stance differs	Hannah Arendt's core notion of the banality of evil to acknowledge that everyday choices contribute to creating monstrous systems	Life is based on a continuum of connections across past, present and future We depend on the earth and we return to it The seed is a synecdoche of life and cannot be commodified Water is another example of why we need to protect the land and water through a transformation in the way we live and relate to one another. Not merely through measuring things in a different way

[a]The argument begins with a critical consideration of the capabilities approach of Martha Nussbaum who is a social democrat whose understanding of property in some ways undermines the capabilities approach that she so carefully proposes, because she does not explore the implications for 'a fairness' underpinning current frames of what property entails. In this regard the work of Butler comes closer to engaging with the economic system that is the root of the problem (along with the notion of the nation state and the limited social contract). The work of Judith Butler who holds the Hannah Arendt chair stresses the right to voice our needs and to be listened to and heard. Her starting point is a focus on *the other*, vulnerability and a recognition that we are vulnerable and should demand the right to be heard and cared for. She starts from the position of vulnerability, not from strength or resilience, because her focus is on interdependency. A rights focus can be very individualistic and this can lead to dualism and this is the problem with the capabilities approach of Martha Nussbaum. However, it does provide the liberative potential as a guideline for human rights. But it does not go far enough

The work of Hannah Arendt has also helped to shape my thinking about the question: how should we live? Her notion of the 'banality of evil' is particularly relevant as we need to consider the broad context and the structures that lead people to make unethical choices at a personal, interpersonal and planetary level. The values that shape our current social and economic choices are the root cause of a way of life *that benefits a minority at the expense of the majority* in this generation and the next.

Currently, my practical concern that there are more refugees and asylum seekers now than ever before and the current form of democracy and governance does not work. I strive to focus on the need to work with many forms of knowledge and across conceptual and spatial boundaries to address the big issues of the day, namely peace, gender mainstreaming and social, economic and environmental security within increasingly densely populated cities that need to respond to increasing poverty and climate change.

This has implications for the quality of life from a human, non-anthropocentric and environmental perspective. We need education that enables co-determination of our future and that protects both human and environmental well-being.

Her notion of the 'banality of evil' is particularly relevant as we need to consider the broad context and the structures that lead people to make unethical choices at a personal, interpersonal and planetary level.[13]

5.1.3 *Multilevel Engagement Through Mindfulness and Critical Systemic Praxis*

Testing out ideas requires preserving freedom, space for doubt, diversity and disagreement to the extent that diversity does not undermine the right to freedoms of others. An open approach to design and to research could be extended through acknowledging the praxis knowledge associated with reading the environment and living systems of which we are part.

Research needs to span consciousness, cultural studies and systemic praxis to link the notion of relationships between humans and the land as a source of well-being and the broader societal need for environmental protection and effective ecosystem management. The following Design of Inquiring Systems Approach provides guideline for designing a public education response. They are based on Ulrich (1983, 1994, see: http://wulrich.com/csh.html).

---

[13]The banality of evil evident in media coverage spanning values, freedom of movement and violence expressed in the language of religion or politics:

A Christian pastor (from South Africa) who lives in Nigeria encouraged his followers 'to drink petrol which they could turn into pineapple juice if they had enough faith'. The SABC (26 and 27th of Sept 2015) covered the event showing people drinking the petrol and in some cases passing out. He declaimed that they could demonstrate their faith through this act. This is one of several acts of violence perpetrated by this pastor in the name of Christianity.

The public deaths of noncombatants captured and portrayed online by ISIS. Large flows of refugees from ISIS have raised concerns about the role of the nation states in Europe and the extent to which training in democracy could help to redress the concerns about the so-called clash of cultures.

The injustice towards asylum seekers who arrive by boat to Australia remains an area of total silence in Australia and frontiers of justice remain closed to asylum seekers. No questions are asked, because no information is provided.

In a recent edition of the Catholic Herald, Pope Francis stressed the need to show hospitality to the displaced.

The ISIS claims that the refugees contained sleeper cells is one tactic to polarise groups and to deter the acceptance of those fleeing from Syria. The massacre in Paris on Friday the 13, 2015 has resulted in France declaring that this is an act of war.

Ulrich poses 12 boundary questions. He stresses that they should be asked in terms of ‘what is the case now?’ (Describes the current situation by using the DIS) and ‘what ought to be the case’? (Making judgments informed by ethical considerations). The questions drawn from Ulrich, Werner and Reynolds, Martin (2010) are as follows (Table 5.1):

The groups of questions cover sources of motivation, control, knowledge and legitimacy. Ulrich uses questioning to unfold the implications of particular approaches: (1) How do we know what to do? (2) What is the best approach? (3) Why is it the best approach?

We need to work with many forms of knowledge and across conceptual and spatial boundaries to address the big issues of the day, namely poverty and climate change within increasingly densely populated cities. Critical Systems thinking and if then questions (heuristics) can help to address these concerns in learning organisations and learning communities to create bridges across diversity.

Table 5.1 The boundary categories and questions of CSH

Sources of influence	Boundary judgments informing a system of interest (S)			
	Social roles (Stakeholders)	Specific concerns (Stakes)	Key problems (Stakeholding issues)	
Sources of motivation	1. *Beneficiary* Who out of be/is the intended beneficiary of the system (S)?	2. *Purpose* What ought be/is the purpose of (S)?	3. *Measure of improvement* What ought to be/is S’s measure of success	The involved
Sources of control	4. *Decision maker* Who ought to be/is in control of the conditions of success of S?	5. *Resources* What conditions of success ought to be/are under the control of S?	6. *Decision environment* What conditions of success ought to be/are outside the control of the decision maker?	
Sources of knowledge	7. *Expert* Who ought to be/is providing relevant knowledge and skills for S?	8 *Expertise* What outfit to be/are relevant new knowledge and skills for S?	9. *Guarantor* What ought to be/are regarded as assurances of successful implementation?	
Sources of legitimacy	10. *Witness* Who ought to be/is representing the interests of those negatively affected by but net involved with S?	11. *Emancipation* What ought to be/are the opportunities fur the interests of those negatively affected to have expression end freedom from the worldview of S?	12. *Worldview* What space ought to be/is available for reconciling differing worldviews regarding S among those involved and affected?	The affected

Adapted from Ulrich (1996, p. 44)
Source Ulrich, Werner and Reynolds, Martin (2010, Chap. 6, p. 245)

When people are affected by the shared health concerns associated with food or water contamination they can strengthen bonds of friendship through enabling people to feel that they have a voice and a right to a voice. Participation in vertical democracy through elections, whilst important needs to be extended to include other forms of participation when discursive discussion and direct votes on policy ideas could help to foster connections and a shared ideology for a sustainable socially and environmentally just future. What sectors could be included and on what basis? In this section I will explore the potential for multicultural education and critique the hypothesis by Huntington (1993: 22) in 'the clash of civilizations', namely,

> It is my hypothesis that the fundamental source of conflict in this new world will not be primarily ideological or primarily economic. The great divisions among human kind and the dominating source of conflict will be cultural. Nation states will remain the most powerful actors in world affairs, but the principal conflicts of global politics will occur between nations and groups of different civilizations. The clash of civilizations will dominate global politics. The fault lines between civilizations will be the battle lines of the future. Conflict between civilizations will be the latest phase in the evolution in the modern world…

The following are amongst the greatest challenges facing humanity and that they potentially pose an 'existential risk': representation of increasingly diverse populations within nation states (b) accountability to ensure that resources are used fairly and equitably and sustainably and (d) engagement to enhance the capability of people to understand that attitudes to consumption have profound implications for social and environmental justice issues. Thus the participatory action research aims to contribute to discovering and piloting new forms of architectures for public education and engagement to support user-centric public policy and to enable multilevel, multi-agency approaches for social and environmental justice.

Geographies of place in an era of displacement require new forms of democracy and engagement to take into account cross cutting identities.[14]

[14]This section develops an argument based on considering the empirical consequences of social, economic and environmental decisions on the quality of life of current and future generations by applying critical heuristics to the question 'how should we live, in order to respond to the growing challenge of meeting the needs of people living in cities'? What do we value and why? What are the implications for human capacity and human capabilities? These research questions are inspired in part by Stuart Hall who asks what is the social, cultural, economic and environmental context that shapes who gets what, when, why and to what effect? I have been privileged to explore the issue of sustainability and mindfulness at the Schumacher Institute in Bristol, a sea port town that made its fortune from trade with through the Royal African Company. The history of slavery is not openly acknowledged. South Africa, Japan and Indonesia where the widening gap between rich and poor is evident in large cities from Bristol to Cape Town, Gauteng, Yokohama and Jakarta. In South Africa I visited a range of University departments, research institutes and NGOs including University of South Africa, University of Cape Town, the Mindfulness Institute in Stellenbosch, South Africa where I attended the Mindfulness Matters Conference, workshops and conversations with colleagues at Living Hope, Embrace Dignity and Africa Tikkum in Cape Town where the concerns for food security, educational opportunities and unemployment are central and at the World Futures Institute, Cyprus Institute of Technology where the ravages of war in nearby Syria, the socio, economic and environmental problems of a society that is familiar with conflict was expressed in conversations with symposium members.

How democratic is democracy if it does not foster human security? Democracy and governance are in need of improvement (Hulme 2009; Giddens 2009) and lessons learned about regeneration and sustainable living could help to revitalise our approaches to ethics, representation and accountability. It seems to me that diverse ways of knowing hold the key to human and environmental security.

As stressed previously the notion of national sovereignty cannot address the entangled challenges we face today associated with growing numbers of displaced people,[15] poverty, conflict, food insecurity, misuse of agricultural land, water shortages, growing urbanisation and energy shortages associated with using non-renewables and nuclear disasters on the scale of Fukushima. These are examples raised by the United Nations. The issues need a new approach to the architectures of governance on a planetary scale.

Just as bricolage can be used for reshaping found objects and turning them into art and usable objects, we can do the same with policies and forms of government that are not working. Cities from Jakarta to Cape Town[16] are becoming food deserts where the voiceless and powerless are preyed upon. It is probably more widespread than readily acknowledged and if nothing is done, then we are complicit in perpetuating the 'banality of evil'. We evolved through the ability to show empathy to others and to reciprocate and not only through our ability to compete. The good

[15]The number of people who have been displaced as a result of war and natural disasters has increased dramatically. The global refugee figure passes 50 million for first time since second world war" and yet the needs of the displaced are not addressed through the current architectures of democracy. Displaced people who are non-citizens are considered to be outside the frame of reference of state protection. It is estimated that this figure could rise to 150 million by 2050 (Rusbridger 2015: 13). Climate change refugees and climate change displacement needs to be addressed through democracy, governance and capacity building through public education. Offshore processing in states whose economy is reliant on aid leads to the development of oppressive camps along the lines of convict colonies. The propensity for riots has already been demonstrated along with the reaction of increased control and increased distancing of displaced people as 'other'. The IPCC formula suggests that the privileged lives of some could lead to 'existential risk' for all forms of life on the planet (Bostrom 2011). The 'price of inequality'—national and global—has escalated. The notion that this is ethically or politically sustainable is also debunked in this paper. Currently, my practical concern is that there are more refugees and asylum seekers now than ever before and the current form of democracy and governance does not work. Policy researchers, practitioners and educators need to work with many forms of knowledge and across conceptual and spatial boundaries to address the big issues of the day, namely peace, gender mainstreaming and social, economic and environmental security within increasingly densely populated cities that need to respond to increasing poverty and climate change. This has implications for the quality of life from a human, non-anthropocentric and environmental perspective. We need education that enables co-determination of our future and that protects both human and environmental well-being. My emphasis in social theory on the ecological footprint has been to theorise the need to shift from merely building the natural and social capital thesis to more organic interventions based on 'being the change' and mobilising with people to redesign the way we live through the appreciation of living systems of which we are part. Reforming economics and accounting and accountability needs to be driven systemically (see Systemic Ethics 2014).

[16]Whilst undertaking a 3-day workshop with the Ministry of Social Affairs in Jakarta (MOSA) one of the key areas of concern was food security. Conversations with staff at NGOs in Cape Town confirmed their concerns about food security.

global neighbourhood programme could be a way forward to evolve in ways that respect others (including the voiceless) and the environment on which we depend. We are the land and by recognising our hybridity and vulnerability we could become more resilient too. Public education needs to address human capacity and capabilities to live in ways that regenerate the land 'through caring for country.'

The chapter addresses ways to address the challenges and opportunities through multicultural education that responds to current challenges, namely the need to live with diversity and differences of opinion on a daily basis. It stresses the need for public education to protect the environment and to develop our capabilities to think across conceptual and spatial boundaries. In a globalising word we are faced with diversity on a daily basis. I will compare South Africa and Indonesian challenges for education. In both contexts populations are growing in increasingly urbanised areas where the divides between rich and poor are widening.[17]

What follows is a suggestion based in part on the work and suggestions I have gained through working with Ph.D. students and academics in Indonesia and South Africa doing research on a range of policy concerns[18] that result from the impact of increased urbanisation on quality of life. Food deserts are the likely scenario if more emphasis is not placed on balance, greening cities and supporting small farmers. I advocate a transformation in the developing curriculum.

The UN Sustainable Development Goals and targets do not go far enough.

Education needs to:

[17]As stressed above we face challenges that will need to be urgently addressed by current and future generations if the global climate change agreement is to have any real impact on mitigating: Climate change—through addressing food, energy and water in increasingly urbanised world living in unsustainable cities. Poverty and food insecurity. Urbanisation along with, poverty (hunger, unemployment and crime).

[18]How do you hold in mind more than one big idea at the same time and why is this important for addressing policy paradoxes? Addressing wicked problems requires choice architectures to nudge people towards better decisions to address complex health, housing and social inclusion, mitigation and adaptation to climate change by enabling people to engage in thinking through options in ways that enhance democracy and governance. Strategic capacity for addressing complex, wicked challenges of poverty and protecting the planet are needed to mitigate and adapt to climate change. Leadership skills are needed to work across sectors and across disciplines which requires knowledge management, communication skills and research skills. Challenges for both South Africa and Indonesia include growth of cities. Urbanisation poses the challenge of people living in food deserts where they have to compete for resources. In South Africa and Indonesia, for example at least 65 and 75% of the population, respectively, will be living in urban areas. Urbanisation poses a systemic threat to the survival and sustainability of culture as we know it today. What is the point of raising these concerns? Opportunities exist for both South Africa and Indonesia include creating greater balance between urban and rural areas, greening cities and striving for better co-determination across regions. This requires a new architecture for governance democracy and ethics—learning mindfulness, setting up and co-creating. The protection of diversity is stressed in Australia through Aboriginal notions of custodianship of the land and their dreaming sites and through Ecological Citizenship endeavours at the local government level. Protecting biodiversity in Indonesia is also a priority. The challenges pose opportunities to provide education and leadership. What is the way forward?

- Address ways to regenerate the rights of farmers (the majority of whom are women) to make decisions about the way they farm and to give opportunities for them to achieve their educational goals on the land, rather than encouraging excellent students to leave the land and to pursue a career in the city.
- Emphasise social inclusion through building the strategic capacity of women to address food, energy and water security.

The emphasis is to address healthy lives and promote well-being for all at all ages through inclusive and equitable public education through promoting life-long learning opportunities for all to address gender equality and to empower all women and girls. It strives to address ways to ensure access to affordable, reliable, sustainable and modern energy for all in cities and human settlements that are more inclusive, safe, resilient and sustainable. The emphasis is on addressing sustainable consumption and production choices across state, market and civil society with the emphasis on the most marginalised and focusing on ways to protect the poor and the affected communities, focusing on women, youth, local and marginalised communities to understand mitigation and adaptation processes. I make the case for:

- Co-determination in regions—new architecture for governance democracy and ethics[19]

[19]Nussbaum (2006, 2011) stresses the need to address human capabilities within and beyond nation states. She stresses that the needs of the voiceless including asylum seekers, young people and the disabled as well as sentient non-human animals need to be respected. She stresses the need to extend the social contract to support the capabilities approach. In discussing human capabilities she stresses health, safety, bodily integrity, education, standard of living, quality of social interaction, productive valued activities, environment, play and basic needs (Nussbaum 2011: 33–34). Introducing a way to think about what we value and as a first step to reframing economics and accounting in response to the plea made by Joseph Stiglitz et al. in 'Mis-measuring our lives', but it also underlined that for transformation to occur, that being the change involves working across many areas of life, social, economic and environmental and that this could lead to quantum change in the sense used by Turok. More importantly this approach to accountability could help to address the size of our ecological footprint by thinking about cultural flows and how our decisions shape the land and the life chances of future generations.

The architectures for governance and democracy need to protect citizens and non-citizens within and beyond the nation state.

Thus my work on systemic ethics is on discussing ways to extend the social contract by enabling people to think about the consequences of their decisions in the short, medium and long terms for themselves, others (including non-sentient beings) and the environment.

But more importantly it needs to contribute to cultural transformation that goes beyond sustainability as it is currently framed. It does not go far enough as it remains uncritical of the economic systems that is so entangled in the creation of the banality of evil. Instead we need to build in a learning organisations approach that could help to enhance the understanding of the UN sustainability development goals and enable implementation and monitoring. The need to protect food and biodiversity requires a public education policy that develops the capability of people to join up the dots so that they understand the systemic flow on effects of decisions that affect social and environmental justice for current and future generations of life. This is more than creating sustainable development goals that respect social, economic and environmental factors that support well-being. The decision as to what to include or exclude in order to make policy that is socially and environmentally just needs to be based on careful ethical considerations. Dobson (2003)

- Co-learning and Mindfulness that support building communities of practice
- Supporting learning communities that help us to think across disciplines and cultures, in order to support co-determination of our future within the region
- Developing new curricula that are supported by architecture for democracy and governance
- Extending solidarity and protection to all forms of life within a region, rather than limiting protection and thus limiting human security which is dependent on biospheres not national boundaries.

It has been (wrongly) assumed that we can continue to develop and modernise the planet. It has been assumed that growth in the economy will sustain a growing population. But this is mistaken. We need to understand that the current way of life is unsustainable and that we need to rethink many of the dimensions of modern culture. Culture after all is simply a way of life and a response to the challenges that we face as human beings. The way we think shapes who we are and the sense we make of our daily experiences.

Neuroscientists such as Cliff Sarron and Al Kasniak stress that mindfulness research has shown how thinking affects the material body that we inhabit and the way we think shapes the body and the environment on which we depend. The environment affects the body and mind. For example, the research by Sarron and Kasniak has found that people who exercise in a natural environment have higher levels of concentration than people who exercise in a built-up environment. Thus the more we decimate the environment the worse it will be for our ability to recreate ourselves when we take a break from learning, working and teaching. Another key

(Footnote 19 continued)

distinguishes between environmental citizenship versus ecological citizenship. The former relates to the role of the state through law and control, the latter refers to the personal choices made by citizens for the common good as they relate to the ecological footprint. This research also draws on the work of Buckminster Fuller on governance of 'Space Ship Earth' and Boulding's (1966) extension of the work to extending domains of knowledge and co-evolution of living systems. The quality of life approach, sufficiency approach and the human security approach (on which this research will be based) strives to emphasise both dimensions. The chapter addresses the extent to which the engagement in public education helps to promote greater resilience through managing food, energy and water consumption and to what extent participatory approaches enhance governance outcomes that support social and environmental justice.

Six principles of sustainable governance underpin the Lisbon principles (1998) to cite and paraphrase these principles are 'responsibility (individual and collective)', 'scale matching (at multiple levels)', 'precaution' rather than narrow minded profit, 'adaptive management' based on integrating as much social, economic and environmental information as possible, 'full cost allocation' based on considering the inter connected dimensions of social and environmental considerations and 'markets should be adjusted' and 'participation based on engagement' using discursive dialogue. Previously the work of Beck (1992, 2005, 2009) discussed the so-called 'boomerang affect' of poverty and pollution, Beck (2009: 96) stresses the need for sub-politics 'from above' and 'below'. An alternative future needs to be modelled and co-determined by with and for people of good will (Dobson 2003). The evolutionary direction is shaped by values, emotions and the ability to think critically and rationally (Bateson 1972; Commoner 1971, 1976; Hulme 2010).

finding from neuroscience is that the telomeres or parts of the cell that protect us from ageing are protected when we have a sense of purpose, when we use all our capabilities each day in an environment that is rich in nature and not degraded.

5.1.4 Critical Systemic Praxis

I am proposing the need to develop participatory action research[20] on education for current and future generations living in sustainable rural and urban environments. I propose the following needs to be researched:

Comparative case studies of the curricula at leading secular and non-secular secondary and tertiary institutions in Jakarta and the surrounding agricultural region could explore policy implications for sustainable development opportunities. It makes the case that the following are amongst the greatest challenges facing humanity and that they potentially pose an 'existential risk': representation of increasingly diverse populations within nation states (b) accountability to ensure that resources are used fairly and equitably and sustainably and (d) engagement to enhance the capability of people to understand that attitudes to consumption have profound implications for social and environmental justice issues. We need to do participatory action research that aims to contribute to discovering and piloting new forms of architectures for public education and engagement to support user-centric public policy and to enable multilevel, multi-agency approaches for social and environmental justice.

5.1.5 Developing an Alternative Vision for Education and Employment

Through striving to reveal 'in the small new ways of seeing the whole' (Adelman 2013: 9) we could co-create new ways of engaging adults and young people in learning for a sustainable South Africa. This could provide lessons for cosmopolitan approaches to rediscovering ways to live through recycling and using resources creatively. It could also foster a sense of stewardship rights and responsibilities (Flannery 2012) that are translated into new approaches to economics and business.

[20]Participatory and systemic approach to case studies will be applied in order to address social, cultural, political, economic and environmental dimensions (see Carspecken 1996, on educational ethnography and McIntyre-Mills 2006; Outhred et al. 2013 on systemic governance approaches). In line with the Paris Declaration, the research will be framed together with key stakeholders in public and private secular and non-secular secondary and tertiary institutions. The Paris Agenda of 1997 stresses the importance of involving all those party to research and development to be part of the design process and evaluation of the initiative. The approach will be to engage with students and staff over a 3-month period through on-site observation, participation, focus groups and interviews.

Where success is valued not in terms of growth and profit, but in terms of distribution and living sustainably by valuing quality of life supported by:

- Socially inclusive communities in which people feel safe, appreciated, respected and happy and in which they are accountable for the governance of sustainable resources (see Ann Florini and Carol Graham of the Brookings Institute).
- Economically sustainable communities where people work and live in ways that create and recreate an environment that will continue to support the current and future generations through caring for food, energy and water supplies.[21]

This community of practice research approach is significant in so far as it contributes to developing insight and foresight into many different ways of knowing and the importance of appreciating diversity in the tertiary education and training sector. It needs to strive to foster and provide pathways for diverse forms of knowledge including '*verbal, visual, physical, musical, mathematical, introspective and interpersonal*' (Bounfour and Edvinssen 2005; Gardner 2008) and thus develop human capacity to address complex socio-environmental challenges (Ehrlich and Ehrlich 2013).

Food security is also being challenged in South Africa as a result of the move to develop biofuels from sugar cane. In an article entitled 'There is a gold rush to invest in ethanol', Carnie (2014: 5)[22] expresses concerns about the way in which arable land will be used for the production of sugar for fuel. He stresses that the Jozini Dam in Pongolo does not have adequate water supplies. Sugar may not be such a loss to the South African diet but if other arable land and other crops are used for fuel when they could be used for food it will become problematic. There is an

---

[21]Resilience is defined as the adaptive capacity of the physical environment, of an individual or of a group. It concerns factors such as the capacity of members of a community to act together and to be able to modify or even transform, existing ways of life (Hulme 2009; Rose 2005). Research needs to draw on diverse ways of knowing (Cruz et al. 2009) that could support regional governance (Wear 2012) to support effective environmental management. We know that environmental sustainability and human well-being are intimately linked, but there is little knowledge about how this linkage can be built upon to improve both areas. Attempts to address climate change are often based on policy information that is not grounded in citizen experiences and fails to address what we do know about human behaviour or choices. The inherent link between engagement in civil society and community well-being (rather than the economic bottom line) needs to be the focus of research. The concepts of stewardship and resilience are important notions through which to explore the nexus across well-being, consumption choices and the environment. We need to develop a deeper understanding of how the intangible aspects of perceived well-being can be measured. But we also need to measure them in relation to the link between well-being and sustainability (Stiglitz et al. 2010). The discourse above remains rooted in the notion that bounded nation states are inevitable. The argument developed in this article is that the nation state needs to take as a given that the cosmopolitan human rights come first and that nation states serve regional needs. They need to be seen as part of an overlapping region with responsibility at a regional level for food security and human security. The costs of climate change will place a heavy burden not only on the state but on the surrounding regions. Benhabib (2007) and Archibugi (2010) stress that instead we need a form of federalist republicanism. But how do we move from the containerist approach that undermines justice through limiting compassion to citizens at the expense of human rights?

[22]Carnie, T 2014 'There is a gold rush to invest in ethanol', Cape Times August 20.

inadequate emphasis on food production for an ever-growing population and there needs to be more emphasis on planning renewable sources of energy.

The Food Security Network (Crush and Fayne 2010; Battersby and Crush 2014; Frayne et al. 2014) has concluded that the reliance on growing food in urban areas whilst a step in the right direction is not enough to address the needs of an increasingly urbanised population in South Africa. Instead the emphasis needs to be on supporting agriculture and arable land.

5.1.6 Participatory Policy Research Needs to Support Curriculum Development and Co-determination for Social and Environmental Justice

The rationale for working at the local, national and international level to support mindful critical and systemic thinking is that with growing urbanisation there will be a need for actively engaged young people to address democracy, climate change, food security and sustainable living in urban and regional environments. This has implications for the big issues of the day, namely representation, accountability and sustainability. Learning to live sustainably and well requires knowledge spanning social, cultural, economic and environmental issues. This requires social engagement and the fostering of the capacity of members of a community to act together and to be able to modify or even transform, existing ways of life (Rose 2005; Hulme 2009). This is the premise on which this approach is based.

Arendt (1972) emphasises both potential and risks through drawing attention to evidence of 'the banality of evil'.[23] She considered the broad context and the structures that lead people to make unethical choices. Arendt's (1972) work 'Crises of the Republic' on how lies and spin contribute to undermining democracy remains relevant as does her report on Eichmann's trial. Arendt stressed the implications of being part of an unquestioned monstrous system that becomes an

[23]The book by Jonasson (2014) a Swedish novelist (and former journalist) entitled the 'Girl who saved the King of Sweden' stressed that the difference between genius and the idiot is that the genius knows where to draw the line. This requires careful judgement of many dimensions. In this dark comedy he conveys in deceptively simple prose 'the banality of evil' in South Africa and in world politics. It tells the story about a young South African girl from Soweto, who works in a sanitation department. She is a self-taught genius who is given no credit and who is exploited by a corrupt system and complicit people. Their cruel thoughtless lives are based on an assumption of their god given rights and a denial that they have stolen the land on which they live. The indigenous people are treated as 'hewers of wood and drawers of water under the system of Apartheid', as designed by the architect, Verwoerd. The existence of a nuclear programme in Pelindaba was kept secret from the world, except for the collaboration of the Israelis who benefited from South Africa's programme. The wry, dark humour addresses some of the issues that remain off limits in world politics. The silencing of those who have tried to raise difficult issues such as nuclear weapons is no secret. But the power to determine the agenda for discussion is held by the powerful. The morality of this agenda and the potential for mishap is the topic of Jasonoff's novel.

unquestioned culture and a taken-for-granted system of bureaucracy. To avoid taking things for granted it is important to develop critical thinking based on the capability to think at a meta-level about the implications of choices (Van Gigch 2003), but also to have in place constitutional structures that protect social and environmental justice for this generation and the next (Jessop 2009).

The next phase of research needs to explore what the UN development goals mean in day-to-day lives. How do culturally diverse participants perceive ecological rights and responsibilities at home, at work and at leisure? To what extent are the choices made on the basis of conscious ethical deliberation on how we house to live? To what extent do participants understand their rights and responsibilities to limit the size of their personal ecological footprints? Can public engagement make a difference to the way in which diverse cultural participants in urban and rural regions in Australia and Indonesia perceive ecological citizenship and the way in which they manage ecological footprints? Growing awareness of the vital role of biodiversity for human well-being has led governments and civil society to set targets to reduce species loss. In line with the Paris Declaration (1997) the research conversations will be framed together with co-researchers with local lived experience. We will facilitate cross-cultural conversations on sustainable living and species diversity. The focus will be on how these aspects relate to food, energy and water.

The conundrum this next phase of the project addresses is that nation states need to find ways to enhance a sense of shared ecological citizenship and responsibility to address food, water and energy security:

First, we need to explore the perceived implications of the stewardship of biodiversity for well-being and resilience. The concepts of stewardship and resilience are important notions through which to explore the relationships across biodiversity, environment, human food, energy and water security and stewardship. Stewardship is based on awareness that the land and biologically diverse ecosystems are a cultural heritage (Flannery 2012) on which the well-being of current and future generations depend. Resilience is defined as the adaptive capacity of the physical environment, of an individual or of a group. It concerns factors such as the capacity of members of a community to act together and to be able to modify or even transform, existing ways of life (Rose 2005; Hulme 2009).

Second, we need to explore whether collaborative approaches drawing on diverse ways of knowing (Cruz et al. 2009) and citizen-centred self-governance of ecological footprints (Podger et al. 2012) could support regional governance (Wear 2012) in line with the UN Sustainable Development Goals. We will try to learn more about diverse ways of knowing (Cruz et al. 2009) that could support effective environmental management by taking into consideration: (1) Non-anthropocentric knowing drawing on the environment as well as knowing through all our senses—including empathy and intuition, (2) Respect for 'know how' developed through empirical trial and error, (3) Creativity in maintaining sustainable relationships with the land, (4) Respect for spiritual awareness, the wisdom of experience as well as the social and natural sciences, Artistic expression of feelings, perceptions and emotions, (5) Caring for future generations of life, and (6) 'Phronesis' or

Aristotelian wisdom to match the right knowledge in context. Thus the research explores the following:

- A priori norms for ecological citizenship
- A posteriori performance measures for ecological citizenship.

The will to make a difference is the motivation for change. We cannot rely on the 1% to make the difference and we cannot expect that the powerful will want to make vast changes to the status quo. The temptation to continue to focus on critique of the 1% along with pleas for transformation is great, but we also need to recognise the 'banality of evil' in the sense used by Arendt. I discussed[24]

- The complicity evident in entire societies that do not focus on the elephants in the room, namely that our way of life is unsustainable.
- Three basic discourses on the notion of truth along a continuum from one truth based on modernist enlightenment to multiple or no truths, based on extreme versions of post-modernist thinking and I explore a middle ground based on co-creating meanings within context. Each discourse is unpacked in terms of policy discourses and the implications for social and environmental justice elsewhere (McIntyre-Mills 2000).

5.1.7 Capacity Building Through Scenarios at the Personal, Interpersonal and Organisational Level

This requires respectful engagement across conceptual and spatial boundaries. Open interfaith and cross-cultural dialogue within and across cultures and to think in terms of transdisciplinarity to address social and environmental justice. The wisdom of many religions can be found in Islam, Buddhism, Hinduism and Christianity. The notion that a belief in the responsibility of each human being to others, the voiceless and the environment is recognised by all indigenous religions and is retained in all mainstream religions.

I propose critical participatory action research on learning organisations and learning communities for public education in post-national regions as a way forward.

---

[24]The paper raised questions about the way in which 'clash' is defined and the way it plays out. I returned from South Africa to South Australia to find that a bill had been proposed to stress that women wearing burkas be requested to sit in an enclosed part of parliament. This bill proposed by the speaker has not been supported by Prime Minister Abbott (ABC news 3 October 2014). Today a bill will be presented in the Australian parliament as to whether Australia will follow Ireland in supporting equality in marriage. But the injustice towards asylum seekers who arrive by boat to Australia remains an area of silence and frontiers of justice remain closed to asylum seekers. No questions are asked because no information is provided.

This is needed to protect current and future generations so as to achieve sustainable rural and urban living environments that balance individual and collective needs. What is a learning organisation within a learning community and post-national region? It is an organisation that is open to the social, economic and environmental context and responds to the needs of the people it purports to serve. It helps to develop opportunities to test out ideas with others. It considers the consequences for self, others and the environment by teaching the ability to think about the consequences of decision-making through applying critical heuristics (see Systemic Ethics, McIntyre-Mills 2014). Future perceived scenarios necessarily encompass complex social, political, economic, cultural and environmental dimensions (see Kahane 1992: 3). Drawing on the Mont Fleur Scenarios, I ask workshop participants in Australia, Indonesia or South Africa to think about how scenarios about how we can think differently about food security:

- What are the implications of denial (or being an ostrich)? Denial leads to 'business as usual' Please give examples in terms of the area of concern, namely food, jobs and gender mainstreaming
- What are the implications of being halfhearted (a lame duck)? This means we do too little too late and so we are stuck with the results! Please give examples
- What are the implications of flying too close to the sun (Icarus)? This is an extreme response which could be counterproductive! Please give examples
- What are the implications of coordinating and working together/flying together for the long haul (being flamingos)? Coordinated long-haul flight to achieve goals.

Box 1

Thinking exercise to support capacity building through facilitating action research for social and environmental justice was adapted from user-centric policy design (see McIntyre-Mills and De Vries 2008: 308). The research aims to explore the extent to which people understand the implications of their choices as to what they should eat. To what extent do people understand that whether they decide to eat has an impact on our ecological footprints? Some food requires more energy and water, depending on whether the food locally grown and locally consumed. The growth of cash crops for marketing at a distance can add costs to people and the planet. When the fortunate eat a punnet of strawberries out of season flown in from elsewhere it makes a difference to the planet. Particularly if it is sold in a plastic container with a polystyrene base, as is often the case.

- How does food security/food insecurity help other aspects of life? For example food enables one to concentrate at school
- How does the bubble hinder other aspects of life? Having food at lunch time means you spend time socialising with others

- How important is the bubble? (Use scale like one below) not at all a bit doesn't matter quite important very important. If I solve this problem or have this asset first, does it make solving other problems easier?
- Do these things always happen together? Or one after another?
- How do I achieve it?
- How do I avoid it?
- Where can I get help for it?
- Who can I help and how, if they need this or have this problem?
- Is it sometimes good and sometimes bad—in what situations?
- Are there other names/terms for the same thing?
- What can stop me from (or make it really hard) getting/achieving it?
- Are there conditions I have to meet to achieve/get it? (Such as age, sex, children, income, employment, etc.)
- Is this a smaller or larger part of another issue? (like Physical Health is parent of diabetes)
- If one thing happens, does another thing usually follow? Both good and bad.

TASK

- Please map out all the factors that influence food security/in security and how it relates to well-being/lack of well-being in your opinion.
- Please write under the headings and draw your own pathways about how the bubbles connect.
- Add issues/needs/solutions as you discuss and think about things.
- You can work as a group or alone—people can choose.
- Add as many more questions/descriptions as they/you want.
- Add bubbles as well, if you want to.

- Please give examples and relate these to the areas of concern by reading/listening to the scenarios and working through the bubble/factor maps, below adapted from the map created by De Vries (2008):
- To what extent participatory approaches enhance an understanding of ecological citizenship rights and the responsibility to self-manage ecological Footprints
- Can Participatory Public Education in Australia and Indonesia on managing the size of our ecological footprint at the local government level nudge people towards changing their attitudes towards their rights and responsibilities as ecological citizens?

5.2 Co-determination in Regions—Challenges for Education: Policy Proposal to Develop Learning Organisations and Learning Communities

> What I feel about the future of cybernetics is that the ethical implications of cybernetics are so profound that we need to continue to do all we can to influence people's thinking to be more systemic. Any change in the understanding of causation and intentionality will have ethical implications. The obvious example is climate change. We have to do more to think about Earth's systems and our dependence on them, and we have to learn to communicate with one another about them in ways that will affect behavior (Mary Catherine Bateson 2016: 102)

Public education is needed to develop learning organisations and learning communities for public education in post-national regions. This is needed to protect current and future generations so as to achieve sustainable rural and urban living environments that balance individual and collective needs. A learning community within a post-national region is an organisation that is open to the social, economic and environmental context and responds to the needs of the people it purports to serve. It helps to develop opportunities to test out ideas with others. It considers the consequences for self, others and the environment by teaching the ability to think about the consequences of decision-making through applying critical heuristics (see Systemic Ethics, McIntyre-Mills 2014). Future perceived scenarios necessarily encompass complex social, political, economic, cultural and environmental dimensions (see Kahane 1992: 3). Drawing on the Mont Fleur Scenarios, I ask workshop participants in Australia, Indonesia or South Africa to think about how scenarios about how we can think differently about food security.

This research explores perceptions on ecological citizenship and rights and responsibilities to manage the size of our ecological footprint before and after running a PE curriculum on ways to manage the size of our ecological footprint and governing the Anthropocene to assess the impact of participation at the local level (Evans 2014) through awareness-raising about well-being and 'well-being stocks', sustainability (Stiglitz et al. 2010), resilience and biodiversity. It examines the value placed on ecosystems (Fisher et al. 2009) and how this impacts on beliefs and intentions.[25] The conundrum this book addresses is that nation states need to find

[25]This research explores the perceived implications of the stewardship of biodiversity for well-being, resilience. Well-being according to Stiglitz et al. (2010: 15) who use a multidimensional measure of well-being spans (1) material living standards (income, consumption and wealth), (2) health, (3) education, (4) personal activities including work, (5) political voice and governance, (6) social connections and relationships, (7) environment (present and future conditions) and (8) insecurity, of an economy as well as a physical nature. Leisure should also be valued. According to Stiglitz et al., the essence of the commissioned work's findings is that wealth needs to include stocks for the future—social, economic and environmental. The way to achieve this is in part through public education. Resilience is defined as the adaptive capacity of the physical environment, of an individual or of a group. It concerns factors such as the capacity of members of a community to act together and to be able to modify or even transform existing ways of life (Rose 2005; Hulme 2009). Biodiversity means "Current rates of species loss exceed those of

ways to address the biggest challenge of the decade, namely to find a way to engage large groups of people effectively and to enhance a sense of shared responsibility for environmental management and consumption. Significantly the research seeks to deepen an understanding of how people perceive local challenges and experiences and develop a new understanding of how to match the processes and governance structures to diverse interest groups—with different life chances and different socio-demographic characteristics. The research will assess the extent to which participation per se could help to promote greater understanding of rights and responsibilities to self-manage the size of our ecological footprints through combining cycles of open and structured dialogue that encourage people to think in terms of the consequences of their values about how we live our lives and how we ought to live our lives and the implications of these choices on the size of our ecological footprint.

The research will explore the extent to which people with diverse demographics, life chances and values show changes in attitudes towards before and after engaging in PE on rights, responsibilities and the UN Development Goals. The protection of biodiversity is stressed in Aboriginal notions of custodianship of the land and their dreaming sites. It is also stressed as a priority for Indonesia to protect the habitat of indigenous animals and to preserve agricultural land for food security at the Annual Council of Islamic Studies Conference, Balikpapan, East Kalimantan in 2014

(Footnote 25 continued)

the historical past by several orders of magnitude and show no indication of slowing. Major drivers of biodiversity loss on a global scale are land-use changes and agricultural intensification. These processes are threatening ecosystem functioning and services on which humans depend.... In consequence, there is a deep concern that a loss of biodiversity and deteriorating ecosystem services contribute to worsening human health, higher food insecurity, increasing vulnerability of ecosystems to natural disasters, lower material wealth; worsening social relations by damage to ecosystems highly valued for their aesthetic, recreational or spiritual values; and less freedom .... Growing awareness of the importance of biodiversity for human well-being has led governments and civil society to set targets to reduce ...loss" (see Convention on Biological Diversity; www.cbdint) (Lindemann-Matthies 2014: 195–6). The research draws on and extends Hulme's research on why we disagree about climate change (2009) and Cornelius' research on systemic approaches to understanding emotions (1996). Values, perceptions and emotions could be better understood from a systemic viewpoint that draws on interdisciplinary theories, in this case such as to 'why more equal societies almost always do better' (Wilkinson and Pickett 2009). The 'best–worst' option, namely democracy is in need of improvement (Hulme 2009; Giddens 2009). The specific challenge that this research addresses is to match forms of appropriate participation that are accessible to diverse groups and not viewed with cynicism (Cooke 2004). Significantly these questions assess whether participatory democracy and governance enhance sustainable living and well-being. It involves public education through local governments, schools, businesses and NGO organisations. It will identify the following: the decision-making context; constraints to achieving outcomes; elements of three scenarios (denial of the need to change, too little action too late, sustainable long-term adjustments); and key factors (variables) in facing up to the risks. It could contribute to the SD21 policy initiative 'Sustainable development in the twenty-first century'. The ability of governments to secure the environmental regions—on which the cities depend—to deliver services and resources will impact on the livability of our communities and human security within our region.

(see McIntyre-Mills 2015) and at the Food, Energy and Water Symposium (Indonesian Research Consortium, Jakarta 2015). To sum up the planetary passport advocates for the careful use of resources within national regions and post-national biospheres need to be governed by means of tiers of government from local to national to post-national. The resource use of each person in terms of haves, needs, what they are willing to add or discard from their lives, what the turning points are for better or worse, what the barriers are to achieving change all need to be carefully documented and updated as a diary entry that is logged on a monthly basis as detailed in Chap. 4 on the planetary passport.

These shared logs could become new identity documents that become a form of passport that controls the greatest consumers. The entries for resources could be verified by stewards connected with local neighbourhoods, local governments who give narrative accounts of the way of life of their local residents based on checking water, energy and food consumption. People who have light footprints should expect the right to live elsewhere, because their own habitats have been affected through the carbon choices made by powerful nation states and powerful elites. Those who are illiterate and innumerate ought to be exempt from this process until they receive education and ought to be protected by the closer governance of those who are privileged and living carbon heavy life styles by stewards responsible to Post-National Regional Governments that operate as a federation to uphold a co-determined global covenant. Producing one's own energy and food could ensure that people begin to live in ways that meet the requirements of the so-called Tuvalu test based on considering what if I were to make this decision or that, what would be the carbon consequences for others? A programme based on the multiple social, cultural, political, economic and environmental consequences of choices could be developed based on life choices measured by being, doing, having and interacting indicators that could enable urban populations to monitor their lives as we find ways to move away from a carbon intensive economy and society one that re-generates resources in local tiered communities that are supported by regional systems of agriculture. Instead of those who are suffering the worst effects having to give evidence to qualify for assistance, we should be applying the Tuvalu test to limit the size of the carbon footprints of emitters. In Australia the social contract protects those within the boundaries of the state but not those who seek asylum from elsewhere. The issue of bounded governance and decision-making that does not consider the consequences of social, economic and environmental choices for our neighbours has been raised by Joy Murray in the so-called 'Tuvalu test' that demonstrates how our social economic choices affect our neighbours through rising sea levels. For example, in Tuvalu and Samoa, agriculture has been affected by rising sea levels and increased levels of salinity. The control of each local community could be based on the principle of subsidiarity. Digital data on the distribution or redistribution of resources would need to be transparent and a basis for a form of taxation that rewards right living and penalises profligate living.

As Pikkety stressed in the Mandela lecture (2015) the data on money trails and wealth could be managed to ensure fairness and reciprocity. In some instances this would require some transfers to restore the balance.

- All public education on global citizenship ought to support understanding the notion of stewardship for current and future generations of life.
- All global citizens ought to be required to be accountable to others through a transparent system of local auditing to ensure that consumption is monitored and that regeneration and sustainable living becomes the norm for the urban elites not just the majority of the world's poor.
- The economy ought to be changed from measuring productivity to valuing the quality of the environment.
- A use value ought not to be placed on the environment, it needs to be considered beyond measure.
- People and their government need to be seen as agents to support the environment as principal, not as a resource.
- But instrumental values should not be placed on people, sentients or the environment. The example set by the new forms of constitution is that (potentially) it helps to achieve a cultural shift towards non anthropocentricism.

Box 2

Four scenarios are adapted from the Mont Fleur exercise conducted in South Africa (Kahane 1992). The four scenarios are as follows:

1. Scenario of the Ostrich in Denial leads to 'business as usual' and participants do not consider the consequences of their choices or deny the need to make any changes in their decisions about their everyday choices.
2. Scenario of the lame duck means we do too little too late and so we are stuck with the results! An example would be to make limited cuts that do not make enough impact on climate change emissions. Another would be accepting to do the minimal to protect displaced or marginalised people.
3. Scenario of Icarus flying too close to the sun considers the counterproductive implications of making too many changes too quickly. The scenario refers to making dramatic changes that could lead to the abandonment of changes for the better and could instead result in draconian form of conservatism. In contrast to the lame duck approach the Icarus approach makes dramatic changes that could lead to social instability. This symbolism is perhaps ironic given the subject matter!
4. Scenario of the flamingos who engage in a coordinated long-haul flight to achieve their goals. This scenario refers to making continual adjustments towards regenerating the social and environmental fabric of life.

- What are the implications of denial (or being an ostrich)? Denial leads to 'business as usual'. Please give examples in terms of the area of concern, namely food, jobs and gender mainstreaming
- What are the implications of being halfhearted (a lame duck)? This means we do too little too late and so we are stuck with the results! Please give examples

- What are the implications of flying too close to the sun (Icarus)? This is an extreme response which could be counterproductive! Please give examples
- What are the implications of coordinating and working together/flying together for the long haul (being flamingos)? Coordinated long-haul flight to achieve goals.

Please give examples and relate these to the areas of concern by reading/listening to the scenarios and working through the bubble/factor maps, below adapted from the map created by De Vries (2008):

- To what extent participatory approaches enhance an understanding of ecological citizenship rights and the responsibility to self-manage ecological footprints
- Can Participatory Public Education in Australia and Indonesia on managing the size of our ecological footprint at the local government level nudge people towards changing their attitudes towards their rights and responsibilities as ecological citizens?

5.3 Representation, Accountability, Regeneration and Sustainability Are Challenges for Education

The traditional economic model argues that growth in population helps to sustain the economy. But it is unsustainable and needs to be reframed (Stiglitz et al. 2010; McIntyre-Mills et al. 2014).[26] According to Turok (2014) we need to understand that boundaries in nature can be understood by reflecting on the smallest and largest scales. These provide lessons for policymakers. Just as a cell has pores that open or close to allow in nutrients and to expel pollutants, we need to enable people to understand how

[26]'Transformation from Wall Street to Wellbeing' discusses current research on an alternative architecture for governance. The concept 'Wall Street' is used to symbolise economic rationalism and 'well-being' to symbolise a reframed approach to 'being interconnected' and through treading lightly (McIntyre-Mills and De Vries 2012; McIntyre-Mills et al. 2014). It is an attempt at introducing a way to think about what we value and as a first step to reframing economics and accounting in response to the plea made by Joseph Stiglitz et al. (2010) in 'Mis-measuring our lives', but it also underlined that for transformation to occur, that being the change involves working across many areas of life, social, economic and environmental and that this could lead to quantum change in the sense used by Turok (2012). The companion volume, Systemic Ethics and Non-anthropocentric Stewardship (McIntyre-Mills 2014b), explores the need to develop the capacity to 'join up the dots' through participatory democracy and governance and through working across disciplines while preserving space for difference, based on the axiom that freedom and diversity need to be fostered to the extent that they do not undermine the rights of others in this generation or future generations of life.

to make decisions on what and who to include in the processes of decision-making. The decision as to what to include or exclude in order to make policy that is socially and environmentally just needs to be based on careful consideration, based on questions that address who, what, why, who, how and so what are the consequences. The following are amongst the greatest challenges facing humanity and that they potentially pose an 'existential risk': representation of increasingly diverse populations within nation states (b) accountability to ensure that resources are used fairly and equitably and sustainably and (d) engagement to enhance the capability of people to understand that attitudes to consumption have profound implications for social and environmental justice issues. Thus the participatory action research aims to contribute to discovering and piloting new forms of architectures for public education and engagement to support user-centric public policy and to enable multilevel, multi-agency approaches for social and environmental justice.

They exposed the abuses of power, namely surveillance and the way in which food is being commodified and controlled and the threats to our food security. One is a physicist come green campaigner and the other is a self-trained computer geek with a genius for hacking. They are both big picture thinkers who are capable of seeing how to transform at the local level. The third underlines the suffering of those outside the social contract.

Addressing diversity in perceptions is a central concern of my praxis. My role is one of building the capacity of graduate students in a range of cultural contexts including Southeast Asia, Canada, New Zealand, for example, and undertaking my own programme of research by working with diverse stakeholders in a range of policy contexts to enhance social, economic and environmental well-being through addressing, in particular:

Economically sustainable communities where people work and live in ways that create and recreate an environment that will continue to support the current and future generations through caring for food, energy and water supplies.[27]

[27]Resilience is defined as the adaptive capacity of the physical environment, of an individual or of a group. It concerns factors such as the capacity of members of a community to act together and to be able to modify or even transform, existing ways of life (Hulme 2009; Rose 2005). Research needs to draw on diverse ways of knowing (Cruz et al. 2009) that could support regional governance (Wear 2012) to support effective environmental management. We know that environmental sustainability and human well-being are intimately linked, but there is little knowledge about how this linkage can be built upon to improve both areas. Attempts to address climate change are often based on policy information that is not grounded in citizen experiences and fails to address what we do know about human behaviour or choices. The inherent link between engagement in civil society and community well-being (rather than the economic bottom line) needs to be the focus of research. The concepts of stewardship and resilience are important notions through which to explore the nexus across well-being, consumption choices and the environment. We need to develop a deeper understanding of how the intangible aspects of perceived well-being can be measured. But we also need to measure them in relation to the link between well-being and sustainability (Stiglitz et al. 2010). The discourse above remains rooted in the notion that bounded nation states are inevitable. The argument developed in this article is that the nation state needs to take as a given that the cosmopolitan human rights come first and that nation states serve regional needs. They need to be seen as part of an overlapping region with responsibility at a regional level

In growing cities, competition for resources will have an impact on human security. We aim to enable public education through primary, secondary and tertiary levels to work with local government to raise awareness about the implications of increased rate of urbanisation in Indonesia and the implications for human security as the competition for resources grows. Jobs, food, energy and water security will be the focus of the engagement process. The research consortium aims to extend and test ways to enable better decision-making to enhance representation, accountability and sustainability and extends research on cosmopolitan approaches to governance and accountability (McIntyre-Mills 2015a, b).

Professor Pratikno (2014) stressed at the presentation for the inaugural meeting of at Flinders that Indonesia and Australia need to foster regional partnerships through building on friendship networks and soft diplomacy. In 2014 Professor Victoria Farrar Myer (Fulbright Distinguished Chair) echoed this plea for greater emphasis on rebuilding relationships and emphasised the need to foster collective determinism through public engagement processes McRae (2014) of the Lowy Institute also echoes this concern:

> Finally, people-to-people ties remain a persistent challenge. Abbott's announcement of a new Australian Centre for Indonesian Studies is a positive initiative, as is the Coalition's signature New Colombo Plan. Neither though is a replacement to committing the resources required to promote Indonesian literacy in Australia. This requires both maintaining teaching infrastructure for Indonesian language in Australia, as well as promoting career options that would encourage Australians to attain Indonesia literacy.[28]

In South Australia and Australia education and development needs to support language skills so that cross-cultural communication is aided. Education also needs to foster an understanding of the challenges of living in one of the driest continents. The recent IPCC reports stressed that South Australia will be affected by climate change and that rainfall levels will decrease. In terms of public policy, Australia has announced the importance of 'studying Indonesian and fostering regional ties'. The following statistics were presented by Praktino and Akbar Susamto (2014) at a seminar at Flinders University stressed that the size of the Indonesian economy would increase from the sixteenth largest to the seventh largest economy in the world, that it would grow from 45 to 145 million members of the so-called 'consuming class'. 55 million skilled workers in the Indonesian economy to 135 million workers in the Indonesian economy who generate 0.5 trillion Australian dollars and this will grow to 1.8 trillion in consumer services, agriculture, fisheries and education".

(Footnote 27 continued)

for food security and human security. The costs of climate change will place a heavy burden not only on the state but on the surrounding regions. Benhabib (2007) and Archibugi (2010) stress that instead we need a form of federalist republicanism. But how do we move from the containerist approach that undermines justice through limiting compassion to citizens at the expense of human rights?

[28]http://www.abc.net.au/news/2013-10-02/mcrae-challenges-ahead-for-the-indonesia-relationship/4993720.

According to Rosyada (2014) the Ministry of Religious Affairs administers only about 24%, of madrasah, while the rest are students of private schools. Thus the role of the government in relation to the provision of non-secular education very low compared with private-owned institutions. According to Rosyada, education strives to address the curriculum and to develop the capacity of teachers as mandated in the constitution.[29] The opportunity exists to do the following:

- Teach learners to think systemically and to join up the dots so that they understand the policy implications of their choices.
- Help them to realise that decisions have consequences in the short, medium and long term.
- Teach them systemic ethics in both secular and non-secular places of learning.
- Help places of learning at all levels to connect with communities.

5.3.1 Co-learning and Mindfulness that Support Building Communities of Practice

- Supporting learning communities that help us to think across disciplines and cultures, in order to support co-determination of our future within the region.
- Developing new curricula that are supported by architecture for democracy and governance.
- Extending solidarity and protection to all forms of life within a region, rather than limiting protection and thus limiting human security which is dependent on biospheres not national boundaries.

It has been (wrongly) assumed that we can continue to develop and modernise the planet. It has been assumed that growth in the economy will sustain a growing population. But this is mistaken. We need to understand that the current way of life is unsustainable and that we need to rethink many of the dimensions of modern culture. Culture after all is simply a way of life and a response to the challenges that we face as human beings. The way we think shapes who we are and the sense we make of our daily experiences. Neuroscientists such as Cliff Sarron and Al Kasniak stress that mindfulness research has shown how thinking affects the material body that we inhabit and the way we think shapes the body and the environment on which we depend. The environment affects the body and mind. For example, the research by Sarron and Kasniak has found that people who exercise in a natural environment have higher levels of concentration than people who exercise in a built-up environment. Thus the more we decimate the environment, the worse it will be for our ability to recreate ourselves when we take a break from learning,

[29]"Control school curriculum, school text book, and the capacity of the teacher, because Indonesia still on a serious efforts to develop moral capacity of the students for being a pious professional, plurals, inclusive, and behaving as a multicultural society, as was mandated in the constitution 1945".

working and teaching. Another key finding from neuroscience is that the telomeres or parts of the cell that protect us from ageing are protected when we have a sense of purpose, when we use all our capabilities each day in an environment that is rich in nature and not degraded.

The will to make a difference is the motivation for change. We cannot rely on the 1% to make the difference and we cannot expect that the powerful will want to make vast changes to the status quo. The temptation to continue to focus on critique of the 1% along with pleas for transformation is great, but we also need to recognise the banality of evil, the complicity evident in entire societies that do not focus on the elephants in the room, namely that our way of life is unsustainable.

With growing urbanisation Indonesia faces challenges such as ensuring that the future economy responds to the need for highly skilled and democratically engaged citizens with a focus on living well and sustainably in city environments that will need to be supported by sustainable agricultural regions. The basis for a democratic and sustainable region rests on regional networks to support human security resting on meeting the quality of life concerns of the current and future generations through securing sustainable cities within sustainable agricultural land. This rests on developing appropriate education and participatory democracy and governance. The research builds on recent publications that consider whether a change in the architecture of democracy and governance could balance individual and collective needs more effectively through participation, guided by the principle of subsidiarity and the axiom that people ought to be encouraged to be free and diverse to the extent that their freedom and diversity does not undermine the rights of others.

This requires a learning organisations approach that could help to enhance the understanding of the UN Sustainability Development Goals and enable implementation and self-monitoring based on Dobson's (2005) notion of ecological citizenship.

5.3.2 Learning Organisations Span Sectors to Support Non-anthropocentric Designs that Do not Commodify Others or the Environment

Policy on education needs to:

- Develop the capability of people to join up the dots so that they understand the systemic flow on effects of decisions that affect social and environmental justice for current and future generations of life. This is more than creating sustainable development goals[30] that respect social, economic and environmental factors

[30] http://sustainabledevelopment.un.org/?menu=1300

"First session of the Open Working Group on Sustainable Development Goals

The intergovernmental Open Working Group on sustainable development goals called for in the Rio+20 Outcome Document convened its first meeting on Thursday and Friday, 14–15 March

that support well-being. Public education needs to encourage respectful dialogue across conceptual and spatial boundaries. Open interfaith dialogue needs to occur within and across cultures and to think in terms of transdisciplinarity to address social and environmental justice. The wisdom of many religions can be found in Islam, Buddhism, Hinduism and Christianity. The notion that a belief in the responsibility of each human being to others, the voiceless and the environment is recognised by all indigenous religions and is retained in all mainstream religions.
- Understand that boundaries in nature can be understood by understanding the smallest and largest scales. These provide lessons for policymakers. Just as a cell has pores that open or close to allow in nutrients and to expel pollutants, we need to enable people to understand how to make decisions on what and who to include in the processes of decision-making. The decision as to what to include or exclude in order to make policy that is socially and environmentally just needs to be based on careful consideration, based on questions that address who, what, why, who, how and so what are the consequences.

The conundrum for public education is to enable the nation state to find ways to address the biggest challenge of the decade, namely to find a way to engage large groups of people effectively and to enhance a sense of shared responsibility for environmental management and consumption. Significantly the book seeks to deepen an understanding of how people perceive local challenges and experiences and develop a new understanding of how to match the processes and governance structures to diverse interest groups—with different life chances and different socio-demographic characteristics. The research will assess the extent to which participation per se could help to promote greater understanding of rights and responsibilities to self-manage the size of our ecological footprints through combining cycles of open and structured dialogue that encourage people to think in terms of the consequences of their values about how we live our lives and how we ought to live our lives and the implications of these choices on the size of our ecological footprint.

The research needs to explore the extent to which people with diverse demographics, life chances and values show changes in attitudes towards before and after engaging in PE on rights, responsibilities and the UN Development Goals. The protection of biodiversity is stressed in Aboriginal notions of custodianship of the land and their dreaming sites. It is also stressed as a priority for Indonesia to protect the habitat of indigenous animals and to preserve agricultural land for food security at the Annual Council of Islamic Studies Conference, Balikpapan, East Kalimantan in 2014 (see McIntyre-Mills 2015) and at the Food, Energy and Water Symposium (Indonesian Research Consortium, Jakarta 2015) to support a learning organisation and learning community approach that could help to enhance the understanding of

(Footnote 30 continued)

2013, in the UN General Assembly Hall in New York. The meeting was addressed by the Secretary-General and the President of the General Assembly".

the UN Sustainability Development Goals and enable implementation and self-monitoring based on Dobson's (2005) notion of ecological citizenship.

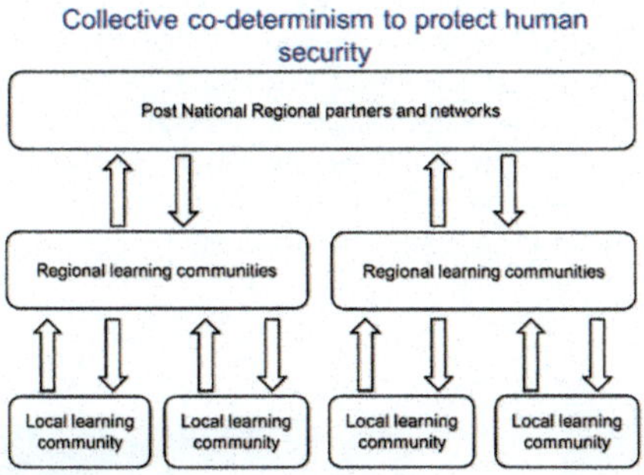

Stewardship for future generations underpins the philosophy of Aboriginal custodians of the land and their dreaming sites. However, there is evidence that many non-Aboriginal urban citizens wish to spend more time living slower lives, walking wherever possible, riding bikes, growing local food, recycling and reusing and consuming less, instead of living stressful, competitive lives that save time, but waste resources as they are reliant on fast food in 'throw away' containers, fast travel and a 'time is money attitude' (McIntyre-Mills 2012a, b, c; McIntyre-Mills and De Vries 2012; McIntyre-Mills et al. 2014). These findings clearly link environmental health and human well-being and raise the question of what can we learn from mobile and egalitarian, place-attached people (Vaske and Kobrin 2001) whose history is recorded in the landscape (Guddemi 2006; Rose 2004). Education needs to rediscover stewardship rights and responsibilities (Flannery 2012) and to become accountable. These goals are considered essential for the governance of sustainable resources by Florini (2003) and Graham (2011) of the Brookings Institute. More applied research is needed at the interface of capacity building, knowledge and biodiversity management, consciousness studies and systemic governance (McIntyre-Mills 2006b, 2010, 2014b) in order to contribute to new ways to address the systemic social, economic and environmental challenges we face today.

We are the land and by recognising our hybridity and vulnerability we could become more resilient too. We need to protect biodiversity through learning organisations. Education policy needs to support the development of curricula that enables people to join up the dots so that they understand the systemic flow on effects of decisions that affect social and environmental justice for current and future generations of life. The rights of sentient creatures were discussed as part of the Asian Conference of Islamic Studies (ACIS) but needs to be translated into practice. The Islamic ban or fatwa on the destruction of endangered animals is one way to try to prevent their demise. Awareness of the need to create sustainable development goals (United Nations 2013) that respect social, economic and environmental factors is vital. Despite the Dyak's cultural emphasis of the iconic hornbill and its use as a logo for the ACIS conference it faces extinction. The attempts to raise awareness within the region through newspaper articles and airport posters need to be embedded in public education and sanctions to protect the natural heritage on which current and future generations depend:

> It is very important to start a movement for the protection of this near extinct bird. Social organizations like the Dayak Custom Communal Council (DAD) must be engaged in the campaign…. NGOs, activists and academics, working together to arouse public sentiment and make people understand the significance of protecting helmeted hornbills. ….Apart from hunting, population decline is caused by environmental degradation and habitat damage. Forest Watch Indonesia in its 2011 report said the forest destruction in Kalimantan during 2000–2009 was over 36 percent, higher than any other island in Indonesia.

Source http://fatbirder.com/links_geo/asia/indonesia_kalimantan_selatan.html. Accessed 8/06/2015

Author's photos taken in Balikapan airport

Authors photos taken in Balikapan airport

5.4 Summing Up: Co-determination Within Biospheres: Rethinking Architectures for Teaching and Learning Based on the Pillars of Morality: Empathy and Reciprocity[31]

The tendency to think in terms of specific disciplines rather than with respect for other ways of thinking has resulted in so-called 'enlightened', ethnocentric and anthropocentric thinking and practice that has led to policies that have affected the quality of agriculture and water management. My area of concern is to introduce a more responsive and open, transdisciplinary and cross-cultural approach to teaching and learning. Instead of making decisions in a top-down manner we need to practice democracy from below so that decisions match the needs of local people. The Ghandian moment has arrived. We need to do and use what we can to make a

[31]The social contract extends rights and expects responsibilities to be fulfilled in return. But what about those who are voiceless, disabled, too young or without citizenship rights (displaced, asylum seekers or refugees)? In order to address this area of concern it is necessary to with a range of different theories spanning disciplines and to take inspiration from transdisciplinarity as well as cross-cultural inspiration on cultural flows. Stanford research on non-anthropocentric approaches to fairness and unfairness shows that primates and other animals understand the concept of the fair distribution of resources and that a sense of morality and reciprocity guides the behaviour of primates and other animals (including human animals). I also draw on Frans De Waal who stresses the need to recognise that we evolved not only through our ability to compete but through our ability to cooperate and to show empathy to others. 'Community rights are necessary for both ecology and democracy. Bureaucratic control by distant and external agencies and markets control by commercial interests and corporations [can] create disincentives for conservation…' (Shiva 2002: 30–31). I explained that the use of water could be used as a synecdoche for discussing participatory governance and democracy. In South Australia the flow of water has been affected by the introduction of dams and weirs. The upstream users benefit at the expense of down steam users, but this has been raised as a problem by Major Sumner, a Narrinjari caretaker who has stressed that we need to think in terms of cultural flows. This means that our thinking and practice needs to be carefully considered so that we consider the consequences of our policy decisions for ourselves, others and the environment.

systemic difference. It could be saving seeds, growing food, walking or perhaps finding ways to geo engineer change. It could be speaking respectfully to your neighbour and creating national and post-national regional neighbourhoods.

- Let us develop the good neighbourhood hood programme
- Learn to talk over the fence
- Learn accept diversity and freedom to the extent that it does not undermine the rights of others
- Support our local green spaces and protect the small farmer on the land.

We could set up regional agoras through communities of practice that support learning communities. We need to enable educators to develop the capacity to consider the value of many ways of knowing.[32]

[32]Educate primary-, secondary- and tertiary-level students through a curriculum that supports their ability to think in terms of this design of enquiry approach so that they will be able to sustain their well-being through joining up the dots. The DIS enables people to think about the consequences of their choices. The curriculum needs to encourage public education to address the question: how should we live? What are the implications for our choices in the public, private and civil society sectors? How do our policy choices impact on this and future generations at the local, national, post-national and regional level? We are all part of one planet and living at the expense of others (next door, or in future generations) is based on the mistaken notion that the zero-sum approach is viable. It is not. There is a groundswell of democracy that seeks to save the planet from our consumption choices, but little is known about those consumption choices and what motivates people to consume excessively. It has been argued that after a certain point, increased consumption does not lead to more happiness (Pretty 2013). Thus, increased consumption, driven by increased economic growth, does not make people happier. Indeed, in more egalitarian societies, all people consume less and are less status conscious (Wilkinson and Pickett 2009). Thus, the link between greater equality or well-being and the prevention of global warming appears to involve 'limiting consumerism' (Wilkinson and Pickett 2009: 221). It is assumed that greater social and economic equality will provide the key to reducing the cultural pressure to consume. These are policy challenges locally and regionally that need to be addressed if this agenda is to be advanced. Specifically, it will require balancing individual and community needs to achieve sustainable human rights (Faist 2009; Layard 2005; Nussbaum 2006) within and beyond nation states (Etzioni 2004, 2012).'Transformation from Wall Street to Wellbeing' (McIntyre-Mills et al. 2014) discusses current research on an alternative architecture for governance. The concept 'Wall Street' is used to symbolise economic rationalism and 'well-being' to symbolise a reframed approach to 'being interconnected' and through treading lightly (McIntyre-Mills and De Vries 2012; McIntyre-Mills et al. 2014). Participatory action research explores whether the proposed architecture for engagement in participatory democracy and governance (Archibugi 2010; Gaventa 2001) could enable people to gain greater awareness of the implications of excessive consumption (Pretty 2013; Urry 2007, 2010) for stewardship (see Australian Government Caring for our Country 2013–2018). The book (McIntyre-Mills et al. 2014) argues that social, economic and environmental accounting could be assisted by means of participation to make this so-called triple bottom line accountability viable across the boundaries of nation states. Thus the participatory action research grasps the nettle to address the challenge posed by the Earth Charter: 'Everyone shares responsibility for the present and future well-being of the human family and the larger living world' (cited by Hayden 2010: 368). The challenge is to move beyond the rhetoric of cosmopolitan citizenship and to address both justice and sovereignty. The nation state needs to be held to account by an Earth Charter that is supported by overlapping regional institutions (supported by inclusive regional research institutions, policies, the rule of law including parliaments and courts).

The global financial crisis and the environmental melt down seem to be seen as two problems—not one. Human beings cannot achieve new growth through increased extraction of profit. Instead, they need to live elegantly and well through redistribution and simpler lifestyles supported by green economies, reciprocal sharing of resources and skills in green communities, supported by energy efficient systems. The hypothesis is that participation through awareness and consciousness-raising (McIntyre-Mills 2010) will influence the way in which people think about boundaries. It could help them to remake connections with others and the environment through valuing the environment and engaging in healthy relationships (Wynne 1996).

This hypothesis is based on the notion of neural plasticity in that the brain shapes the environment and, in turn, is shaped by the environment (Bateson 1972; Beer 1994; Capra 1996; Greenfield 2000). The research tests the principle of subsidiarity, namely that decisions need to be taken at the lowest level possible through the well-being and environmental stewardship score card that adapts and extends the Max-Neef Human Development Index. The latter provides the architecture for applying the Aarhus Convention (1998) based on the right to participate and the right to freedom of information on issues pertaining to local environmental well-being concerns. The engagement architecture that we develop and test enables diverse opinions to be mapped and scored in terms of social, economic and environmental indicators of what works why and how to support personal and environmental well-being through exploring perceptions. These perceptions will help policymakers narrow the gap between haves, wants and needs when addressing service delivery. The companion volume, Systemic Ethics and Non-anthropocentric Stewardship (McIntyre-Mills 2014b), explores the need to develop the capacity to 'join up the dots' through participatory democracy and governance and through working across disciplines while preserving space for difference, based on the axiom that freedom and diversity need to be fostered to the extent that they do not undermine the rights of others in this generation or future generations of life. Education needs to enable all people to understand our dependence on one another and the land and to protect the well-being of current and future generations of life.

The theoretical perspective developed by Nussbaum (2006) in 'Frontiers of social justice' is applied to address complex challenges within and across boundaries in order to address quality of life and capabilities for all.

There is evidence that many desire more environmentally sustainable lives, but little is known about the influences on choices around the management of land, water, food and energy supplies that affect the environment. Government responses to human well-being are often based on economic development, which inadvertently increases consumerism, resulting in greater environmental degradation and a heavier carbon footprint.[33] The implementation of coal plants and the lack of local

[33]My own reflection spanned social, economic and environmental considerations. Climate change is a significant problem in Australia. There is little doubt that accelerated climate change will adversely affect well-being and sustainability in Australia (Flannery 2005; Pretty 2013; Stiglitz et al. 2010)—particularly if we continue to consume at current rates (Davies and World Institute 2008)—resulting in significant devastation and a compromised quality of life. The impact of climate change

government support for green energy sources are a case in point for both Australia and South Africa.

We know, therefore, that environmental sustainability and human well-being are intimately linked, but there is little knowledge about how this linkage can be built upon to improve both areas or how we can draw on the lived experience of people to create and co-create pathways to sustainable living through valuing different ways of living and sustainable knowledge that has stood the test of time. Attempts to address climate change are often based on policy information that is not grounded in people's experiences and fails to address what we do know about human behaviour or choices. How can we increase engagement in the environment that balances individual and collective well-being and simultaneously protects the environment when we know little about this nexus and when we do so little to provide opportunities for the recognition of prior learning and do not draw on cultural case studies of sustainable living?

Research needs to draw on diverse ways of knowing (Cruz et al. 2009) that could support regional governance (Wear 2012) to support effective environmental management. There is evidence that many desire more environmentally sustainable lives, but little is known about the influences on choices around the management of land, water and food that affect the environment. Government response to human well-being is often based on economic development, which inadvertently increases consumerism, resulting in greater environmental degradation. We know, therefore, that environmental sustainability and human well-being are intimately linked, but there is little knowledge about how this linkage can be built upon to improve both areas through enhancing capacity building for sustainable living and sustainable employment. Attempts to address climate change are often based on policy information that is not grounded in citizen experiences and fails to address what we do know about human behaviour or choices. The inherent link between engagement in civil society and community well-being (rather than the economic bottom line) needs to be the focus of research. The concepts of stewardship and resilience are important notions through which to explore the nexus across well-being, consumption choices and the environment.

We need to develop a deeper understanding of how the intangible aspects of perceived well-being can be measured and how different kinds of knowledge can be applied to support social and environmental justice. But we also need to measure them in relation to the link between well-being and sustainability (Stiglitz et al.

(Footnote 33 continued)

has been underestimated (Lovelock 2009; Rockström et al. 2009) and local solutions have been overlooked. Aboriginal cultures teach us about stewardship and relationships with the land, but these relationships have been lost in non-Aboriginal cultures. As Major Sumner, an Ngarrindjeri Aboriginal elder from the periodically drought-ravaged lower Murray River in South Australia and custodian of the river stresses, we are the land and the land is us. Re-establishing relationships with the land is at the heart of effective cultural ecosystem management (see http://www.mdba.gov.au/what-we-do/working-with-others/aboriginal-communities/ringbalin).

2010). Our conversation then explored the shared area of concern, namely that lived experience and alternative ways of knowing are not sufficiently appreciated.

The Ngarrindjeri caretaker Major Sumner teaches that learning to appreciate water flows is based on years of experience of testing out ideas and drawing on the empiricism of learning from the land. But we need to also appreciate that non-anthropocentric care taking requires learning from narratives of what works, why and how. We need to 'join up the dots' and appreciate our interdependency on many forms of life.

Shiva for example stresses the need to learn from nature and from non-human animals. The wisdom of the earthworm to sustain organic material and the wisdom of the kangaroo, an environmentally evolved creature that suits the semi-arid environment and that fertilises it with manure that sustains the bush on which it depends (Flannery). A post-colonial approach to agriculture could recognise appropriate farming in Australia and South Africa.

The ability to live sustainably will require valuing different ways of knowing such as the ability to be an entrepreneur and to have the skills to create artworks and functional items out of recycled material. The ability to mend and recycle has been forgotten in a consumerist world.

Thus the issue of creating barriers to university entry, rather than pathways, needs to be explored widely. Best practice examples of not merely recognising prior learning through competency tests set by the university or other colleges for vocational education have already been implemented in Australia. For example, in Alice Springs the Centre for Desert Knowledge and CAT demonstrate working in partnership with Indigenous Australians.

Thus the notion of participatory action research with those who have experience could be a methodology for learning to live sustainably.

The group then explored specific questions that flowed from their engagement with our shared area of concern, namely the need to respond to the employment needs of South Africans through enabling learners to achieve pathways to sustainable living. Many of the issues faced by South Africans are issues faced by those in other parts of Africa. The droughts in Tunisia, for example, have led to protests and the flow of asylum seekers to Europe. Many also seek asylum in South Africa from other parts of Africa.

In Australia the social contract protects those within the boundaries of the state but not those who seek asylum from elsewhere. The issue of bounded governance and decision-making that does not consider the consequences of social, economic and environmental choices for our neighbours has been raised by Joy Murray in the so-called Tuvalu test that demonstrates how our social economic choices affect our neighbours through rising sea levels. For example, in Samoa, agriculture has been affected by rising sea levels and increased levels of salinity.

Thus the issue of employment and unemployment is one that needs to be placed within the context of the environment of the problem, namely a sustainable future in which learning supports social, economic and environmental well-being.

After sharing my own concerns about the implications for climate change locally and regionally and showing the interconnectedness of social, economic and environmental considerations by means of a soft systems map which I reproduce below as follows, each participant was asked to raise their own areas of concern. To facilitate a shared process I asked participants to reflect on their concerns by considering how they draw the framework of the research, how they will approach the research in terms of methodology to address their area of concern. The following slide was shown summarising a set of basic questions as a precursor to applying Churchman's (1979) Design of Inquiring Systems Approach.

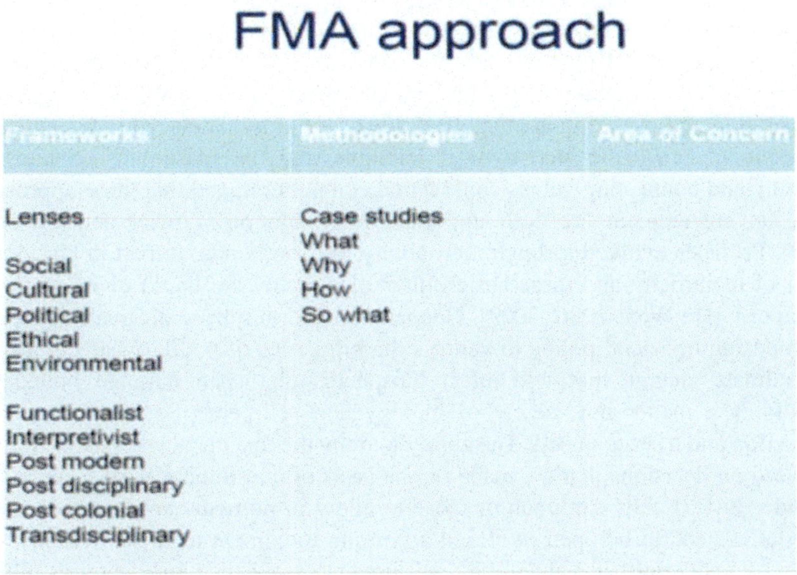

Boundaries are sociological constructions. In biological systems the most basic boundary is the membrane of a cell which opens and closes according to context. In the hierarchy of systems (Boulding 1956) biological systems are above inorganic systems because they are (a) open and responsive to their environment and (b) able to adapt and learn as self-evolving systems (see McIntyre-Mills 2014a, b for more details). The webs of relationships that are fostered across all forms of inorganic and organic life are recognised in physics (Gunaratne 2009; Turok 2012). Each particle is in motion and it is the movement and flows of energy that make life possible. The transfer of information flows across all living systems. A core capability for sociologists who wish to respond to the complex interconnected social, cultural, political and economic challenges will be the ability to transcend disciplinary boundaries and to work with diverse perspectives. Thus those who inform the argument for this entry include De Waal and Dawkins (primatology and philosophy), Hirschman and Hannah Arendt (on economics and politics), Amartya Sen (on economics and morality), Stuart Hall (on identity) and Martha Nussbaum (on social justice). The work of Stiglitz on well-being stocks is extended through drawing on Vandana Shiva (on the intersections spanning economics, politics and the environment). Sociologists need to work together with many stakeholders who can contribute diverse ways of knowing, including non-anthropocentric approaches informed by an understanding of nature and other life forms. In terms of personal praxis as sociologists it will require our learning the ability to be inclusive by enabling everyone to feel that they are respected and heard. This requires giving time and developing the ability to communicate respectfully across diverse participants in sociological conversations that span diverse cultures, interests, experiences and discipline-based knowledge. The focus on anthropocentric humanism and human rights has led to an unethical divide or boundary between the human and the animal. The human being is seen as the controller of nature. But the divided nature of control and compete is only one part of the story. The continuum of relationships with nature and with animals needs to be seen as co-evolving. Cooperation and nurturing are the other side of the story. We need to recognise the connections across the stereotypical feminine and the masculine principles of empathy and connecting versus conflict and competition, and that these approaches to social engagement are both important for evolution (Fowler and Christakis 2010). Primates evolved through their ability to form bonds of trust in groups as a result of reciprocity and these basic pillars of morality are based on empathy and reciprocity (De Waal 2006, 2009). Cooperation and nurturing is equally important when designing social policy to address the big issues of the day, namely poverty and climate change that will affect biospheres that span national boundaries. Identity is a matter not merely of biology and self-perception, but also social interaction and a sense of self. The ability to draw the line in social systems needs to be based on decisions that are made on the basis of questioning what to include or exclude. Just as cells can open or close to allow in nutrients and to expel toxins, boundaries need to be open or closed according to context in social systems. This requires considering our values and realising that we see the world in terms of these values. We are the boundaries according to Haraway (1991, 1992, 2010). She exhorts those without a voice to be the designers. We can choose to destroy the environment or to be stewards based on rewiring our brains through thinking about our thinking, cooperating and nurturing. Space, time and identity are changing as our awareness of interconnectivity is rediscovered. The work of Hall (1992, 1997), Hall and Du Gay (1996) on cultural identity and representation discussed the diversity within the modern state and stressed the need to consider the way resources are distributed and consumed within nation states by asking What is the social, cultural, economic and environmental context? Who gets what, when, why and so what? The implications of greed and lavish consumption on others and the environment. The will to make a difference is the challenge, according to Held et al. (2005: 33–4), who calculate the amounts in American dollars required to make a difference to the life chances of others:

We need to be able to think about our thinking—need to be open to testing out ideas with others through a new form of public engagement that protects human rights and the rights of the planet.

Science needs to serve needs through being conducted in a democracy that respects open testing of ideas and that is guided by systemic ethics and supported by governance structures that support a strong cosmopolitan approach. Democracy and governance needs to be based on the ethical axiom that we can be free and diverse to the extent that we do not undermine the rights of others. Science cannot be disconnected from society it is about designing responses in service to both human beings and the environment on which all life depends. It is not about commodification of gadgets for profit. It needs to be based on testing out ideas in conversation and dialogue with others through creating rapport based on many ways of knowing and by learning from all aspects of nature.[34]

5.4.1 Making Policy Recommendations on the Basis of the Research to Create an Appropriate Curriculum that Helps Students and Members of the Public Appreciate Their Interconnectedness Functions

According to Bourgon (2004) the challenge for the public sector is to shift from responding to a predictable social, economic and environmental context to being innovative and responsive. This is equally relevant for the private and non-government sectors. Compliance in terms of the rules of law now needs to consider reframing and recasting the future of democratic governance to take into account our increased vulnerability and interconnectedness.

Policy and governance outputs and outcomes need to be responsive; one size does not fit all in increasingly diverse nation states. Currently, the challenge is that on the one hand the nation state is too big to address the needs of diverse voters and too small to address the collective needs and common good of people in regional areas facing food energy and water challenges that will need a multilevel and multi-agency approach to ensure social and environmental justice within and beyond the boundaries of the nation state.

The Open Democracy Forum was the starting point for research in response to Held et al.'s (2005) proposal shared through Open Democracy that the core challenge of the day is to address the vast differences in standard of living between rich and poor through reforming world trade and using the market to address global warming by means of a global covenant that prevents unilateral interventions.

In response to his critics Held claims that the resources exist for a reformed United Nations to make a difference, if there is a will. But the UN according to Scruton (2005: 47–8) is corrupt, the bureaucrats are overpaid and he argues that national sovereignty is the only way to make a difference through controlling corruption. But nationalist responses fall short of correcting the market failures and Held did not anticipate the extent of the market failures that have actually occurred. The problem is not only concerning externalities that are not factored into

---

[34]Reciprocity and empathy can be extended through encompassing more ways of knowing to achieve.

increasingly impoverished communities in cities without adequate resources to maintain a decent quality of life. We face challenges that will need to be urgently addressed by current and future generations:

- Climate change—food, energy and water in increasingly urbanised world living in unsustainable cities.
- Poverty and food insecurity.
- Urbanisation along with poverty (hunger, unemployment and crime) is a way of life.
- Public education needs to address human capacity and capabilities to live sustainably.

This book has tried to develop a discussion on ways that could protect the interest of those who have been displaced and marginalised. I make the case that if we are able to accept the capabilities approach for all human and sentient beings then we can bypass the challenges of 'us versus them' thinking and thus will not have to work so hard on repairing divides, because we will appreciate that our starting point is our interconnectedness. Projects such as the NGO providing integrated care expresses the spirit of Ubuntu by making an effort to being the change on a daily basis. The vulnerability of children was recognised by this NGO but with rising numbers of people coming to the cities the threat of hunger will become even greater.

The spirituality of Africa is expressed in many ways. Through a sense of being a person through others as Tutu (1995) emphasised in his preface to the 'African Prayer Book' by Bishop Desmond Tutu but it is also recognised through a sense of learning from and being connected to nature, expressed by this prayer which he cites entitled a 'A Blessing' which he included by Samburu from Kenya:

> May god raise you up
> Above everything
> Spread out like water of a lake.
> Be abundance that never ends,
> That never changes.
> Be like a mountain.
> Be like a camel.
> Be like a cloud –
> A cloud that brings rain always.
> And God promised that it would be so

We could add Camagu, 'let it be so', if we realise that we are the change and our spirituality could evolve towards more inclusive ways of knowing that become more like pure physics—a realisation that 'we are indeed the boundaries', as Haraway suggested—and that we can evolve and change as we develop greater insights into the nature of science, spirituality, Relationships and presence.

Chapter 6
Executive Summary

Representation and Accountability in Glocal Governance and the 2030 Development Agenda: Narrowing the Gap Between Perceived Needs and Outcomes

6.1 Introduction: Representation, Accountability and Re-Generation

The challenge is to face up to our interconnectedness and to appreciate that the current siloed or containerist approaches are no longer relevant to current challenges. Representation and accountability responses are needed that are transformational in their scope and integrated in their responses. We need to transform the architectures of both democracy and governance.

Most people are interested in their ***personal*** wellbeing and safety and what they can do to enhance the wellbeing and safety their family and community. This is the starting point for engaging people to think about the implications of their choices, rather than beginning a conversation about ethical choices, sustainability or stewardship. This has been detailed in: '*Transformation from Wall Street to Wellbeing*' (McIntyre-Mills et al. 2014) in which a case is made for pparticipatory democracy to protect local citizen's right to a safe environment. Ethical decision making about consumption, in order to enhance wellbeing through living and consuming in a sustainable manner has been addressed in detail in this volume and in its companion volume, *Systemic Ethics* (McIntyre-Mills 2014) in the *Contemporary Systems Series*. Together they develop an argument based on considering the consequences of social, economic and environmental policy decisions for current and future generations of life—based on conversations with informants about how we ought to live.

J. McIntyre-Mills, *Planetary Passport*, Contemporary Systems Thinking,
DOI 10.1007/978-3-319-58011-1_6

The addiction[1] by the powerful to consume luxury goods is at the expense of future generations of life. The book explores the inherent link between wellbeing and engagement in the environment and discusses research on ways to express personal rights and responsibilities as ecological citizens (Dobson 2003, 2012; Dobson and Eckersley 2006) concerned about both *people and the planet*. It has two main objectives:

Firstly, to make the case for social change through exploring post disciplinary and post materialist frameworks to address greed, zero sum competition for resources, the commodification of the powerless and the environment.

Secondly, to develop a reframed approach to measuring wellbeing—not productivity—as a sign of economic success.

Thus the paper considers the challenge posed by Stiglitz et al. (2010) to the Australian Productivity Commission, namely to foster an understanding that the wellbeing of humanity is dependent on protecting the global commons and ensuring that some do not live at the expense of others and future generations of life.

In so doing it makes a case for cosmopolitan approaches that scales up local engagement to:

- Enable monitory democracy 'from below' (Kareiva 2016; Kitcher 2016; McIntyre-Mills 2014a, b) and
- Enhance accountability (Ocampo and Gomez-Arteaga 2016) through empowering the strategic rights of the marginalized.

The capabilities approach is conceptualised by the Development Economist Amartya Sen (2000) and extended by Martha Nussbaum who address this issue of population growth by stressing that—only through enabling women in the developing world to achieve literacy and numeracy—would it be possible to address their life chances, thus enabling the demographic transition through '*Development as Freedom*' and ensuring that women are agents who control their own bodies and are capable of meeting basic needs and able to voice their concerns. Women with increased agency and control over their lives have fewer children:

> Indeed, the ecological footprint facilitates viewing ecological sustainability as a numbers game, since a reduced population would allow the expansion of the 1.8 gha per capita

[1]The desire of the rich to consume and the desire of the majority of the poor to leave the ghetto are expressed through emulating the rich. Socio-demographic research shows that debt and bankruptcy in developed nations—such as the USA and nations within the EU—are driven by advertising, easy credit and the desire to 'keep up appearances' by emulating the standards of the very rich (Frank 2007; Wilkinson and Pickett 2009). Making and invoking treaties and conventions that control commodification needs to be a priority and needs to be ongoing if justice is to be maintained. The approach could be buttressed through federations supported by, for example the Lisbon Treaty (Horvath and Odor 2010) which requires that social, economic and environmental legal considerations be met. Unfortunately the structural and process mechanisms of the EU are not able to manage the distribution of power or funding (Rhodes 1997). Clearly if the EU is to survive it needs to be supported by a means to balance individual and national interests with the collective good of the union. More and more members of the union are disenchanted for a number of reasons.

currently at our disposal. In any case, the numbers game will not go away easily, especially considering the population will continue to grow until it reaches some 9 billion (Cao 2015: 246).

Instead of Big Brother controlling *from above*, the people are encouraged to monitor the use of resources *from below* at a local level. The process of monitoring needs to be supported by means of the principle of subsidiarity and buttressed by international law and UN policy spanning post national regions such as the Millennium Goals, the UN Sustainable Development Goals and the 2030 Development Agenda.[2]

6.2 Area of Concern and Statement of the Problem

Archibugi (2008) and Benhabib (2007) stress that we need a form of federalist republicanism. But how do we move from the containerist approach that undermines justice through limiting compassion to citizens at the expense of human rights? Bohman (2005p 102, 111) stresses Arendt's concept of 'the capacity' as a way to 'begin a democratic dialogue':

> For democracy to promote justice, it must already be just. Call this the democratic circle. While it can never be said to disappear, the circle can become virtuous through what I refer to as the 'democratic minimum': the achievement of a democratic arrangement sufficient for citizens to exercise their creative powers to re-shape democracy according to the demands of justice—that is the capacity to initiate democratic deliberation (2014 p111).

The paper addresses the potential for new architectures for democracy and governance through addressing:

- **A priori norms** and a posteriori measures for transformation towards non-anthropocentric, re-generative living that is protected by an expanded social and environmental contract
- **Matching** social, cultural, economic and environmental decisions to perceived needs with a focus on food, energy and water security
- **Narrowing** the gap between perceived needs and the way resources are distributed

The increased complexity of challenges faced by the state and civil society is that poverty, climate change, movement of people and international crime span boundaries. The nation state is becoming increasingly diverse and increasingly ineffective in meeting the perceived needs of marginalized citizens and voiceless residents who are without the protection of the social contract, because they are asylum seekers or refugees. More people than ever have been displaced by disasters that are linked with war, famine and unemployment and the hope of a better future.

[2]https://sustainabledevelopment.un.org/post2015/transformingourworld.

All three projects test out the principle of subsidiarity and Ashby's (1956) Rule of Requisite Variety[3] as a means to balance individual and collective needs. The areas of concern are:

- What combinations of social, cultural, political and environmental factors make a difference to wellbeing, resilience and our capability to adapt to climate change at the local level and to mitigate its effects by scaling up local interventions?
- Who decides on the norms, values and indicators that should structure so-called good governance within and beyond the boundaries of the nation state?
- What a priori norms and values ought to guide participatory governance to support sustainable development goals?[4]

The focus is on the need to balance individual and collective needs in the interests of social and environmental justice for this generation and the next. Ashby's Rule refers to socio-cybernetics research which makes the case that the complexity of the decision should be matched by the complexity of the decision makers. Subsidiarity refers to the notion that decisions should be taken at the lowest level possible in society and with as much freedom and diversity of opinion as possible (to the extent that local freedom does not compromise the freedom and diversity of the whole).

Some scholars (Christakis and Flanagan 2010; Dryzek 2010; Erlich and Erlich 2010, 2012) argue that a lack of cultural understanding about our interlinked and entangled lives affects human rights in so far as it shapes access to water, energy and food security. Social movements seek to save the planet from our consumption

---

[3]As a non-Indigenous researcher, I build on 18 years of research with Indigenous Australians including an ARC linkage grant and several publications including 5 monographs. The process of participation is the subject of her recent publications and other monographs, entitled '*User-centric policy design*' (McIntyre-Mills 2008), based on research conducted in South Australia and funded by an Australian Research Council Linkage Grant and '*Critical Systemic Praxis for Social and Environmental Justice*' based on research conducted in the Northern Territory and funded by Local Government (McIntyre-Mills 2003). These publications, respectively made the case that participation enhances attachment to policy ideas and narrows the gap between perceived needs and service outcomes. In '*Transformation from Wall Street to Wellbeing*' (based on the South Australian Local Government Assoc. funded research) McIntyre discusses in detail public engagement to enable people to make better social, economic and environmental consumption choices and thus to think about wellbeing and sustainability. Specifically, this research builds on the findings of three of my previous projects: (1) a Participatory Action Research (PAR) project with Indigenous Australians focused on wellbeing (ARC linkage LP0560406) and (2) a PAR project on decision-making to promote adaptation to climate change (Local Government Grant https://wellbeing.csem.flinders.edu.au/) and (3) ethical non-anthropocentric approaches to decision making. The research builds on research monographs and articles published in Consciousness Studies, Systems and Behavioural Science and Cybernetics and Human Knowing. An encyclopedia entry on expanded pragmatism and 'if then heuristics' associated with decision making (Churchman 1971, 1984) used in the research appears in the 'Encyclopedia of Food and Agricultural Ethics', entitled: '*Systemic Ethics*' addresses decision making, ethics, consumption and consumerism.

[4]https://sustainabledevelopment.un.org/post2015/transformingourworld.

choices, but little is known about those consumption choices and what motivates people to consume excessively. It has been argued that after a certain point, increased consumption does not lead to more happiness (Pretty 2013). Thus, increased consumption, driven by increased economic growth, does not make people happier. Indeed, in more egalitarian societies, all people consume less and are less status conscious. Thus, the link between greater equality or wellbeing and the prevention of global warming appears to involve 'limiting consumerism' (Wilkinson and Pickett 2009: 221). Greater social and economic equality offer the key to reducing the cultural pressure to consume at the expense of the majority in this generation and the next. These are policy challenges locally and regionally that need to be addressed if this agenda is to be advanced. Specifically, it will require balancing individual and community needs to achieve sustainable human rights (Layard 2006; McIntyre-Mills, forthcoming).

The IPCC formula suggests that the privileged lives of some could lead to 'existential risk' for all forms of life on the planet (Bostrom 2011). The 'price of inequality'—national and global has escalated. The global figure for displaced people passed 50 million for first time since the Second World War (United Nations High Commission for Refugees, Antonio Guterres (15 January 2014).

It is assumed that climate change is real and that it will impact negatively on food, energy and water security and that forced migration of large groups (Held 2016)[5] will become commonplace unless changes are made.

Addressing norms and values is the starting point for the research. The different views and emotions on the challenges are the starting point for striving to narrow the gap between service users and service providers (government/NGO) to address the size of our carbon footprint.

Shaping norms together can be linked with public education and it provides a basis for developing shared indicators of wellbeing that are used to address social, economic and environmental indicators as part of a Triple Bottom Line (Elkington 1994, 1997) accountability process linked with public education. Participation in public education by ecological citizens is facilitated through the application of the UN local Agenda 21 approach. The paper reflects on user-centric policy design to address:

---

[5]"Despite twenty years of multilateral negotiations under the UN, a global deal on climate change mitigation or adaptation has until recently been elusive, with differences between developed countries, which will drive future emissions, forming the core barrier to progress…Multilateral governance has been grid locked over climate change and, in this context, it is not unreasonable to expect climate change to become an ever more powerful cause of migration (Hale et al. 2013). Climate change is wreaking havoc on the world's diverse species, Biosystems and socioeconomic fabric. Violent storms are becoming more frequent, water access is becoming a battleground, rising sea levels may well, as predicted, displace millions, the mass movement of desperate people will become more common, and death levels from serious diseases in the world's poorest countries will increase rapidly (largely because bacteria will spread more quickly, causing greater contamination of food and water). The overwhelming body of scientific opinion maintains that climate change constitutes a serious threat not only in the long term, but in the here and now…." Held (2016:4–5).

'Wellbeing' (defined in terms of social, cultural, spiritual, economic and environmental dimensions) is now widely located in mainstream transdisciplinary literature that re-frames what we value as a society (Stiglitz et al. 2010; McIntyre-Mills 2010). Well-being is (a) Achieved by making time to connect with people and place; a sense of community, (b) Enhanced through access to sustainable public & private infrastructure for housing such as affordable transport, safe cycle paths and accessible public transport. (c) Agencies/facilitators of wellbeing include; willingness to use & re-use resources carefully; give up greed & waste; sense of community & respect others. (d) Constraints to achieving wellbeing include: lack of time & money; sense of hopelessness; inappropriate infrastructure. (e) Related to/closeness of fit; a sense of community & belonging; connection with family friends and nature.[6]

The size of urban footprints need to be managed through encouraging low carbon living and minimising the size of the ecological footprint. The paper proposes a way to enable this to occur. Urbanisation per se may or may not be problematic depending on the way in which wellbeing stocks are protected. The purpose of this paper is to explain the need to develop the capability to appreciate wellbeing stocks through public education linked with a so-called planetary passport—which provides a way to enhance ecological awareness on ways to manage social, economic and environmental resources for current and future generations of life and the importance of ensuring that when 65–75% of the population are urbanised (as is the case in South Africa and Indonesia) that the balance between rural and urban areas is protected through appreciating Indigenous ways of relating to others and the living and non-living systems on which we depend. South Australia, the driest state in Australia has faced rising water and energy costs and farmers reliant on the River Murray have faced global financialization and privatising of water during the past decade whilst consumers have faced rising living costs.

So we need to think about how to do different kinds of governance and democracy to promote food, energy and water security through minimising the size of the urban footprint. What I have suggested is a way to manage the *wellbeing stocks* locally through enhancing human development capabilities in terms of:

- Public education to 'join up the dots' so that people are aware of the implications of their social, economic and environmental decisions.
- Representation of local people (particularly the Indigenous population, the marginalised and the dispossessed) and accountability in terms of the fair use and distribution of resources.
- Sustainability and in terms of the way resources are used.
- Governance needs to be based on a new architecture to support people and the planet by reframing what accountability means, the policy tools that are used to

[6]This information was provided to local government to help them to plan and deliver services to residents. The administrators are able to access summarized data in excel spread sheets. Axial themes according to Max Neef (1991) resonate with the axial themes identified in McIntyre-Mills et al. (2008, 2011, 2014).

support it, the role of government, the style and the goals of government as detailed below.

The starting point for the planetary passport idea is to adapt the five areas of priority highlighted by Archibugi (2008) as:

(a) Control over the use of resources and equitable sharing enforced through both positive and negative sanctions.
(b) Acceptance of cultural diversity by ensuring internal sovereignty based on constitutional rules that enable freedom and diversity to the extent that the freedom and diversity of others is not undermined. External sovereignty is 'replaced by global constitutionalism'(Archibugi 2008 in Wallace and Brown 2010:321);
(c) Strengthening the self-determination of people based on participatory democracy and the 'absence of domination' over other states based on a notion that each nation state can be free and diverse to the extent that freedom and diversity does not undermine the rights of others by mis-using resources that need to be shared with others;
(d) Monitoring internal affairs, based on a form of monitory democracy and governance;
(e) Participatory management of the global commons as the fabric of life.

Towards Indigenizing a Planetary Passport for Re-generation: knowing our place through recognizing our hybridity'[7] makes a case for, describes and develops a new architecture for governance and democracy to protect both people and the planet through promoting social and environmental justice for all through (a) promoting public education and (b) enabling accountability for low carbon living.

Designing public education requires enabling participants to address representation, accountability, sustainability and re-generation. The planetary passport provides a way to encourage 'if then thinking'. In this respect it draws on the work of West Churchman's (1979) *Design of Inquiring Systems Approach* based on questioning the boundaries of inclusion and exclusion and it examines the so-called 'enemies within' (religion, mortality, politics and aesthetics) by considering the consequences of our choices for ourselves, others and the environment.[8]

[7]The aim of the Human Development Capabilities Approach is to empower the marginalised (particularly women) to have a public voice. The more women are able to demonstrate personal and interpersonal agency in their lives the more likely developing countries will move towards achieving a demographic transition. This type of pyramid is common for developing countries with high birth and death rates. Relatively short life expectancy, as well as low level of education and poor health care also describe this kind of population age distribution model. Food, energy and water insecurity poses a post national threat to international security if water is not shared within and beyond national boundaries in a careful manner that is representative, accountable and sustainable in terms of delivering social and environmental justice.

[8]It is based on questioning boundaries of inclusion and exclusion, it examines the so-called 'enemies within' (religion, mortality, politics and aesthetics) and it considers the consequences of our choices.

The purpose of public education is to help people to question their choices and to rethink what local residents value by enabling them to appreciate our dependency on living systems of which we are a strand. The public education focus is on addressing a core issues such as food, water and energy security. The process enables people to think about their consumption decisions and it demonstrates that current consumption choices are both challenges and opportunities.

The book uses the analogy of flows by building on the work of Vandana Shiva, who is an eco-feminist, trained as a physicist and winner of the Sydney Peace Prize. It also builds on the work of Joseph Stiglitz. The idea is to adapt and apply the Global Covenant idea developed by Held (2004) to protect the global commons and to find a way for this cosmopolitan approach to be applied in practice through drawing on the liberative potential of the Aarhus convention that supports access to information and the right to participate and to be heard on environmental matters.

The notion of a planetary passport (McIntyre-Mills 2016) is a heuristic device to argue for protecting this generation of life and the next through re-thinking rights and responsibilities in terms of governing our human footprints and protecting habitat. Habitat refers to the living spaces for human beings and other living systems which we share and on which we are mutually dependent in overlapping biospheres. Its companion volume 'Balancing Individualism and Collectivism' (McIntyre-Mills, forthcoming) is an edited collection of papers that together speculate on whether participatory democracy and governance could enhance sustainable consumption choices that support wellbeing.[9]

6.4 Human Development, Capabilities and Demographic Transition

Achieving the demographic transition to low population growth is vital today through public education to provide numeracy, literacy and opportunities for women so that they are able to have more control over their own lives. Personal agency is enhanced as educated women will strive to choose the number of children and to become part of the paid workforce and part of the democratic and governance processes so that they can make policy and shape the politics for transformation.

This has been emphasised by Martha Nussbaum (2011) and Amartya Sen (2000, 2002) in their work on human development and human capabilities. This research adapts the Human Development approach and builds on the need to enhance capabilities for women (Crocker 1995). This is one aspect of the challenges facing the planet. The other is to enable the most marginalised people to voice their

[9]The two volumes consider diverse ways of interacting with the environment and extend the work of the Special Integration Group 'Balancing Individualism and Collectivism' that McIntyre has chaired since 2006 for International Systems Sciences conference (see also McIntyre-Mills 2015).

Table 6.1 Re-generation of the environment, greater rural-urban balance and sustaining this balance through enhancing wellbeing stocks

Social changes—growing populations and increased urbanization	Wellbeing stocks need to be developed and protected through recognizing the flow on effects
Economic approaches that focus on growth in debt and growth in profit	The pursuit of profit at the expense of people and the environment is a central problem for democracy and governance
Environmental approaches that focus on profit for some at the expense of others and the environment	Habitat for living systems needs to be the focus for new forms of transparent governance based on positive sanctions that reward the protection, promotion and re-generation of wellbeing stocks. Negative sanctions by organizations for social and environmental justice need to protect living systems and prevent complicit exploitation of people and the planet by big business

concerns about being dispossessed and to be listened to by those who are supposed to represent their interests (Butler 2005, 2011; Butler and Athanasiou 2013; Dobson 2012) (Table 6.1).

The past president of the World Bank, Stiglitz links wellbeing with economics in an address to the Australian Productivity Commission. '*Mis-Measuring our lives*' (Stiglitz et al. 2010) addresses the role of social inclusion in discussions on climate change and building capacity. They link quality of life, perceived wellbeing about what we value and our attitudes towards the environment and living in ways that develop and protect stocks for the future. They stress the need to transform our attitudes towards productivity and consumption. But in valuing the fabric of life, if water, air, and earth are given a price—who will pay for protecting the common good? Surely not the very poor? This falls into the trap once again of commodification. Transboundary democracy and governance for accounting and accountability may be considered problematic by many for diverse reasons, such as: Why should people save resources, so that others can squander them? How can we ensure that everyone has a fair share?

Rights without responsibilities cannot sustain the fabric of life for the next generation. Sen stresses the need to build capabilities to develop education and quality of life for the development not only amongst the illiterate and innumerate—but amongst the profligate bankers who awarded themselves profits and parcelled up the debt for others to bear the burden. The will to do things differently is lacking, not the means to make a difference. Held (2005) cites data cited from the World Bank that three billion people live on less than $2.50 per day. The will to make a difference is the challenge (Held 2005: 33–34).

6.5 Research Approach

Government response to human wellbeing is often based on the notion that growth equals development, which inadvertently increases consumerism, resulting in greater environmental degradation. We know that environmental sustainability, consumption choices, and human wellbeing are intimately linked, but there is little knowledge about how this linkage can be built upon to improve both areas. Consumption based on living simply, ethically and well—versus consumerism to express status are very different. Zavestoski (2002) has stressed that the voluntary simplicity movements (Alexander 2011) such as post materialism, slow living, eating local food, reducing energy usage, re-using, recycling and re-generating tend only to occur when status needs have already been met. Paradoxically some of the changes become status driven consumerism that can appear to be simple, but often lead to change for the sake of appearance (Binkley 2008) rather than substantive change and greater happiness. The gap between rich and poor has grown wider than ever before in human history (Hay and Beaverstock 2016; Stiglitz 2012; Oxfam 2016). Now the wealthiest 1% own more than the poorest 50% of people on the planet. But empirical research shows paradoxically that more equal societies do better in terms of achieving a range of social, economic and environmental indicators (Wilkinson and Pickett 2009). Higher levels of unhappiness lead to higher levels of addictive habits spanning food choices and what people tend to buy and why. It has been argued that when people have a sense of low status they tend to consume status items for appearance sake and to alter the mind through drugs and food choices (Wilkinson and Pickett 2009).

The IPCC concluded that the goal of reducing the carbon footprint should be 387 parts per million of carbon, but Rockström et al. (2009) have argued that it should be lower at 350 parts per million. But to halt the growth in greed, the growth of the gap between rich and poor and shift to a demographic transition (from high growth rates and high infant mortality—this is typical of developing countries) we need more participatory democracy and governance not more top down control of the size of the ecological footprint. This is an ongoing debate that rages between 'the population bomb' supporters and those (like Amartya Sen and Martha Nussbaum who say that literacy and the right to a strategic voice will make a difference to achieving the demographic transition.

The book discusses ways to enhance capabilities and public education so that they understand the implications of their choices socially, culturally, politically economically and environmentally. Significantly, the Mixed Method research applies Participatory Action Research to address the need for a different approach to the way in which we live our lives in terms of our consumption of energy resources, fresh water, transport and food consumption choices and the implications for wellbeing and sustainable living (Pretty 2013). It seeks to deepen an understanding of how people perceive relationships and how they impact local challenges and experiences, in order to develop a new understanding of environmental responsibility and the stewardship of shared habitats—for both current and future generations.

Democracy is in need of improvement (Hulme 2009). The specific challenge is to match forms of appropriate stewardship participation that are accessible to diverse groups and not viewed with cynicism (Cooke 2004).

The application of policies discussed (McIntyre-Mills 2016, forthcoming) such as the UN Local Agenda 21 and the Aarhus convention provide potential for a scaled up approach to managing resources through considering what people have, what they need and what they are prepared to add or discard from their lives, in order to use resources carefully and to feed forward resources to future generations. This works along the lines of the so-called 'veil of ignorance approach' underpinning the social contact but instead, this contract is post national and inter-generational in its approach to distributing resources for social and environmental justice. It provides a new approach to representation and accountability as suggested in column 6 (Table 6.2).

The research explores the discourses of Indigenous and non-Indigenous Australians on consumption and our relationships with one another and the land. A common Participatory Action Research (PAR) design and mixed qualitative and quantitative methodology was used in across the three studies (two completed, one based on secondary date and one PAR in progress) to compare and contrast thinking and practice pertaining to wellbeing and sustainable living. The research develops and tests an engagement process to enable people to develop social, economic and environmental indicators of wellbeing for current and future generations known collectively as 'wellbeing stocks' (Stiglitz et al. 2010). It does this by exploring the suggestion made by Florini in 'The Coming New Democracy' that the Aarhus Convention (1998) on 'Freedom of Environmental Information and Participation could be usefully extended to support the nexus between sustaining human and environmental wellbeing and resilience through making careful social, economic and environmental decisions pertaining to habitat, multiple species, food, energy and water consumption. It tests the hypotheses that:

1. The greater the level of participation, the better the match between service users and providers. Major challenges in local communities include the need to achieve or maintain: (a) access to safe domestic housing (including energy and water), appropriate education and employment; within (b) livable cities with liminal biodiverse spaces; that are in turn supported by (c) sustainable regions.
2. The greater the level of participation by the marginalized the greater the support for policy decisions, the narrower the gap between perceived needs and policy outcomes. Participation through awareness and consciousness-raising (McIntyre-Mills 2010) will influence the ways in which people value the environment and engage in healthy relationships with the land. This hypothesis is based on the notion of neural plasticity in that the brain shapes the environment and, in turn, is shaped by social and environmental factors (Bateson 1972; Greenfield 2000). The challenge for governing the Anthropocene ethically and wholesomely is one of moving away from disciplinary and functional differentiation, in order to span biological, psychological, social, cultural, spiritual, political, economic and environmental dimensions to support living

Table 6.2 New architectures for governance

	1. Ancient public administration	2. Traditional public administration	3. New public management	4. New public governance	5. New architectures for people and the planet
Accountable to	Subjects	Voters	Customers	Citizens	Living systems
Policy tool	Minimal	Direct	Contracting out	Co-production	Co-determination
Role of government	Rule	Row	Steer	Facilitate	Steward ecofacturing
Style	Royal	Post Autocratic Bureaucratic	Post Bureaucratic Competitive	Collaborative	Post national Cascade economies
Accountability	Leader	Hierarchy	Market	Network	People and the planet
Goal	Loyalty	Law	Results/performance	Relationships	Protection

Source Column 6 is added as new architectures to extend governance approaches outlined by Zeger Van der Wal, Flinders Symposium, April 2016

ethically in ways that redress the worst aspects of modernisation (Berger 1966; Berger and Luckman 1974). Indigenous cultures teach us about stewardship and relationships with the land, but **these relationships have been oversimplified** (Langton and Longbottom 2012; Langton 2015) **or lost** in non-Indigenous cultures that tend to caricature the notion of stewardship without understanding the social and environmental justice implications for current and future generations. As Major Sumner, an Ngarrindjeri Indigenous elder from the periodically drought-ravaged lower Murray River in South Australia and custodian of the river stresses, we are the land and the land is us. Re-establishing relationships with the land is at the heart of effective cultural ecosystem management (see http://www.mdba.gov.au/what-we-do/working-with-others/aboriginal-communities/ringbalin) and sustainable employment. There is evidence that many non-Indigenous people desire more environmentally sustainable lives, but little is known about the influences on choices around the management of land, minerals, water and food that affect the environment.

This research explores whether collaborative approaches drawing on diverse ways of knowing (Cruz et al. 2009) and user-centred governance of resources (McIntyre-Mills 2006; McIntyre-Mills et al. 2008, 2012; Podger et al. 2012) could support regional governance (Wear 2012) to support effective environmental management based on the UN Sustainable Development Goals.

The Max-Neef (1991) Human Development Index (HDI) has been applied in a project funded by the Local Government Grant to promote wellbeing and better management of resource consumption to address sustainable development goals. The Attorney General's Department funded the Resilience Score card (Gebbie et al. 2012) to enable preparedness and public education to prepare for and promote resilience to disasters such as droughts, floods, fire and earthquakes. The purpose of the research is to: explore the wider potential of Max-Neef's index as it relates to the UN Sustainable Development Goals (2014) and consumption choices. The '*Being, Having, Doing and Interacting Index*' was used in focus group conversations on consumption, sustainability and wellbeing. The conversations will continue and extend to others via a face to face and digital community of practice to meet the different engagement needs of residents (Wenger 2010). The engagement process, prompted by the index enables people to think about the way they live in terms of what they have, what they need, what they are prepared to add or discard from our lives and the turning points for the better or worse in terms of wellbeing, sustainability and resilience. Their responses were used as the basis for developing indicators.

6.6 Multilevel Engagement Through Mindfulness and Critical Systemic Praxis

Ulrich (1993) poses 12 boundary questions. The questions drawn and adapted from Ulrich, Werner and Reynolds, Martin (2010: 245) are asked in terms of: 'what is the case now?' (Describes the current situation by using the DIS) and 'what ought to be the case'? (Describes judgments informed by ethical considerations).

The groups of questions cover sources of motivation, control, knowledge and legitimacy. Ulrich uses questioning to unfold the implications of particular approaches: 1. How do we know what to do? 2. What is the best approach in this context and why, who says so? Our approach to building capability to think critically and systemically is to:

- Develop software to enhance decision-making.
- Enable service users, policy makers and providers to develop clear priorities in response to contextual challenges through careful consideration of the consequences of social, economic and environmental decisions.
- Understand the diverse perceptions on mitigation and adaptation. Mitigation means reducing our impact on climate change. Adaption involves changing our daily decisions and behavior.
- Enhance awareness of our interconnectedness and dependency on nature.
- Build the capacity of participants to realize that they can live elegantly and well and that they do not need to polarize productivity versus the planet (Charlton 2011) by building wellbeing stocks for the future.

6.7 The Engagement Process

Personal stories based on addressing the question 'how should we live in order to achieve wellbeing' reveal patterns in the data. The software is populated with the social, economic and environmental indicators of wellbeing as perceived by the participants. Patterns of adaptation to the effects of climate change and mitigation of causes provide residents and service providers the opportunity to explore pathways to wellbeing. The new architecture for democracy and governance (McIntyre-Mills et al. 2014) enables participants to build on other people's experiences. The software summarises personal experiences and perceptions for service providers based on collecting people's responses and perceptions to the following:

- *I have* the following things in my life
- *I need* in my life
- *I will add* to my life
- *I will discard* from my life

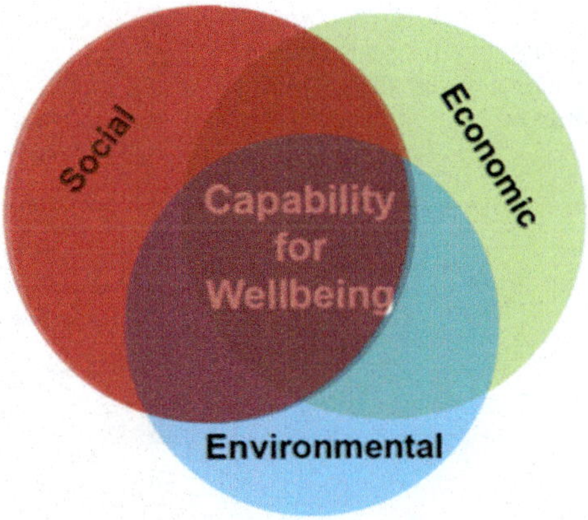

Diagram 6.1 Enhancing capabilities for transformation. *Source* Details can be found at: https://wellbeing.csem.flinders.edu.au/ and now http://wirasoft.com/pathways-to-wellbeing/

Plus, self-reflection on:

- *The turning points*—for better or worse
- *Consideration of the barriers*—for self and society

In 'Towards Indigenizing a Planetary Passport' it is suggested that the de-identified data could help glocal regional governments to plan and deliver services to residents. Human security would thus no longer be tied to nation states. Instead regional overlapping regions could undertake to ensure representation, accountability and sustainability. For service users a pathway to well-being that can be further developed after reflection. For service providers it provides a spread sheet, listings and cross tabulations of the inputs from residents providing information to assist in developing strategies to deal with poverty and climate change. The core point made by the participants to date is that people need to take time or find time to make a difference to the way they live. The architecture for engagement could have potential if applied as a means to implement monitoring 'from below'. These publications (McIntyre-Mills et al. 2014; McIntyre-Mills 2016) discuss the testing of the software to develop raft of systemic wellbeing indicators to address the notion of wellbeing stocks and 'being the change' glocally.

The aim of the participatory action research was to test and prefigure a means to hold the market to account—to ensure that the use of resources that are necessary for the common good are accessible and equitable. This has implications for food, energy and water security in terms of distribution and sharing (Diagrams 6.1 and 6.2).

Diagram 6.2 A proposed architecture for post national regional governance post Brexit

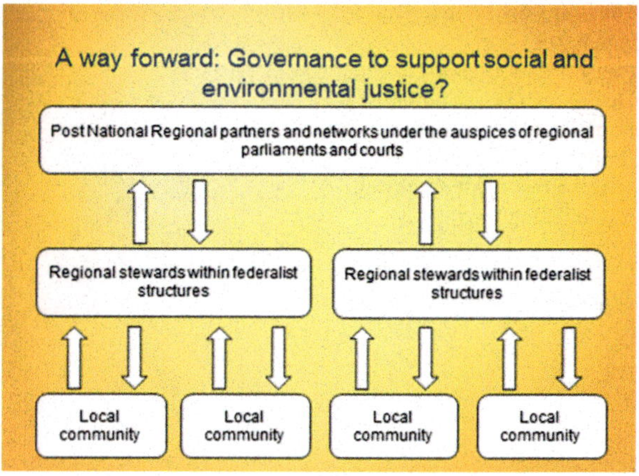

6.8 Water Flows: A Synecdoche for Planetary Democracy and Distributive Governance

Pauli (2010) stresses the need to protect so-called ‘ecofacturing’ in ‘cascade economies’ where waste becomes a resource.

This approach uses water flows as a way to explain a new approach to living ethically and well by protecting living systems of which we are a strand. As we do not know where the next disaster will be located within the region, human security and hospitality needs to be extended across the region. Resources need to be managed carefully so that resource rich areas share with those in need. Stocks are not stored and used within contained areas or interest groups. They are shared according to a system of accountability that measures resources in terms of a planetary passport and score card. This has implications for complex decision making within and across regions. Rivierère (2015) stresses that water security is a regional concern. Water management is considered in this paper as a synecdoche for new architectures and processes for governance and democracy. It has implications of building *stocks for the future* (Stiglitz et al. 2010) and re-framing not only economics but post national governance to protect social and environmental justice in cascade economies which provide opportunities from sources of waste that is re-valued (Pauli 2010).

Shiva (2002) stresses the importance of enhancing the strategic needs of women, not just their basic needs and giving women a voice in decisions at the domestic level, community level as well as a strong voice in the public and private sectors. Polly Higgins[10] also suggests that a new planetary law should be passed to protect

[10]See https://www.youtube.com/watch?v=QPUmN88htCo and Earth democracy http://m.youtube.com/watch?v=Bq0lrbznsjc. http://www.theguardian.com/world/2015/may/31/the-heat-and-the-death-toll-are-rising-in-india-is-this-a-glimpse-of-earths-future.

the viability of the planet and that current systems of law are inadequate to protect people and the planet Stiglitz has made the same point in relation to economics, whilst the need to reframe boundaries across a range of disciplines has made similar points as I stressed in a recent entry in Sociopedia called: 'Reconsidering boundaries' (McIntyre-Mills 2014). Whilst the proposal was made by Higgins and others to the UN, Bolivia has succeeded in leading the way by passing a law to recognize the need to protect the earth. This could help to ensure that elected leaders remain more connected to the people they are supposed to represent. Human development needs to be fostered through the capabilities approach to address poverty and carbon emissions by using the Max-Neef Human Sustainability Index (1991) applied to valuing, measuring and managing environmental and human assets. The index that supports the engagement process to help people think about their well-being is called the *Being, having, doing and interacting Index (*McIntyre-Mills with Binchai 2013. It enables people to think about the implications of what we have, what we need, what we are prepared to add or discard from our lives and the turning points for the better or worse. These indicators are assessed by the interactive, web based *Pathways to Wellbeing and Resilience Score Card* that could serve as way to develop a regional passport that becomes a basis for demonstrating ecological citizenship that is informed by indigenous praxis. The research explores whether engagement in thinking through 'if then scenarios' could aid discursive dialogue within communities of practice to support post materialism and wellbeing.

The community approach to water governance underpins this notion of glocally interconnected living choices that underpin the passport to protect the global commons. The laws for water management highlight the challenges, but also the way forward for water management. I will discuss the four theories of water rights raised by Shiva (Shiva 2002: 77):

> The first law is the Harmon Doctrine which protects the sovereignty of states and is derived from British private property law. Sharing of water for downstream users is only moderated by the suggestion that good neighbourliness is preferable.
>
> The second approach is the **natural water flow** theory based on protecting the rights of up and down stream riparian users. …
>
> The third is the **Helsinki principle** based on the equitable apportionment of water, but as Shiva (2002:78) stresses: 'equitable utility is defined as the maximum benefit accruing to all the riparian states, in light of their differing economic and social needs. **Community of interest is closely related to the Helsinki Rules on the Uses of the Waters** of International Rivers, "Each basin state is entitled, within its territory, to a reasonable and equitable share in the beneficial uses of the waters of an international drainage basin (Shiva 2002: 79).

This reasonable and equitable distribution needs to be based on what we have, what we need, what we are prepared to add or discard, turning points for the better and worse (in terms of food, energy and water security).

6.9 Conclusion

The paper explores whether the architecture for engagement in participatory democracy and governance could enable people to gain greater awareness of the implications of excessive consumption (Pretty 2012; Urry 2010) for stewardship. To sum up, this research made an effort to address axiology, ontology, and epistemology as follows:

- **Axiologically** the research emphasis is on representing the views of the service users and providers as a basis for social and environmental justice that empowers service users to have a say in shaping wellbeing for representation, accountability and re-generation.
- **Ontologically** it is assumed that the nature of reality can in fact be shaped when people speak out about the way in which resources ought to be used.
- **Epistemologically** it draws on deductive, inductive, retroductive[11] and abductive[12] logic informed by the perceptions of the service users and providers on what works why and how and what does not work why and how.

The research is relevant to re-discovering stewardship rights and responsibilities (Flannery 2012) and to being accountable through better decision making about consumption choices. New local forms of engagement and governance (Held 2004) are needed to protect residents against environmental hazards and the subsequent economic and social consequences. Major challenges in local communities include the need to achieve or maintain: (a) access to safe housing (including energy and water), appropriate education and employment; within (b) livable cities[13]; that are in turn supported by (c) sustainable regions. The paper discusses the potential for rebuilding relationships and the need to foster collective determinism through public engagement processes. The accountability framework for the 2030 Development Agenda could be supported by the policy framework of UN local Agenda 21 and the Aarhus convention which suggests the need for engagement in setting policy indicators, the right to access to information and to participate. Social engagement is important for public education, representation and accountability within and across national boundaries, despite the increasing trend towards inward looking policies that do not foster post nationalist agendas, as demonstrated by the recent Brexit decision.

The practical outcome of a new architecture for participation could be to enable participating citizens and organisations to balance individual and collective needs. It is hoped that the ongoing research in Indonesia will:

---

[11]Retroductive logic is applied to find patterns in the data.

[12]Abductive logic leaps beyond the taken for granted and develops a new way of thinking and in this case encourages the development of an 'ecological mindset' (Bateson 1972).

[13]Opportunities exist to include creating a better balance between rural and urban areas and greening cities (Dryzek 1999, 2000, 2010) through the creation of urban agriculture in open spaces, on top of, attached to and within buildings.

- **Contribute to the theory** on democratic engagement and how to address re-generative and sustainable living through governance and also extend and integrate theory on wellbeing and consumption in relation to cross species habitat protection, water, energy and food.
- **Enable stakeholders** in a range of locations to contribute their ideas and preferred choices, in order to manage contextual development and consumption choices, such as natural resource management, building codes and alternative energy options in their local and regional areas.
- **Inform responsive policy decisions** and service delivery to address destruction of habitat, rising temperatures and risks pertaining to water, energy and the associated rise in food and living costs.

The new architecture for democracy and governance could provide a means to enabling this to occur. The sticking point is that weak cosmopolitanism argues for equality of all human beings without realizing that recognition of human rights, requires securing the basic conditions for life. The research is relevant to re-discovering the stewardship rights and responsibilities (Flannery 2012) and accountability that are considered by Ann Florini and Carol Graham of the Brookings Institute to be essential for the governance of sustainable resources. It strives to foster and manage diverse forms of knowledge including verbal, visual, physical, musical, mathematical, introspective and interpersonal (Gardner 2008) and thus develop human capacity to address complex socio-environmental challenges (Ehrlich and Ehrlich 2013). The challenge for governance and public administration is to match processes and governance structures to diverse interest groups—with different life chances and different socio-demographic characteristics.

To sum up, the approach to social justice applies Critical Systems Thinking, a Meta level approach to consciously thinking about ways to protect living systems. This has implications for personal and public ethics, education, democracy and governance. The systemic focus of the book is on being mindful of the need to design policy that supports living systems. 'Ecological citizenship needs to ensure that the requisite variety of habitat is protected to support living systems as a planetary concern. It is important to realize that the wellbeing of individual citizens cannot be protected unless the global commons is protected through governing the size of our shared carbon footprints that have a planetary impact in this so-called caused by human beings in this era of the Anthropocene. The exploitation of sentient beings (workers and animals) and the environment for profit is no longer sustainable. The way in which diverse people understand the implications of urbanisation, loss of territory, loss of species and the implications for living systems of which we are a strand (Rose 1996, 2005; Atkinson 2002; Shiva 1988, 2005, 2007) needs to be addressed.

Bibliography

A frontier-free Europe. 1988. *Commission of the European Communities.* Brussels and Luxembourg: CEC.

Aarhus Convention. 1998. *On access to information. Public participation and access to justice in environmental matters*, Denmark, 25 June. http://ec.europa.eu/environment/aarhus/.

Ackoff, R.L., and J. Pourdehnad. 2001. On misdirected systems. *Systems Research and Behavioral Science* 18: 199–205.

Africa Recovery, Vol. 18 1 April 2004. Watch page, 'Southern Africa, Another year of hunger' Website: www.africarecovery.org, africa_recovery@un.org.

Akomfrah, J. 2013. 'The Stuart Hall Project'. Documentary produced by David Lawson & Lina Gopaul and directed by John Akomfrah.

Alexander, S. 2011. The voluntary simplicity movement. The International Journal of Environmental, Cultural, Economic and Social Sustainability, 7. http://www.Sustainability.

Alexander, P. 2013. Interview: South Africa after Marikana. *International Socialism: Quarterly Journal of Socialist Theory*, 8th January.

Ali, Ayaan Hirsi. 2015. *Heretic: Why Islam Needs a Reformation Now.* New York: Harper.

Ali, M., J. Kos, P. Lietz, D. Nugroho, A. Furqon Zainul, and E. Emilia. 2014. Quality of Education in Madrasha. World Bank Report.

Almond, P. 2015. Paris attacks: Why Islam and Christianity are twin religions of war and peace. *The Conversation*. November 16th.

Arendt, H. 1972. Crises of the Republic. Orlando: Harcourt Brace.

Ariely. D. 2008. Predictably irrational: The hidden forces that shape our decisions. New York: Harper Collins. www.e-reading.club/bookreader.php/138702/Ariely_-_PredictablyIrration.

Archibugi, D. 2008. The architecture of cosmopolitan democracy. In *The Global Commonwealth of Citizens: Towards Cosmopolitan Democracy*. Princeton University Press and also Archibugi, D. 2010. The architecture of cosmopolitan democracy. In *Cosmopolitan Reader*, ed. Brown, G.W., and Held, D. Cambridge: Polity Press.

Arendt, H. 1959. *The Human Condition*. New York: Doubleday Anchor.

Arendt, H, 1963. *Eichmann in Jerusalem: A Report on the Banality of Evil*. New York: Viking Press.

Arendt, H. 1967. *The Origins of Totalitarianism*. London: Allen and Unwin.

Arendt, H. 1972. *Crises of the Republic*. Harcourt Brace, Orlando. First published in 1969.

Arnstein, S. 1969. A ladder of citizen participation. *Journal of American Institute of Planners* 35 (4): 216–224.

Article Stable URL: http://www.jstor.org/stable/307253410.2307/3072534.

Ashby, W.R. 1956. *An Introduction to Cybernetics*. London: Chapman and Hall.

Aspinal, E., and G. Fealy (Eds). *Local Power and Politics in Indonesia*. Decentralisation and democratisation Indonesian update series. Research School of Pacific and Asian Studies, ANU, 259–274.

J. McIntyre-Mills, *Planetary Passport*, Contemporary Systems Thinking,
DOI 10.1007/978-3-319-58011-1

Athanasopoulos, C. 2014. Christian Mysticism and Food. *Encyclopedia of Food and Agricultural Ethics.*

Atkinson, J. 2002. *Trauma Trails*. London: Recreating Song Lines. Spinefex.

AusAid, 2009. Disaster risk reduction. Investing in a safer future.

Australian Government Department of Infrastructure and Transport. 2011. 'Our cities our future. A national urban policy for a productive, sustainable and liveable future', Department of Infrastructure and Transport. GPO Box 594, Canberra ACT 2601, Australia. ISBN 978-1-921769-26-9INFRA–1118/May 2011c, Commonwealth of Australia 2011.

Australian Government State of Australian Cities. Department of Infrastructure and Transport. 2011. Major cities Unit, GPO Box 594, Canberra ACT 2601, Australia, Internet: http://www.majorcities.gov.au.

Australian Government Caring for our Country 2013–2018

Australian Government Caring for Country. http://www.nrm.gov.au/system/files/resources/706caf49-5dab-44e8-b89f-d2ace54cee58/files/achieve-report-appendix.pdf.

Australian Public Service Commission. 2007. *Tackling wicked problems*. http://www.apsc.gov.au/publications. Accessed January 8, 2013.

Bacchi, C. 2009. *Analysing Policy. What is the Problem Represented to be?* New South Wales: Pearson.

Bakker, K. 2007. The commons versus the Commodity. *Antipode* 39 (3): 430–458.

Balibar, E. 2006. *Strangers as enemies. Further reflections on the aporias of transnational citizenship*. Globalization and Autonomy On line Compendium. http://www.globalautonomy.ca/global1/article.jsp?index. Accessed March 29, 2011.

Bammer, G. 2005. Integration and implementation sciences: Building a new specialisation. *Ecology and Society* 10 (2): 6.

Banathy, B. 1996. *Designing Social Systems in a Changing World*. London: Plenum.

Banathy, B. 2000. *Guided Evolution of Society: A Systems View*. London: Kluwer/Plenum.

Barber, B. 2005. Global governance from below. In *Debating Globalization*, ed. Held, D. et al. Cambridge: Polity.

Bartley, T. 2011. Transnational governance as layering of rules: Intersections of public and private standards. *Theoretical Inquiries in Law* 12 (2) Article 6.

Baruma, I., and A. Margalit. 2004. *Occidentalism: The West in the Eyes of Its Enemies*. New York: Penguin.

Bateson, G. 1972. *Steps to an Ecology of Mind*. New York: Ballantine.

Bateson, M.C. 2016. Living in cybernetics—Making it personal. *Cybernetics and Human Knowing* 23 (1): 96–102.

Battersby, J., and J. Crush. 2014. Urban Forum 25: 143. June 2014 25 (2): 143–151. doi:10.1007/s12132-014-9225-5.

Baubock, R. 2006. *Who are the citizens of Europe?* (http://eudocitizenship.eu). Rainer.Baubock (at) eui.eu.

Bauerlein, M. 1998. Richard Rorty. Achieving our country: Leftist thought in twentieth-century America. *Walt Whitman Quarterly Review*, 104–108.

Bauman, Z. 2011. From agora to the market place, and where to from here? *Journal of Globalisation Studies* 2 (1): 3–14.

Bausch, K. 2006. Be your enemy: The contributions of West Churchman to doing better governance and international relations. In *Rescuing the Enlightenment from Itself: Critical and Systemic Implications for Democracy*, ed. McIntyre-Mills, J., vol. 1, C. West Churchman Series. New York, Boston, London: Springer.

Bausch, K., T. Flanagan, J. McIntyre-Mills, T. Made, K. Mackenzie, C. Morse, and G. Underwood. 2012. Striving for sustainable global democracy through a group decision-making process: A critical review of an online course to model transformative praxis. *Journal of Globalization Studies*, 3 (1), May 2012. http://www.sociostudies.org/journal/jogs/archive/2012_1/. Accessed August 15, 2012.

Beardsworth, R. 2011. *Cosmopolitanism and International Relations Theory*. Cambridge: Polity.

Beck, U. 1992. *Risk Society Towards a New Modernity*. London: Sage.
Beck, U. 1998. *Democracy Without Enemies*. Cambridge: Polity.
Beck, U. 2005. *World Risk Society.*. Cambridge: Polity.
Beck, U., and N. Sznaider. 2006. Unpacking cosmopolitanism for the social sciences: A research agenda. *The British Journal of Sociology* 57 (1): 23.
Beck, U. 2009. *World at Risk*. Cambridge: Polity.
Beck, U. 2010. Climate for change, or how to create a green modernity. *Theory, Culture and Society* 27 (2–3): 254–266.
Beer, S. 1974. *Designing Freedom*. London: Wiley.
Beer, S. 1975. *Platform for Change*. New York: Wiley.
Beer, S. 1994. Governance or government. In *Beyond Dispute: The Invention of Team Syntegrity*. London: Wiley.
Bell, G. 2009. *Getting Connected, Staying Connected*. South Australia: Department of Premier and Cabinet.
Benhabib, S. 2002. *The claims of culture: Equality and diversity in the global era*. Princeton, NJ: Princeton University Press.
Benhabib, S. 2004. *On the existential question that moves her scholarly and political life*. http://globetrotter.berkleu.edu/people4/Benhabib. Accessed July 27, 2012.
Benhabib, S. 2007. Twilight of Sovereignty or the emergence of cosmopolitan norms? Rethinking citizenship in volatile times. *Citizenship Studies* 11 (1): 19–36.
Benhabib, S. 2008. *On the public sphere, deliberation, journalism and dignity*. http://www.resetdoc.org/Benhabib-interview-media.php. Accessed July 27, 2012.
Berger, P.L. 1974. *Homeless Mind: Modernization and Consciousness*. London: Penguin.
Berger, P., and T. Luckmann. 1966. *The social construction of reality*. London: Penguin.
Berlin, I. 1953. *The Hedgehog and the Fox: An Essay on Tolstoy's View of History*. London: Weidenfeld & Nicolson.
Berlin, I., and H. Hardy. 1979. *Against the Current: Essays in the History of Ideas*. London: Hogarth Press.
Best, S. 2013. *Zygmundt Bauman: Why Good People Do Bad Things*. Farnham: Ashgate.
Bignall, S., S. Hemming, and D. Rigney. 2016. Three ecosophies for the anthropocene: Environmental governance, continental posthumanism and indigenous expressivism. *Deleuze Studies*, 455–478.
Binkley, S. 2008. Liquid consumption anti-consumerism and the fetishizedde-fetishization of commodities. *Cultural Studies* 22 (5), September. doi:10.1080/09502380802245845.
Bogue, R. 1989. *Deleuze and Guattari*. London: Routledge.
Bohman, J. 2004. Expanding dialogue: The Internet, the public sphere and prospects for transnational democracy. *Editorial Board of the Sociological Review* (Oxford: Blackwell).
Bohman, J. 2004. Republican cosmopolitanism. *The Journal of Political Philosophy* 12 (3): 336–352.
Bohman, J. 2005. The democratic minimum: Is democracy a means to global justice? *Ethics and International Affairs* 19 (1): 101–115.
Bohmann, J., and Lutz-Bachman (Eds). 1997. *Perpetual Peace: Essays on Kant's Cosmopolitan Ideal*. Cambridge: MIT press.
Bonanno, G. 2004. Loss trauma, and human resilience. *American Psychologist* 59 (1): 20–28.
Borradori, G. 2003. *Philosophy in a Time of Terror*. Dialogues with Habermas and Derrida: University of Chicago Press.
Bostrom, N. 2011. *Existential risk prevention as the most important task for humanity*. Faculty of Philosophy & Oxford Martin School University of Oxford. http://www.existential-risk.org/.
Bottici, C., and B. Challand. 2006. Rethinking political myth: The clash of civilizations as a self-fulfilling prophecy. *European Journal of Social Theory* 9 (3): 315–336.
Boulding, K. 1956. General systems theory—The skeleton of science. *Management Science* 2: 197–208.

Boulding, K.E. 1966. The economics of the coming spaceship earth. *Environmental Quality in a Growing Economy*, 3–14. Baltimore: Johns Hopkins Uni. Press.

Bourdieu, P. 1977. *Outline of a Theory of Practice*. Cambridge: Cambridge University Press.

Bourdieu, P. 1999. *The Weight of the World*. Oxford: Polity.

Brown, G.W., and D. Held (eds.). 2010. *Cosmopolitan Reader*. Cambridge: Polity Press.

Brown, L., and Strega, S. 2005. *Research as Resistance: Critical, Indigenous and Anti-Oppressive Approaches*. Canadian Scholars Press.

Brundtland, G. 1987. *The report "Our Common Future"*. http://www.worldinbalance.net/agreements/1987-brundtland.html.

Buckminster Fuller. 1969. *Operating manual for spaceship earth.*http://designsciencelab.com/resources/OperatingManual_BF.pdf.

Burns, D. 2007. *Systemic Action Research. A Strategy for Whole System Change*. Bristol: Policy Press.

Butler, J. 2005. *Giving an Account of Oneself.* Fordham University Press.

Butler, J. 2011. Precarious life: The obligations of proximity, May 28, 2011. *The Neale Wheeler Watson Lecture 2011*, given by Professor Judith Butler. Location: Nobel Museum, Svenska. http://www.youtube.com/watch?v=KJT69AQtDtg.

Butler, J., and A. Athanasiou. 2013. *Dispossession: The Performative in the Political*. Cambridge: Polity.

Cao, B. 2015. *Environment and Citizenship*. London: Routledge.

Capra, F. 1996. *The Web of Life: A New Synthesis of Mind and Matter*. Harper Collins.

Carens, J. 1995. Aliens and Citizens: The case for open borders. In *The Rights of Minority Cultures*, ed. Kymlicka, W. Oxford: Oxford University Press.

Carspecken, P.F. 2014. *Critical Ethnography in Educational Research*. New York: Routledge.

Castells, M. 1998. *The Information Age: Economy, Society and Identity*. Blackwell: Malden Mass.

Chakraboritty, A. Hidden money corrupts. *The Guardian Weekly*, 15.0416.

Chambers, R. 1983. *Putting the Last First*. New York: Routledge.

Chambers, R. 1997. *Whose Reality Putting the First Last*. Sussex: Intermediate Technology.

Charlton, A. 2011. Man-made world: Choosing between progress and the planet. *Quarterly Essay*, Issue 44.

Checkland, P., and J. Scholes. 1991. *Soft Systems Methodology in Action*. London: Wiley.

Christakis, A. 2006. A retrospective structural inquiry into the predicament of Humankind: Prospectus of the Club of Rome. In *Rescuing the Enlightenment from Itself. Critical and Systemic Implications of Democracy*, ed. McIntyre-Mills, J., Volume 1 of the 'C. West Churchman and Related Works Series'. Van Gigch, J (series editor). London: Springer.

Christakis, A., and T. Flanagan. 2010. *The Talking Point: A Collaboration Project of 21st Century Agoras*. Information Age Publishing.

Christakis, A., and K. Bausch. 2006. *How People Harness Their Collective Wisdom and Power to Construct the Future in Co-laboratories of Democracy*. Greenwich: Information Age.

Churchman, C.West. 1971. *The Design of Inquiring Systems*. New York: Basic Books.

Churchman, C.West. 1979. *The Systems Approach and its Enemies*. New York: Basic Books.

Churchman, C.West. 1982. *Thought and Wisdom*. Californian: Intersystems Publications.

Cobbett, E., and R. Germain. 2012. 'Occupy Wall Street' and IPE: Insights and implications. *Journal of Critical Globalisation Studies* (5): 110–113.

Collins, R. 2010. Geopolitical conditions of internationalism, human rights and world law. *Journal of Globalisation Studies* 1: 29–45.

Comaroff, J. 2011. Theory from the South: Or how Europe is evolving towards Africa. Key note address at Knowledge and value in a Globalising world. Disentangling dichotomies, querying unities. University of Western Australia. Perth 5-5th July.

Commoner, B. 1976. *The Poverty of Power*. New York: Knopf.

Commonor, B. 1971. The Closing Circle: Nature, Man, and Technology Knopf, 1971, digitalised 17 August 2006.

Cooke, B. 2004. Rules of thumb for participatory change agents. In *Participation from Tyranny to transformation?*, ed. Hickey, S., and G. Moham. New York: Zed Books.
Costanza, R, J. Cumberland, H. Daly, R. Goodland, R. Norgaard, I. Kubiszewski, C. Franco. 2015. Examples of policies, instruments and institutions. In *An Introduction to Ecological Economics*. London: Taylor and Francis.
Crocker, D. 1995. Functioning and capability: The foundations of Sen's and Nussbaum's development ethic', Part 2. In *Women, Culture and Development: A Study of Human Capabilities*, ed. M. Nussbaum, and J. Glover. Oxford: Clarendon Press.
Crossen, T., and V. Niessen. 2007. NGO standing in the European Court of justice—Does the Aarhus regulation open the door? *RECEIL* 16 (3): 332–338. ISSN 09628797 at 332.
Crowder, G. 2003. Hedgehog and fox. *Australian Journal of Political Science* 38 (2) (July 2003): 333–77.
Crowder, G. 2003. *Pluralism, relativism and liberalism in Isaiah Berlin*. Refereed paper presented to the Australasian Political Studies Association Conference University of Tasmania Hobart. 29 September–1 October.
Crush, J., and B. Fayne. 2011. *Pathways to Insecurity: Urban Food Supply and Access in Southern African Cities*. Urban food security series. African Food Security Urban Network, 1–44. http://www.alnap.org/resource/6452.
Cruz, I., A. Stahel, M. Max-Neef. 2009. Towards a systemic development approach: Building on the human scale development paradigm. *Ecological Economics* 68: 2021–2030.
Davies, J.B., and World Institute. 2008. *Personal Wealth*. Oxford University Press.
Dawkins, R. 1996. *1976 (updated preface) The Selfish Gene*. Oxford: Oxford University Press.
Dawkins, R. 2006. *The God Delusion*. Black Swan Berkshire.
De Lang, K. 2016. http://www.theguardian.com/world/2015/may/31/the-heat-and-the-death-toll-are-rising-in-india-is-this-a-glimpse-of-earths-future.
De Tocqueville, A. 1945. *Democracy in America*. New York: Knopf.
De Waal, F. 2006. Morally evolved: Primate social instincts, human morality, and the rise and fall of Veneer theory. In *Primates and Philosophers: How Morality Evolved*, ed. Macedo, S., and J. Ober. Princeton University Press. Princeton Science Library.
De Waal, F. 2009. *The Age of Empathy Nature's Lessons for a Kinder Society*. Harmony.
Derrida, J. 2005. The principle of hospitality. *Parallax* 11 (1).
Desai, M. 2005. Social democracy as world Panacea. In *Debating Globalization*, ed. Held, D. et al. Cambridge: Polity.
Dobson, A., and R. Eckersley. 2006. *Political Theory and the Ecological Challenge*. Cambridge University Press, 10 Aug 2006—Political Science.
Dobson, A. 2007. Environmental citizenship: Towards sustainable development. *Sustainable Development* 15, 276–285. www.interscience.wiley.com. doi:10.1002/sd.344.
Dobson, A. 2012. Listening: The new democratic deficit. *Political Studies* 60: 843–859. doi:10.1111/j.1467-9248.2012.00944.x.
Dobson, A., and D. Bell. 2005. *Environmental Citizenship*. Cambridge: MIT Press.
Donaldson, S., and W. Kymlicka. 2011. *Zoopolis: A Political Theory of Animal Rights*. Oxford: Oxford University Press.
Douglas, M. 1978. *Purity and Danger. An Analysis of the Concepts of Pollution and Taboo*. London: Routledge and Kegan Paul.
Drahos, P. 2005. *An alternative framework for the global regulation of intellectual property rights*. Accessed from http://cgkd.anu.edu.au/menus/workingpapers. Working paper forthcoming Austrian Journal of Development Studies.
Dryzek, J. 1999. *Democratising Rationality in Discursive Democracy*. Cambridge: University Press.
Dryzek, J. 2000. *Deliberative Democracy and Beyond*. Oxford: University Press.
Dryzek. J. 2005. *The Politics of the Earth. Environmental Discourses*, 2nd edn, 156–157. New York: Oxford University Press.
Dryzek, J. 2010. *Green Democracy*. Cunningham Lecture Series. ANU, Occasional Paper.

Eder, K. 1996. Part 3: The politics of the environment: Exhaustion or renewal. The institutionalisation of Environmentalism: Ecological discourse and the second transformation of the public sphere. In *Risk, environment and modernity. Towards a New Ecology*, ed. Lash, S., B. Szerzynski, and B. Wynne. London: Sage.

Ehrlich, P., and A. Ehrlich. 2013. *Can a collapse of global civilization be avoided?*, 1–9. http://rspb.royalsocietypublishing.org/.

Elkington, J. 1994. Globalisation's reality checks. In *Debating Globalization*, ed. Held, H et al. Cambridge: Polity.

Elkington, J. 1997. *Cannibals with Forks*. Oxford: Capstone.

Elliot, A. 2002. Beck's sociology of risk: A critical assessment. *Sociology* 36: 293–315.

Elliot, A. 2012. *Life on the Move: Mobilities in a Global Age*. Public Lecture, Flinders University 19th June. Victoria Square.

Elliot, A., and J. Urry. 2010. *Mobile Lives*. New York: Routledge.

Etzioni, A. 2004. *From Empire to Community: A New Approach to International Relations*. New York: Palgrave Macmillan.

Etzioni, A. 2010. Unmanned aircraft systems: The moral and legal case. JJFQFQ/issue 57, 2d quarter 2010: 77-71.

Etzioni, A. 2012. The domestic sources of global adhocracy. *Social Change Review* 10 (2), December 2012.

Evans, M. 2013. Confronting insecure employment. *Advocate* 20 (3): 28.

Evatt, H.V. 1984. *'Introduction' The Eureka Stockade*, by Carboni, R. 1855. First edition printed by Atkinson and Co, Melbourne.

Eyerman, R. 1981. False consciousness and ideology in Marxist theory. *Acta Sociologica*. http://www.sagepublications.com.on behalf of the Nordic Sociological Association. Downloaded Flinders University July 31, 2001.

Fairlie, 2009. http://www.thelandmagazine.org.uk/articles/short-history-enclosure-britain. This article originally appeared as 'A short history of enclosure in Britain' in *The Land*, Issue 7.

Faist, T. 2009. The transnational social question. *International Sociology* 24: 17–36.

Farrell, H. 2012. The consequences of the internet for politics. *Annual Review of Political Science* 15: 35–5 online.

Fenby, J. 2013. Scandals undermine Chinas regime. *The Guardian Weekly* 17.05.13, p. 20.

Fisher, B., R. Turner, and P. Morling. 2009. Defining and classifying ecosystem services for decision making. *Ecological Economics* 68: 643–653.

Fiske, A. 1992. The Four Elementary Forms of Sociality: Framework for a Unified Theory of Social Relations Psychological Review, 99 (4): 689–723.

Flanagan, T., and K. Bausch. 2011. A democratic approach to sustainable futures. A workbook for addressing the global problematique. Ongoing Emergence Press. Riverdale.

Flannery, T. 2005. *The Weather Makers. The History and Future Impact of Climate Change*. Melbourne: Text Publishing.

Flannery, T. 2010. *Here on Earth. An Argument for Hope*. Melbourne: Text Publishing.

Flannery, T. 2012. After the future: Australia's new extinction crisis. *Quarterly Essay*, Issue 48.

Flood, R., and E. Carson. 1993. *Dealing with Complexity: An Introduction to the Theory and application of Systems Science*, 2nd ed. London: Plenum.

Flood, R., and N. Romm. 1996. *Diversity Management: Triple Loop Learning*. Chichester: Wiley.

Florini, A. 2003. *The Coming Democracy*. Washington DC: Island Press.

Flyvbjerg, B. 1998. Habermas and Foucault. *British Journal of Society* 49 (2): 200–232.

Folke, C. 2006. Resilience: The emergence of a perspective for social-ecological systems analyses. *Global Environmental Change* 16: 253–267.

Follesdal, A. 2006. Subsidiarity, democracy and human rights in Europe. *Journal of Social Philosophy* 37 (1): 61–80.

Foote, P. 1967. The problem of abortion and the doctrine of the double effect. *Oxford Review*, no 5.

Foucault, M., and C. Gordon (eds.). 1980. *Power/Knowledge*. Harvester: Brighton.

Fougere, G. 2007. *Wellbeing is an Idea Whose Time Has Come. New Opportunities for Health Impact Assessment in New Zealand Public Policy and Planning*. Wellington, New Zealand: The Public Health Advisory Committee.

Fourade, M., and J. Savelsburg. 2006. Global processes, national institutions, local bricolage: Shaping law in an era of globalisation. *Law and Social Inquiry* 31 (3): 515–519.

Fowler, J.H., and N. Christakis. 2010. Cooperative behavior cascades in human social networks. *Proceedings of the National Academy of Sciences* 107 (9): 5334–5338.

Fox, D. 2011. *The limits of intelligence: The laws of physics may well prevent the human brain from evolving into an ever more powerful thinking machine*. Scientific American. http://www.scientificamerican.com/article.cfm?id=the-limits-of-intelligence on line 14 June. Accessed August 29, 2013.

Frank, R.H. 2007. *Falling Behind: How Rising Inequality Harms the Middle Class, Berkley*. CA: California Press.

Frankl, V. 1955. *The Doctor and the Soul: From Psychotherapy to Logotherapy*. New York: Vintage.

Frayne, B., C. McCordic, and H. Shilomboleni. 2014. *Urban Forum*. June 2014, 25 (2): 177–18. doi:10.1007/s12132-014-9219-3.

Freire, P. 1997. *Pedagogy of the Heart*. New York: Continuum.

Friere, P. 1982. Creating Alternative Research Methods: Learning to Do It. In *Creating Knowledge: A Monopoly?*, ed. Hall, B., A. Gillete, and R. Tandon, 29–37. New Delhi Society for Participatory Action Research in Asia.

Friere, P. 1985. *The Politics of Education*. Westport, CT 06881: Begin and Garvey, Inc.

Fukuyama, F. 2002. *Our Post Human Future*. New York: Farrar, Struas and Giroux.

Fukuyma, F. 2004. *State Building: Governance and World Order in the 21st Century*. Ithaca: Cornell University Press.

Gadamar, H.G. 2001, Education is self-education. *Journal of Philosophy of Education*, 35 (4).

Gallhofer, S., and A. Chew. 2000. Introduction: Accounting and indigenous peoples. *Accounting, Auditing and Accountability Journal* 13 (3): 256–267.

Galtung, J. 1990. Cultural violence. *Journal of Peace Research* 27 (3): 291–305. Sage.

Gamble, A. 2000. Economic governance. In *Debating Governance: Authority, Steering and Democracy*, 1st ed, ed. J. Pierre. Oxford: Oxford University Press.

Gandhi, L. 1966. *Affective Communities: Anti-Colonial Thought, fin-de-siècle Radicalism and the Politics of Friendship*. Duke University Press.

Gardner, H. 2008. *Multiple Intelligences: New Horizons in Theory and Practice*. New York Basic Books.

Garnaut, R. 2011. *The Garnaut Review: Australia in the Global Response to Climate Change*. Melbourne: Cambridge University Press.

Gaventa, J. 2001. Global citizen action: Lessons and challenges. In *Global Citizen Action*, ed. M. Edwards, and J. Gaventa. Boulder: Lynne Rienner.

Gebbie, K. et al. 2012. *The Disaster resilience score card for Australian Government National Emergency Management Program.*www.torrensresilience.org.

Gebbie, K., and Arbon, P. Cusack, L. 2014. Developing a model and tool to measure community disaster resilience. Disaster Resilience Australia. Get Ready. *Australian Journal of Emergency Management* 29 (4): 12–16.

Geertz, C. 1973. Thick description: Towards an interpretive theory of culture. In *The interpretation of cultures. Selected essays (3-32)*, ed. Geertz, C. New York: Basic Books.

Geertz, C. 1980. *Negara: The Theatre State in Nineteenth Century Bali*. Princeton: Princeton University Press.

Geertz, C. 2004. What is a state if it is not sovereign? Reflections on politics in complicated places. *Current Anthropology* 45: 577–593.

Getano, Lui. 1993. *Voices from the Land*. Boyer lecture Series.

Gibbons, M., C. Limoges, H. Nowotny, S. Schwartzman, P. Scott, and M. Trow. 1994. *The New Production of Knowledge: The Dynamics of Science and Research in Contemporary Societies*. London: Sage.

Giddens, A. 2009. *The Politics of Climate Change*. Cambridge: Polity.

Girarde, H. 2015. *Sustainability is unhelpful: We need to think about regeneration*. http://www.theguardian.com/sustainable-business/blog/sustainability-unhelpful-think-regeneration.

Goff, S. 2007. Participatory practices: On being our field together. *ALAR Journal* 12 (2): 106–126.

Goldsmith, J., and E. Posner. 2005. *The Limits of International Law*. Oxford: Oxford University Press.

Gore, A. 2007. *Assault on Reason*. London: Bloomsbury.

Gould, C. 2007. Transnational solidarities. *Journal of Social Philosophy* 38 (1): 148–164.

Gouldner, A.W. 1971. *The Coming Crisis of Western Sociology*. London: Heinemann.

Graham, C. 2009. *The economics of happiness*. http://www.brookings.edu/.

Graham, C. 2011. *The Pursuit of Happiness: An Economy of Well-being*. Washington: Brookings Institution Press.

Graver, M. 2007. *Stoicism and Emotion*. Chicago: University of Chicago Press.

Greenberg, G., and E. Tobach. 1987. *Cognition, Language and Consciousness: Integrative Levels*, vol. 2. New Jersey: Hillsdale.

Greenfield, S. 2000. The private life of the brain. *Emotions, consciousness and the secret of the self*. New York: Wiley.

Greenfield, S. 2003. *Tomorrow's People How 21st Century Technology is Changing the Way We Think and Feel*. Penguin.

Greenfield, S. 2008. *ID: The Quest for Meaning in the 21st Century*. London: Sceptre, Hodder and Stoughton.

Greenwald, G. 2012. *New York's top court highlights the meaninglessness and menace of the term 'terrorism'*. http://www.guardian.co.uk/commentisfree/2012/dec/16/court-terrorism-morales-gangs-meaningless.

Grodzki, L. 1995. *Approaching a theory of emotion: An interview with Candace Pert*, Ph.D. http://www.primal-page.com/pert.htm.

Guddemi, P. 2006. Human views of ecological nature. *Cybernetics and Human Knowing* 12 (1): 86–90.

Gunaratne, S.A. 2009. Globalization: A non Western perspective: The bias of social science/communication oligopoly. *Communication, Culture and Critique* 2: 60–82.

Habermas, J. 1984. *The Theory of Communicative Action: Reason and the Rationalization of Society*. Boston: Beacon.

Habermas, J. 1989. *The Structured Transformation of the Public Sphere*. Cambridge: Polity Press.

Habermas, J. 2001. *The Postnational Constellation. Political Essays*. Cambridge: Polity.

Habermas, J., J. Derrida, and G. Borradori. 2003. Philosophy in a time of terror Dialogues with Jurgen Habermas and Jacques Derrida. University of Chicago Press.

Habermas, J. 2008. *Between Naturalism and Religion*. Cambridge: Polity Press.

Habermas, J., and J. Derrida. 2003. February 15, or what Binds Europeans Together: A plea for a Common Foreign Policy, Beginning in the Core of Europe. *Constellations* 10 (3): 291–297.

Haidt, J. 2008. The emotional dog and its rational tail: A social intuitionist approach to moral judgment. In *Reasoning: Studies of Human Inference and Its Foundations*, ed. Adler, J.E., and J. Lance, 1024–1053.

Hall, S. 1992. *Our Mongrel Selves*. New Statesman and Society. 19 June.

Hall, S. 1997. *Representation: Cultural Representations and Signifying Practices*. London: SAGE Publications Ltd presentation.

Hall, S., and P. Du Gay. 1996. *Cultural Identity*. London: Sage.

Hammadi, S., and A. Kelly. 2013. Bangladesh's stark lesson for buyers. *Guardian Weekly* 17.05.13, p. 44.

Hannerz, U. 2003. Being there…and there…and there! Reflections on multi-site ethnography. *Ethnografeast* 4 (2): 201–216 [1466–1381(200306) 4:2; 201–216; 034179]. Downloaded from eth.sagepub.com at Flinders University on November 23, 2016.

Harari, Y.N. 2011. *Sapiens: A Brief History of Humankind*. New York: Harper.

Harari, Y.N. 2015. Homo Deus. *A brief history of tomorrow*. London: Random House.

Haraway, D. 1991. *Cyborgs, Simians, and Women: The Reinvention of Nature*. London: Free Association Books.

Haraway, D. 1992. The promises of monsters: A regenerative politics for inappropriated others. In *Cultural Studies*, ed. Lawrence Grossberg, Cary Nelson, and Paula A. Treichler, New York: Routledge, 295–337.

Haraway, D. 2010. *Speculative fabulations for technoculture's generations: Taking care of unexpected country*. Australian University E press. http://epress.anu.edu.au/guidelines/author_guidelines.html.

Hardin, G. 1968. The Tragedy of the Commons. *Science* 162: 1243–1248.

Hardt, M. 2010. Revolution. In *The Examined Life: Excursions with Contemporary Thinkers*, ed. A. Taylor. New York: The New Press.

Harris, J. 2013. Can Labour transform for 2015? *Guardian Weekly*, 17.05.13, p. 21.

Harris, L.D., and J. Wasilewski. 2004. Indigenous wisdom of the people's forum. *Systems Research and Behavioural Science* 21: 505–514.

Harrison, K. 2002. Policy perspectives: Social capital. In *Democratising Local Government. The South African Experiment*, ed. Parnell, S., M. Swilling, and D. Wooldridge. Landsdowne: University of Cape Town Press.

Hassan, R. 2008. *Inside Muslim Minds*. Melbourne: Melbourne University Press.

Hay, I.E. 2013. *Geographies of the Super-rich*. Cheltenham: Edward Elgar.

Hay, I.M., and J. Beaverstock (eds.). 2016. *Handbook on Wealth and the Super-Rich*. Cheltenham: Edward Elgar.

Hayden, P. 2010. The environment, global justice and world environmental citizenship. In *Cosmopolitanism*, ed. Brown, G.W., and D. Held. Cambridge: Polity.

Hegel, G.W.F. 1977. *Phenomenology of the Spirit*. Translated by A.V. Miller. Oxford: Oxford University Press.

Held, D., A. McGrew, D. Goldblatt, and J. Perraton. 1999. *Global transformations. Politics Economics and Culture*. Stanford California: Stanford University Press.

Held, D. 2004. *Global Covenant: The Social Democratic Alternative to the Washington Consensus*. Cambridge: Polity.

Held, D. 2016. Climate change. *Migration and the Cosmopolitan Dilemma*. doi:10.1111/1758-5899.12309.

Held, D. et al. 2005. *Debating Globalization*. Cambridge: Polity.

Heller, P. 2001. Moving the state: The politics of democratic decentralisation in Kerela, South Africa, and Porto Alegra. *Politics and Society* 29 (1): 131–163.

Hesse-Biber, S. 2010. Qualitative Approaches to Mixed Methods Practice. http://qix.sagepub.com/content/16/6/455.

Hesse-Biber, S. 2014. Thinking outside the box: Applying a feminist praxis lens to austerity policy in Eurozone countries. *Psychology of Women Section Review* 16 (1): 9–23.

Heycox, J., P. Meadows, and B. Vernon. 1997. *Environment and Energy Statistics Section Australian Bureau of Statistics. Measuring environmental expenditures and revenues in Local Government*. Australian Centre for Regional & Local Government Studies Working Paper No 2. University of Canberra.

Hickey, S., and G. Mohan (eds.). 2004. *Participation: From Tyranny to Transformation?*. New York: Zed Books.

Hil, R. 2013. Beyond the cyber cell. *Advocate* 20 (3): 28–29.

Hill, D. 2015. 'Is Bolivia quietly planning to frack' Mother Earth'? *The Guardian Weekly* 6.03.2015.

Hill Collins, P. 2000. *Black Feminist Thought: Knowledge, Consciousness and the Politics of Empowerment*. New York: Routledge.

Hinnells, J.R. 1984. *A Handbook of Living Religions*. London: Penguin.

Hirschman, A.O. 1970. *Exit, Voice, and Loyalty: Responses to Decline in Firms, Organizations, and States*. Harvard University Press.

HM Government. 2005. Securing the Future: Delivering UK Sustainable Development. Strategy Presented to Parliament by the Secretary of State for Environment, Food and Rural Affairs by Command of Her Majesty, March 2005. HMSO: St Clements House, 2-16 Colegate, Norwich.

Hobbes, T. 1651. *Leviathon*. http://www.gutenberg.org/ebooks/3207.

Hobbes, T. 2002. *Leviathan*. http://ebooks.adelaide.edu.au/h/hobbes/thomas/h68l/index.html. Accessed June 25, 2012.

Hobson-West, P. 2007. Beasts and boundaries: An introduction to animals in sociology, science and society. *Qualitative Sociology Review* 3 (1): 23–41.

Hofstede, G., G.J. Hofstede, and M. Minkov. 2010. *Cultures and Organisations: Software of the Mind: Intercultural Co-operation and Its Importance for Survival*. New York: McGraw Hill.

Hoggett, P., and S. Thomson. 2002. Towards a democracy of emotions. *Constellations* 9 (1): 102–126.

Hoggett, P. 2010. *Politics, Identity and Emotion*. Boulder: Paradigm.

Holloway, J. 2011. The economic platform of the Australian Greens, philosophy, Policy and electoral appeal. Flinders University Honours Thesis.

Hooks, B. 1989. *Talking Back*. Boston: South End.

Horvath, Z., and Odor, B. *The Union after Lisbon. The Treaty Reform of the EU*. HVG-ORAC Publishing. Hungary.

http://mg.co.za/article/2012-02-21-youth-unemployment-south-africas-ticking-bomb/.

http://www.eurozine.com/articles/2006-12-23-baubock-en.html. Accessed December 19, 2012.

Hua-Ching, N. 1989. The complete works of Lao Tzu, Tao The Ching and Hua Hu Ching. College of Tao and Traditional Chinese Healing. LA.

Hulme, M. 2009. *Why We Disagree about Climate Change*. Cambridge: University Press.

Hulme, M. 2010. Cosmopolitan climates. Hybridity, foresight and meaning. *Theory, Culture and Society*. 27 (2–3): 267–276.

Hulme, M. 2011. Climate Change and four narratives of growth. AHTC Workshop. University of Cambridge. http://climatehistories.innerasiaresearch.org/wp-content/uploads/2011/01/Mike-Hulme.pdf. Accessed February 8, 2011.

Huntington, S. 1993. The clash of civilisations? *Foreign Affairs*, 22–49.

Huntington, S. 1996. *The Clash of Cultures and the Remaking of the World Order*. New York: Simon and Schuster.

Husserl, E. 1970. *The Crisis of European Sciences*. Translated by David Carr. Northwestern University Press, Evanston.

icps.gwu.edu/"The Good Life: An International Perspective," There's a Future: Visions for a better world. https://docs.google.com/viewer?a=v&q=cache:8rnWNHvfIcAJ:icps.gwu.edu/files/2013/02/The-Good-Life-An-International-Perspective.pdf+the+good+life++an+international+perspective+%2B+etzioni&hl=en&gl=au&pid=bl&srcid=ADGEESgpk7jXTiA7uWJH2nWGJm4gOc8rGDjuAHm-oeyW7IQl.

Ingold, T. 1993. The temporality of the landscape. *World Archeology* 25 (2): 152–174. (Conceptions of Time and Ancient Society, October).

International Commission on the Future of Food and Agriculture. 2007. *Manifesto on the future of seeds*. www.future-food.org.

Irvin, R., and J. Stansbury. 2002. Citizen participation in decision making. *Public Administrative Review* 64 (1): 55–65.

Irvine, L. 2007. The question of animal selves: Implications for sociological knowledge and practice. *Qualitative Sociology Review* 3 (1): 5–22.

Irwin, T. 1985. *Aristotle Nicomachean Ethics Translated*. Cambridge: Cambridge University Press.

Ison, R. 2005. Geoffrey Vickers: Contemporary applications and changing appreciative settings systems. *Research and Behavioural Science* 22 (4) (July–August): 277–284.
Jackson, M. 1991. *Systems Methodology for the Management Sciences*. London: Plenum.
Jackson, M. 2000. *Systems Approaches to Management*. London: Kluwer.
Jamieson, D. 2008. *Ethics and the Environment: An Introduction*. Cambridge: Cambridge University Press.
Jamrozik, A. 2001. *Social Policy in the Post Welfare State: Australians on the Threshold of the 21st Century*. NSW: Longman.
Jasonoff, S. 2003. Technologies of humility: Citizen participation in governing science. *Minerva* 41: 223–244.
Jessop, S. 2009. Children's participation: An Arendtian criticism. *Educational Philosophy and Theory* 43 (9). Article first published online: 8 NOV 2009. doi:10.1111/j.1469-5812.2009.00560.x.
Jones, O., and M. Woodward. 1910. A game keeper's notebook. London: Edward Arnold. https://archive.org/stream/gamekeepersnoteb00jonerich#page/40/mode/2up.
Judt, T. 2010. *Ill Fares the Land*. London: Penguin.
Kabeer, N., and R. Subrahmanian. 1996. *Institutions, relations and outcomes: Framework and tools for gender-aware planning*. Institute of Development Studies Discussion Paper, No. 357, Brighton, UK.
Kahane, A. 1992. The Mont Fleur Scenarios: What will South Africa be like in the year 2002? *Deeper News* 7 (1). Republication of The Weekly Mail and Guardian Weekly.
Kaldor, M. 2003. *Global Civil Society: An Answer to War*. Cambridge: Polity Press.
Kaldor, M. 2004. *What is Human Security? Debating Globalisation*. Cambridge Polity: Open Democracy Debate.
Kanai, R, B. Bahrami, R. Roylance, and G. Rees. 2011. Online social network size is reflected in human brain structure. In *Proceedings of Royal Biological Society B*. Published online 19 October 2011.
Kant, I. 1976. *The Moral Law*. Translated by Paton, H.J. Oxford: Hutchinson.
Karni, A. (Ed). 2006. *A Celebration of Democracy*. Wahid Institute and Asia Foundation.
Kauffman, L. 2006. Virtual Boundaries. *Cybernetics and Human Knowing* 13 (2): 94–104.
Keane, J. 2009a. *The Life and Death of Democracy*. London: Simon and Schuster.
Keane, J. 2009a. *Democracy failure: The root of the global economic crisis is political*. http://scholar.google.com.au/scholar?cites=16220241904322824583&as_sdt=2005&sciodt=0,5&hl=en. Accessed March 31, 2011.
Kelly, L. 2016. *The Memory Code*. Sydney: Allen and Unwin.
Kickert, W.J.M., E.H. Klijn, and J.F.M. Koppenjan. 1999. *Managing Complex Networks: Strategies for the Public Sector*. London: Sage.
Kirby, M. 2011. Animals deserve our protection. *Weekend Australian*, p. 22. October 1–2.
Kirby, M. 2013. *What Would Ghandi Do?* New York: Penguin.
Kitcher, P. 2016. Masking the meaningful. *Global Policy* 7 (Supplement 1). May.
Kivisto, P., and T. Faist. 2007. *Citizenship, Discourse, Theory, and Transnational Prospects*. Blackwell Publishing.
Kolben, K. 2011. Transnational labor regulation and the limits of governance. *Theoretical Inquiries in Law* 12 (2) Article no 2.
Kovach, M. 2005. Emerging from the margins: Indigenous Methodologies. In *Research as Resistance: Critical, Indigenous and Anti–Oppressive Approaches*, ed. Brown, L., and S. Strega. Canadian Scholars Press.
Kymlicka, W. 2010. Citizenship in an Era of Globalization. In *Cosmopolitan Reader*, ed. Brown, G.W., and D. Held, Cambridge: Polity Press.
La Forgia, R. 2009. The politics of international law: Twenty years later: A reply to Martti Koskenniemi. *The European Journal of International Law* 20 (4): 979–984, at 993.
Lakshmi, R. 2016. Battle to move Mumbai's trash mountain. *The Guardian* April 29, 2016.
Lane, R.J. 2000. *Jean Baudrillard*. London: Routledge Critical Thinkers.

Langton, M., and J. Longbottom. 2012. *Community Futures, Legal Architecture*. Oxon: Routledge.

Langton, M.2015. The question of constitutional recognition: Marcia Langton talks to David Leyonhjelm. *Meanjin* 74 (3).

Laouris, Y. 2014. *Reengineering Reinventing Digital Era*. Broadbent, S.,Dewandre, Charles Ess, Luciano Floridi, Jean-Gabriel Ganascia, Mireille Hildebrandt, Yiannis Laouris, Claire Lobet-Maris, Sarah Oates, Ugo Pagallo, Judith Simon, May Thorseth, Peter-Paul Verbeek the Onlife project . European Commission.

Laszlo, A. 2000. The epistemological foundations of evolutionary systems design. In *Proceedings of the World Congress of the Systems Sciences and ISSS 2000 International Society for the Systems Sciences 44th Annual Meeting*, Toronto Canada, ed. Allen, J.K., and J. Wilby.

Lawn, P. 2010. *Ecological Economics: The Impact of Unsustainable Growth*. ed. Lever-Tracey, C. Routledge handbook of climate change and society, 453–472.

Layard, R. 2005. *Happiness*. London: Allen Lane.

Leonard, A. 2011. Governance in the relative when. *Systems Research and Behavioural Science* 28: 431–436.

Lever-Tracey, C. 2008. Global warming and sociology. *Current Sociology* 56 (3): 445–466.

Lever-Tracey, C. 2010. *Alternative scenarios: Varieties of capitalism*. ed. Lever-Tracey, C. Routledge handbook of climate change and society, 274–279.

Levin, S. 2006. Learning to live in a global commons: Socioeconomic challenges for a sustainable environment. *Ecological Research* 21: 328–333.

Lincoln (2001). Engaging sympathies. In *Handbook of Action Research: Participative Inquiry and Practice*, ed. Reason, P., and H. Bradbury. Sage.

Lipton, B. 2007. *Biology of Belief: Unleashing the Power of Consciousness, Matter and Miracles*. India: Hay House Publishers.

Lovelock, J. 2006. *The Revenge of Gaia: Why the Earth Is Fighting Back—And How We Can Still Save Humanity*. Santa Barbara: Allen Lane.

Lovelock, J. 2009. *The Vanishing Face of Gaia: A Final Warning: Enjoy It While You Can*. Allen Lane.

Lyons, S. 2012. *A most eminent victorian: Thomas Henry Huxley* 76 Automne: 85–104

Levin, M. Greenwood, D. 2001. *Pragmatic Action Research and the Stuggle to Transform Universities into Learning Communities*, ed. Reason, P., and H. Bradbury. Handbook of Action Research. Participative Inquiry and Practice. London: Sage.

Maathai, W. 2004. *The Green Belt Movement*. New York: Lantern Books.

Mangcu, X. 2014. *The Arrogance of Power: South Africa's Leadership Meltdown*. Cape Town: Tafelburg.

March, C., I. Smyth, and M. Mukopadhyay. 1999. *A Guide to Gender-Analysis Frameworks*. Oxford: Oxfam.

Mathews, F. 1991. *The Ecological Self*. London: Routledge.

Mathews, F. 2010. Planet beehive. *Australian Humanities Review* 50: 159–178.

Maturana, H., and F. Verela. 1980. *Autopoeisis and Cognition: The Realization of Living*. Boston: Reidel.

Maturana, H.R., and F.J. Varela. 1987. *The Tree of Knowledge: The Biological Roots of Human Understanding*. London: Shambhala New Science Library.

Max-Neef, M. 1991. *Human Scale Development*. London: Apex.

McDermott, S.K. 2012. Clash of paradigms. *The Mail and Guardian Newspaper*, February 01 2012.

McIntyre, J. 1996. Tools for ethical thinking and caring. *Community Quarterly Melbourne*. 190 pp.

McIntyre, J. 2000. *Global Citizenship and Social Movements: Creating Transcultural Webs of Meaning for the New Millennium*. The Netherlands: Harwood.

McIntyre-Mills, J. 2000 *Global citizenship and Social Movements: Creating Transcultural Webs of Meaning*. Netherlands:Harwood.

McIntyre-Mills, J. 2003. *Critical Systemic Praxis. For Social and Environmental Justice: Participatory Policy Design for a Global Age*. The contemporary systems series. London, UK: Kluwer.

McIntyre-Mills, J. 2006. *Systemic Governance and Accountability: Working and re-working the conceptual and spatial boundaries of international relations and governance*. Volume 3 of the 'C. West Churchman and Related Works Series', London: Springer.

McIntyre-Mills, J. 2008. User-Centric Design to Meet Complex Needs. N.Y.: Nova Science.

McIntyre-Mills, J., K. Bausch, A. Christakis, and D. De Vries. 2008. How can we break the mould? Democracy, semiotics and regional governance. *Systems Research and Behav Science* 25: 305–321.

McIntyre-Mills, J. 2009a. Ethics, boundaries and sustainable futures. In *Systemic Development: Local Solutions in a Global Environmen*, ed. Sheffield, J. Litchfield, USA: ISCE Publishing.

McIntyre-Mills, J., and D. De Vries. 2010a. Addressing complex needs. *International Journal of Interdisciplinary Social Sciences* 5 (5): 11–32.

McIntyre-Mills, J. 2010. Wellbeing, mindfulness and the global commons. *Journal of Consciousness Studies* 17 (7–8): 44–72.

McIntyre-Mills, J. 2012a. Anthropocentricism and wellbeing: A way out of the lobster pot? *Systems Research and Behavioural Science*. Published online in Wiley Online Library. (wileyonlinelibrary.com). doi:10.1002/sres.2131.

McIntyre-Mills, J., and D. de Vries. 2012. Transformation from Wall Street to Well-being. *Systems Research and Behavioural Science*. First published online: October 10, 2012. doi:10.1002/sres. 2133.

McIntyre-Mills, J. 2013. Systemic ethics for wellbeing. *Encyclopaedia of Food and Agricultural Ethics*. doi:10.1007/978-94-007-6167-4_342-6. # Springer Science+Business Media Dordrecht.

McIntyre-Mills, J. 2014. *Systemic Ethics' Springer Encyclopedia of Food Ethics*. ISBN 978-94-007-0928-7.

McIntyre-Mills, J. 2014. Systemic Ethics and Non-anthropocentric Stewardship: Implications for Transdisciplinarity and Cosmopolitan Politics. New York: Springer.

McIntyre-Mills, J.J. 2014c. Reconsidering Boundaries. Sociopedia, International Sociological Association.

McIntyre-Mills, J. 2015. Governing the anthropocene through balancing individualism and collectivism. In *59th Annual International Systems Sciences, Berlin Conference*, 2–7th August, Published http://journals.isss.org/index.php/proceedings59th.

McIntyre-Mills, J., and D. De Vries. 2011. *Identity, Democracy and Sustainability*, 380. ISCE, USA, Litchfieldpp: Emergence.

McIntyre-Mills, J., R. Wirawan, and Indonesian Research Consortium. 2017. Chapter 3: Pathways to wellbeing—Low carbon challenge to live virtuously and well. In *Balancing Individualism and Collectivism: Social and Environmental Justice*, ed. McIntyre-Mills, J., N. Romm, Y. Corcoran-Nantes. London: Springer.

McIntyre-Mills, J., and R. Wirawan. 2017. Chapter 4: Governing the anthropocene: Through balancing individualism and collectivism as a way to manage our ecological footprint. In *Balancing Individualism and Collectivism: Social and Environmental Justice*, ed. J. McIntyre-Mills, N. Romm, and Y. Corcoran-Nantes. London: Springer.

McIntyre-Mills, J., D. De Vries, and N. Binchai. 2014. *Wall Street to Wellbeing: Joining Up the Dots Through Participatory Democracy and Governance to Mitigate the Causes and Adapt to the Effects of Climate Change*. New York: Springer.

McKay, V., and N. Romm. 1992. *People's Education in Theoretical Perspective*. Cape Town: Mascew Miller.

McNevin, A., P. Mares, D. Mitric, K. Neumann, and S. Taylor. 2016. *Beyond deterrence*. http://insidestory.org.au/beyond-deterrence-reframing-the-asylum-seeker-debate.

Mead, G. 1995. A royal omission: A critical summary of the evidence given to the Hindmarsh Island Bridge Royal Commission with an alternative report. South Australia.

Meadows, D., and J. Randers. 1992. *Beyond the Limits. Global Collapse of a Sustainable Future.* London: Earthscan Publications.

Mertens, D., F. Cram, and B. Chilisa (Eds). 2013. *Indigenous Pathways into Social Research.* Left Coast Press, Inc.

Mertens, D. 2016. Assumptions at the philosophical and programmatic levels in evaluation. *Evaluation and Program Planning.* Elsevier 27340047. doi:10.1016/j.evalprogplan.2016.05.010.

Meyers, D. 2012. *eBook Australian Universities: A portrait in decline.*https://www.themonthly.com.au/issue/2012/august/1345613277/don-watson/comment-new-dusk.

Midgley, G. 2000. *Systemic Intervention: Philosophy, Methodology, and Practice.* New York: Kluwer.

Miller, D. 1999. Bounded citizenship. In *Cosmopolitan Citizenship.* ed. Hutchinson, K., and R. Dannreuther. London: Macmillan.

Miller McPherson, L., L. Smith-Lovin, M. James, and J.M. Cook. 2001. 'Birds of a feather: Homophily in Social Networks'. *Annual Review of Sociology* 27: 415–444.

Mills, J. 1983. *Health, healing and dis-ease in a South African township.* MA thesis. University of Cape Town.

Mills, J. 1985. Intwaso, an anthropological re-appraisal. *Journal of South African Sociology* 16 (1): 9–13.

Mills, J. 1987. Diviners as social healers within an urban township context. *Journal of South African Sociology* 18 (1): 7–13.

Molefe, T.O. 2014. South Africa's political fires. *New York Times.* May 28th.

Monbiot, .G. 2013. Money just makes the rich suffer. *Guardian Weekly,* 17.05.13. p. 9.

Monbiot, G. 2014. *The pricing of everything.* http://www.monbiot.com/2014/07/24/the-pricing-of-everything.

Moore, G.D. 2009. *Visions of Culture: An Annotated Reader.* New York: Rowman.

Muhammad, Adib Shomad. 2014. A comparative case study of a secular and religious-based university in Indonesia: A critical review of research capacity. Unpublished Ph.D. thesis, Flinders University.

Murray, et al. 2006. *Systems for social sustainability: Global Connectedness and the Tuvalu Test.* Tuvalu at the Common Wealth Games, Melbourne, 2006. Photograph by Geoff Ludbrook. www.isa.org.usyd.edu.au/publications/documents/Tuvalu_Test.pdf.

Murray, J., C. Dey, and M. Lenzen. 2007. Systems for Social Sustainability: Global Connectedness and the Tuvalu Test. *Cybernetics and Human Knowing* 14 (1): 87–105.

Ngarrindjeri Regional Authority Inc. (2014) Appendix 2 Ramsar Ecological Character Description Statement of Commitment between the Ngarrindjeri regional Authority Inc and the partnership and Stewardship Group. Department of Environment, Water and Natural Resources. Goyder Institute for Water Research is a partnership between the South Australian Government through the Department of Environment, Water and Natural Resources, CSIRO, Flinders University, The University of Adelaide and The University of South Australia.

Nakamori, Y., A. Wierzbicki, and Z. Zhu. 2011. A theory of knowledge construction. *Systems Research and Behavioural Science* 28: 15–39.

Nazarea, V. 2006. Local knowledge and memory in biodiversity conservation. *Annual Review of Anthropology* 36: 317–335.

Ng, C. 2011. National resilience: Developing a whole-of-society response: A more robust society begins with clarity about shared priorities, risks, constraints and capabilities. *Ethos* 10: 21–29.

Nguyen, N., O. Bosch, and K. Maani. 2011. Learning laboratories for sustainable development in biospheres. *Systems Research and Behavioural Science* 28: 51–62.

Nussbaum, M. 1986. *The Fragility of Goodness: Luck and Ethics in Greek Tragedy and Philosophy.* Cambridge: University Press.

Nussbaum, M. 2006. *Frontiers of Justice.* London: Harvard University Press.

Nussbaum, M. 2011. Chapter 2. The central capabilities. In *Creating Capabilities: The Human Development Approach.* London: The Belknap Press of Harvard University Press.

Nussbaum, M., and J. Glover. 1995. *Women, Culture and Development: A Study of Human Capabilities*. Oxford: Clarendon Press.

Nussbaum, M., and A. Sen. 1993. *Quality of Life*. New York: Clarendon Press.

Nye, J. 1990. *Bound to lead: The changing nature of American power*. New York: Basic Books.

Ocampo, J., and N. Gomez-Arteaga. 2016. Accountability in international governance and the 2030 development agenda. *Global Policy*. doi:10.1111/1758-5899.12322.

Odum, H.T. 1996. *Environmental Accounting: Energy and Environmental Decision Making*. New York: Wiley.

Official Journal of the European Union. 2009. C115, Volume 51. Accessed at http//europa.eu. July 9, 2009.

Olayiwola, A. 2002. Truth and Reconciliation. *African Studies Review* 45 (3): 31–34. New York: Organization, Basic Books.

Osborn, D. 1997a. *A rationale and approach for applying environmental accounting frameworks in Local Government*, Australian Centre for Regional and Local Government Studies, University of Canberra, Working Paper No 1.

Osborn, D. 1997b. *Interpreting measures of Local Government's Environmental Protection payments and revenues*, Australian Centre for Regional and Local Government Studies, University of Canberra, Working paper no 3.

Osler, A., and H. Starkeyb. 2005. Violence in schools and representations of young people: A critique of government policies in France and England. *Oxford Review of Education* 31 (2), June 2005: 195–215.

Ottawa Health Charter. 1986. http://www.who.int/healthpromotion/conferences/previous/ottawa/en/.

https://www.oxfam.org/sites/www.oxfam.org/files/file_attachments/bp210-economy-one-percent-tax-havens-180116-en_0.pdfOxfam. Davos report https://www.oxfam.org/.../2016.../62-people-own-same-half-world-reveals-oxfam-da.

Ozbekan, H. 1970. On some of the fundamental problems in planning. *Technological Forecasting* 1 (3): 235–240.

Palombi, L. 2007. Trips, bilateralism and patents: How they are failing both the developed and the developing world and what to do about it. *Electronic Journal in Communication, Information and Innovation in Health* 1 (1): 71–81.

Papadimitriou, D. 2014. The coming 'tsunami of debt' The Guardian.com, Sunday 15 June 2014 17.58 BST.

Parker, J. 2012. Transdisciplinary systems approach to paradigm change in sustainable economy. *The Systemist* 34 (3), Winter 2012: 138–149.

Parry, M., C. Rosenzweig, A. Iglesias, G. Fischer, and M. Livermore. 1999. Climate change and world food security: A new assessment. In *Global Environmental Change*. S51–S67. Elsevier. Available online 25 October. Accessed February, 7th, 2011.

Paskal, C. 2007. *How climate change is pushing the boundaries of Security and foreign policy*. Chatham House. URL http://www.chathamhouse.org.uk/files/9250_bp0607climatecp.pdfconsulted, June 2011.

Pauli, G. 2010. *The Blue Economy: Report to the Club of Rome*. New Mexico: Paradigm Publications.

Pearce, F. 2015. Climate's conflict zone. *New Scientist*. December, no 3050.

Pert, C. 1999. *The Molecules of Emotion: Why You Feel the Way You Feel*. New York: Simon and Schuster.

Pidgeon, N., and B. Fischhoff. 2011. *The role of social and decision sciences in communicating uncertain climate risks*. http://www.nature.com/doifinder/10.1038/nclimate1080.

Pikketty, T. 2015. *13th Nelson Mandela annual lecture*. https://www.nelsonmandela.org/news/entry/transcript-of-nelson-mandela-annual-lecture-2015.

Podger, A., J. Wanna, J. Ma, and T. Su. 2012. Putting the citizens at the centre: Making government more responsive. *Australian Journal of Public Administration* 71 (2): 101–110.

Poe, D. 2010. Subsidiarity and the bias for the local. *Journal of Globalisation Studies* 1: 131–140.

Pogge, T. 1992. Cosmopolitanism and sovereignty. *Ethics*, 103: 48–75.

Pogge, T.W. 2001. *Global Justice*. Malden, MA: Blackwell.

Pogge, T. 2002. *World Poverty and Human Rights: Cosmopolitan Responsibilities and Reforms*. Cambridge: Polity.

Polanyi, K. 2001. *The Great Transformation. The Political and Economic Origins of Our Time*. Boston: Beacon Press.

Pratikno, 2014. Fostering Indonesian and Australian friendship: Put education and culture first. Flinders University Seminar on Rumah Budaya, April.

Preston, N., C. Sampford, and C. Connors. 2002. *Encouraging Ethics and Challenging Corruption: Reforming Governance in Public Institutions*. Sydney: Federation Press.

Pretty, J. 2013. The consumption of a finite planet: Well-being, convergence, divergence and the nascent green economy. *Environmental and Resource Economics*. doi:10.1007/s10640-013-9680-9.

Price, C. 2012. http://za.effectivemeasure.net/em_image“ alt=“““style=“position: Absolute; left:-5px;”/>.

Reason, P. 2002. Justice, sustainability and participation. *Concepts and Transformations* 7 (1): 7–29.

Rees, W., and M. Wakernagel. 2008. Urban ecological footprints: Why cities cannot be sustainable. 10.1007/978-0-387-73412-5_35 Springer Link.

Rhodes, R.A.W. 1997. *Understanding Governance*. Maidenhead: Open University Press.

Rigney, D., S. Bignall, and S. Hemming. 2015. Negotiating indigenous modernity. *Alternative* 11 (4): 334–347.

Ritchey, T. 2011. Wicked problems—Social messes risk. *Governance and Society* 17 (2011): 1–6.

Rittel, H., and M. Webber. 1984. Planning problems are wicked problems. In *Developments in Design Methodology*. New York: Wiley.

Rivière, D. 2015. Thirsty Elephant—India's water security challenges: A test for regional relations over the next decade. Colonel Daniel Rivière, CD, PMP, September 2015.

Robins, K. 1996. Interrupting identities: Turkey and Europe. In *Cultural Identity*, Hall, S., and P. Du Gay. London: Sage.

Rockström, J., W. Steffen, and K. Noone. 2009. A safe operating space for humanity. *Nature* 461: 472–475.

Rodenhoff, V. 2002. The Aarhus Convention and its implications. *RECIEL* 11 (3).

Romm, N. 2001. *Accountability in Social Research: Issues and Debates*. London: Kluwer.

Romm, N. 2002. A trusting constuctivist view of systems thinking in a knowledge age. In *Systems Theory and Practice in a Knowledge Age*, ed. Ragsdell et al. New York: Plenum.

Romm, N. 2010. *The New Racism*. London: Springer.

Romm, N.R., and Hsu, C.Y. 2002. Reconsidering the exploration of power distance: An active case study approach. *Omega* 30 (6): 403–414.

Rorty, R. 1989. *Contingency, Irony and Solidarity*. Cambridge: Cambridge University Press.

Rorty, R. 1999. Achieving Our Country: The William E. Massey Sr. Lectures in the History of American Civilization. Harvard University Press, original publication, 1998.

Rorty, R. 2000. *Philosophy and Social Hope*. New York: Penguin.

Rose, D.B. 1996. Nourishing terrains: Australian aboriginal views of landscape and wilderness. Australian Heritage Commission, Canberra, ACT, BCIN Number: 236409.

Rose, D.B. 2004. *Reports Form a Wild Country*. Sydney: University of New South Wales Press.

Rose, D.B. 2005. *Dislocating the frontier*. http://epress.anu.edu.au/dtf/html/frames.php, see http://epress.anu.edu.au.

Rosen, R. 1991. *Life Itself: A Comprehensive Inquiry into the Nature, Origin and Fabrication of Life*. New York: Columbia University Press.

Rosyada, D. 2014. Islamic Education in Indonesia. Flinders University Seminar on Rumah Budaya, April.

Rusbridger, A. 2015. Why we are putting the climate threat to earth front and centre. *The Guardian Weekly*, 13.03.15.

Sahlins, M. 1990. The political economy of grandeur in Hawaii from 1810–1830. In *Culture Through Time: Anthropological Approaches*, ed. E. Ohnuki-Tierney. Stanford: Stanford University Press.

Saul, B., S. Sherwood, J. McAdam, T. Stephens, and J. Slezak. 2012. *Climate Change and Australia: Warming to the Global Challenge*. Sydney: The Federation Press.

Schneider, H. 2013. World Bank opens floodgates: Hydropower projects are back in favour as greener energy solution. *Guardian Weekly*, 17.05.13, 14.

Schumacher, F.F. 1973. *Small is Beautiful. Economics as If People Mattered*. New York: Harper and Row.

Scott, J. 1998. Seeing Like a State. Yale University Press.

Sen, A. 2000. *Development as Freedom*. New York: Knopf.

Sen, A. 2009. *The idea of Justice*. Cambridge: The Belknap Press.

Senge, Peter M. 2006. *The Fifth Discipline: The Art and Practice of the Learning Organization*. Crown Pub.

Shiva, V. 1989. *Staying Alive: Women, Ecology and Survival in India*. London: Zed Books.

Shiva, V. 2002. *Water Wars: Privatization, Pollution And Profit*. London: Pluto Press.

Shiva, V. 2011. Earth Democracy, Portland University. http://www.youtube.com/watch?v=UOfM7QD7-kk/.

Shiva, V. 2012a. *Monocultures of the Mind: Perspectives on Biodiversity and Biotechnology*. Penang: Third World Network.

Shiva, V. 2012b. *Making Peace with the Earth*. Winipeg: Fernwood Publishing.

Silverstein, J., and M. Kaiser. 2014. *Who welcomes who in multicultural Australia?* October 2nd. https://newmatilda.com/2014/10/02/who-welcomes-who-multicultural-australia.

Singer, P. 1981. *The Expanding Circle: Ethics, Evolution and Moral Progress*. Princeton: Princeton University Press.

Singer, P. 2002. *One World: The Ethics of Globalisation*. Melbourne: Text Publishers.

Snyder, S.H. 2011. Mind molecules. *The Journal of Biological Chemistry*, June, 2011. 286 (2): 21023–21032.

Stanescu, J. 2012. Species trouble: Judith Butler, mourning, and the precarious lives of animals. *Hypatia VL* 27: 563–582.

Steffen, W., A. Persson, L. Deutsch, J. Zalasiewicz, M. Williams, K. Richardson, C. Crumley, P. Crutzen, L. Gordon, M. Molina, V. Ramanathon, J. Tockstrom, M. Scheffer, H. Schellnhuber, and U. Svedin. 2011. The anthropocene: From global to planetary stewardship. *Royal Swedish Academy of Social Sciences Springer Ambio* 40: 739–761.

Stern, N. 2007. *The Economics of Climate Change*. Cambridge: Cambridge University Press.

Stiglitz, J.E. 2012. *The Price of Inequality: How Today's Divided Society Endangers Our Future*. WW Norton.

Stiglitz, J., A. Sen, and J.P. Fitoussi. 2010. *Mis-Measuring Our Lives: Why the GDP Doesn't Add Up*. New York: The New Press.

Sunstein, C. 2013. *Simpler: The Future of Government*. New York: Simon and Schuster.

Taylor, A. 2010. *The Examined Life: Excursions with Contemporary Thinkers*. New York: The New Press.

The heat and the death toll are rising in India. Is this a glimpse of Earth's future? *The Guardian*, April 29, 2016.

The UN Declaration of the Rights of Indigenous Peoples (2008). http://www.ohchr.org/Documents/Issues/IPeoples/UNDRIPManualForNHRIs.pdf.

Treaty of Maastricht. 1992. Article 5 on Subsidiarity.

Triandis, H.C. 1995. *Individualism and Collectivism*. Oxford: Boulder.

Triandis, H., and M. Gelfand. 1998. Converging measurement of horizontal and vertical individualism and collectivism. *Journal of Personality and Social Psychology* 74 (1): 118–128 (American Psychological Association).

Turner, V. 1987. The liminal period in Rites de Passage. In *Betwixt and Between: Patterns of Masculine and Feminine Initiation*, ed. Mahdi, L.C., S. Foster, M. Little. Open Court Publishing Company.

Turner, R., and C. Whitehead. 2008. How collective representations can change the structure of the brain. *Journal of Consciousness Studies* 15 (10–11): 43–57.

Turok, N. 2012. The universe within Allen and Unwin. Based on the CBC Massy Lectures.

Uglow, J. 2006. *Nature's Engraver: A Life of Thomas Bewick*. London: Faber.

Ulrich, W. 1983. *Critical Heuristics of Social Planning: A New Approach to Practical Philosophy*. New York: Wiley.

Ulrich, Werner, and Reynolds, Martin (2010). Critical systems heuristics. In: Reynolds, Martin and Holwell, S. eds. Systems Approaches to Managing Change: A Practical Guide. London: Springer, pp. 242–292.

UNESCO. 2007. *Biosphere Reserves: Dialogue in Biosphere Reserves. References, Practices and Experiences*. ISSN 2071-1468. Paris. UNESCO.

United Nations G.A. 2007. Resolution 61/295 of 13 September.

United Nations Climate Change Summit, Durban. 2011. *United Nations climate change summit*. https://unfccc.int/meetings/durban_nov_2011/meeting/6245.php.

United Nations Economic and Social Development Local Agenda 21, 1992.

United Nations Sustainable Development Goals. 2015. www.un.org/sustainabledevelopment/sustainable-development-goals/.

United Nations Human Development Index. 2003. *A compact among nations to end poverty*. New York: UNDP, Oxford University Press.

Van Gennep, A. 1960. *Rites de Passage*. University of Chicago Press.

United Nations. 2011. *The Millennium goals report*. http://www.aidsdatahub.org/en/whats-new/287-all-data-alerts/728-the-millennium-development-goals-report-2011-united-nations-2011. Accessed July 19th, 2011.

United Nations. 1993. Integrated Environmental and Economic Accounting: Handbook of National Accounting: Studies in Methods, Series F. No 61, Department for Economic and Social Information and Policy Analysis, Statistical Division, United Nations, New York.

United Nations. 2014. World Urbanization Prospects. Department of Economic and Social Affairs Revised report ST/ESA/SER.A/352.

Uhr, J. 2005. *Terms of Trust*. Sydney Australia: UNSW Press.

Urry, J. 2007. *Mobilities*. Cambridge: Polity.

Urry, J. 2010. Consuming the planet to excess. *Theory, Culture and Society* 27: 191–212.

Van Gennep, A. 1960. *Rites de Passage*. University of Chicago Press.

Van Gigch, J. 2003. *Metadecisions: Rehabilitating Epistemology*. Plenum: Kluwer.

Vaske, J., and K. Kobrin. 2001. Place attachment and environmentally responsible behavior. *Journal of Environmental Education* 32 (4): 16–21.

Vickers, G. 1983. *Human Systems are Different*. London: Harper and Row.

Wadsworth, Y. 2008. Is it safe to talk about systems again yet? Self-organizing processes for complex living systems and the dynamics of human inquiry. *Systemic Practice and Action Research* 21 (2): 153–170.

Wadsworth, Y. 2010. *Building in Research and Evaluation. Human Inquiry for Living Systems*. Sydney: Allen and Unwin.

Walker, P. 2013. Research in relationship with humans, the spirit world, and the natural world. In *Indigenous Pathways into Social Research*, ed. Mertens, D, F. Cram, and B. Chilisa. Left Coast Press, Inc.

Wallace-Brown, G., and D. Held (eds.). 2010. *Cosmopolitan Reader*. Cambridge: Polity Press.

Watson, D. 2012. A new dusk. *Monthly Review*. https://www.themonthly.com.au/author/don-watson.

Weir, A. 2012. Collaborative approaches to regional governance—lessons from Victoria. *Australian Journal of Public Administration* 71 (4): 469–474.
Wenger, E. 1998. *Communities of Practice*. Cambridge: Cambridge University Press.
Wenger, E., N. White, and J. Smith. 2009. *Digital Habitats: Stewarding Technology for Communities*. Portland: CP Square.
Wheatley, M. 2009. *Turning to One Another: Simple Conversations to Restore Hope to the Future*. San Franciso: Barett-Koehler.
Widianingsih, I. 2014. International donor policy & local government in Indonesia: Establishing good governance? Unpublished Ph.D. thesis, Flinders University, School of Social and Policy Studies.
Wilkinson, R., and K. Pickett. 2009. *The Spirit Level. Why More Equal Societies Almost Always Do Better*. London: Allen Lane.
Wynne, B. 1996. May the sheep safely graze? A reflexive view of the expert—lay knowledge divide. In *Risk, Environment and Modernity: Towards A New Ecology*, ed. Lash, S., B. Szerszynski, and B. Wynne. London: Sage.
Yeatman, A. 2003. Global ethics, Australian citizenship and the 'boat people': A symposium. *Journal of Sociology* 39 (1): 15–22.
Young, I.M. 2000. *Inclusion and Democracy*. Oxford: Oxford University Press.
Younge, G. 2015. The danger of polarized debate is all too real: The truth about the Paris attacks is more complex and more difficult than many would like to accept. *The Guardian Weekly* January 16, 2015, 18–19.
Zavestoski, S. 2002. Psychological bases of anticonsumption. *Attitudes Psychology and Marketing* 19 (2): 149–165.

Index

J. McIntyre-Mills, *Planetary Passport*, Contemporary Systems Thinking,
DOI 10.1007/978-3-319-58011-1

R

S

T

U

V

W